GOVERNMENT SERIES

W9-BQT-909

Smart Grid

Modernizing Electric Power Transmission and Distribution; Energy Independence, Storage and Security; Energy Independence and Security Act of 2007 (EISA); Improving Electrical Grid Efficiency, Communication, Reliability, and Resiliency; Integrating New and Renewable Energy Sources

Compiled by TheCapitol.Net

Authors: Stan Mark Kaplan, Fred Sissine, Amy Abel, Jon Wellinghoff, Suedeen G. Kelly, and James J. Hoecker

TheCapitol.Net, Inc. is a non-partisan firm that annually provides continuing professional education and information for thousands of government and business leaders that strengthens representative government and the rule of law.

Our publications and courses, written and taught by *current* Washington insiders who are all independent subject matter experts, show how Washington works.™ Our products and services can be found on our web site at <*www.TheCapitol.Net*>.

Additional copies of *Smart Grid* can be ordered online: <*www.GovernmentSeries.com*>.

Design and production by Zaccarine Design, Inc., Evanston, IL; 847-864-3994.

∞ The paper used in this publication exceeds the requirements of the American National Standard for Information Sciences—Permanence of Paper for Printed Library Materials, ANSI Z39.48-1992.

v 1

Smart Grid, softbound:
ISBN: 158733-162-4
ISBN 13: 978-1-58733-162-6

Summary Table of Contents

Table of Contents

Chapter 2:

Chapter 3:

Chapter 8:
Testimony of Commissioner Jon Wellinghoff,
Federal Energy Regulatory Commission Before the Energy
and Environment Subcommittee of the Committee on
Energy and Commerce, U.S. House of Representatives,
Hearing on "The Future of the Grid: Proposals for

Chapter 9:
Testimony of Commissioner Suedeen G. Kelly,
Federal Energy Regulatory Commission Before
the Committee on Energy and Natural Resources,

Chapter 10:
Prepared Statement of James J. Hoecker,
Counsel to WIRES, Before the Select Committee
on Energy Independence and Global Warming,
U.S. House of Representatives, Hearing on
"Get Smart on the Smart Grid: How Technology
Can Revolutionize Efficiency and Renewable Solutions,"

xviii

Introduction

Smart Grid:

Modernizing Electric Power Transmission and Distribution; Energy Independence, Storage and Security; Energy Independence and Security Act of 2007 (EISA); Improving Electrical Grid Efficiency, Communication, Reliability, and Resiliency; Integrating New and Renewable Energy Sources

The electric grid delivers electricity from points of generation to consumers, and the electricity delivery network functions via two primary systems: the transmission system and the distribution system. The transmission system delivers electricity from power plants to distribution substations, while the distribution system delivers electricity from distribution substations to consumers. The grid also encompasses myriads of local area networks that use distributed energy resources to serve local loads and/or to meet specific application requirements for remote power, municipal or district power, premium power, and critical loads protection.

The concept of a "smart grid" lacks a standard definition but centers on the use of advanced technology to increase the reliability and efficiency of the electric grid, from generation to transmission to distribution. However, the smart grid does not necessarily replace the existing infrastructure, most of which was installed in the 1970s.

The move to a smart grid is a move from a centralized, producer-controlled network to one that is less centralized and more consumer-interactive.

- It enables informed participation by consumers
- Accommodates all generation and storage options
- Enables new products, markets, and services
- Provides the power quality for the range of needs
- Optimizes asset utilization and operating efficiency
- Operates resiliently to disturbances, attacks, and disasters

The Department of Energy, Office of Electricity Delivery and Energy Reliability is charged with orchestrating the modernization of the nation's electrical grid. The office's multi-agency Smart Grid Task Force (www.oe.energy.gov/smartgrid_taskforce.htm) is responsible for coordinating standards development, guiding research and development projects, and reconciling the agendas of a wide range of stakeholders, including utilities, technology providers, researchers, policymakers, and consumers.

The National Institute of Standards and Technology (NIST), has been charged under the Energy Independence and Security Act (P.L. 110-140, Dec. 19, 2007) with identifying and evaluating existing standards, measurement methods, technologies, and other support services to Smart Grid adoption.

1626SmartGrid.com

Summary

This report provides background information on electric power transmission and related policy issues. Proposals for changing federal transmission policy before the 111th Congress include S. 539, the Clean Renewable Energy and Economic Development Act, introduced on March 5, 2009; and the March 9, 2009, majority staff transmission siting draft of the Senate Energy and Natural Resources Committee. The policy issues identified and discussed in this report include:

Federal Transmission Planning: several current proposals call for the federal government to sponsor and supervise large scale, on-going transmission planning programs. Issues for Congress to consider are the objectives of the planning process (e.g., a focus on supporting the development of renewable power or on a broader set of transmission goals), determining how much authority new interconnection-wide planning entities should be granted, the degree to which transmission planning needs to consider non-transmission solutions to power market needs, what resources the executive agencies will need to oversee the planning process, and whether the benefits for projects included in the transmission plans (e.g., a federal permitting option) will motivate developers to add unnecessary features and costs to qualify proposals for the plan.

Permitting of Transmission Lines: a contentious issue is whether the federal government should assume from the states the primary role in permitting new transmission lines. Related issues include whether Congress should view management and expansion of the grid as primarily a state or national issue, whether national authority over grid reliability (which Congress established in the Energy Policy Act of 2005) can be effectively exercised without federal authority over permitting, if it is important to accelerate the construction of new transmission lines (which is one of the assumed benefits of federal permitting), and whether the executive agencies are equipped to take on the task of permitting transmission lines.

Transmission Line Funding and Cost Allocation: the primary issues are whether the federal government should help pay for new transmission lines, and if Congress should establish a national standard for allocating the costs of interstate transmission lines to ratepayers.

Transmission Modernization and the Smart Grid: issues include the need for Congressional oversight of existing federal smart grid research, development, demonstration, and grant programs; and oversight over whether the smart grid is actually proving to be a good investment for taxpayers and ratepayers.

Transmission System Reliability: it is not clear whether Congress and the executive branch have the information needed to evaluate the reliability of the transmission system. Congress may also want to review whether the power industry is striking the right balance between modernization and new construction as a means of enhancing transmission reliability, and whether the reliability standards being developed for the transmission system are appropriate for a rapidly changing power system.

This report will be updated as warranted.

Electric Power Transmission: Background and Policy Issues

Contents

Figures

Congressional Research Service

Electric Power Transmission: Background and Policy Issues

- Generating plants produce electricity, using either combustible fuels such as coal, natural gas, and biomass: or non-combustible energy sources such as wind, solar energy, and nuclear fuel.

- Transmission lines carry electricity from the power plant to demand centers. The higher the voltage of a transmission line the more power it can carry. Current policy discussions focus on the high voltage network (230 kilovolts (kV) rating and greater) used to move large amounts of power long distances.[2]

- Near customers a step-down transformer reduces voltage so the power can use distribution lines for final delivery.[3]

Figure I. Elements of the Electric Power System
Simplified Schematic

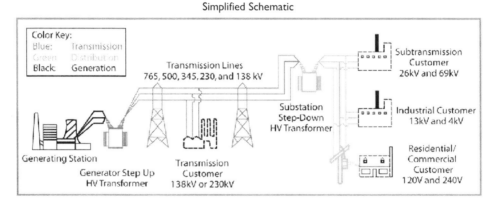

Source: U.S.-Canada Power System Outage Task Force, Final Report on the August 14, 2003 Blackout in the United States and Canada: Causes and Recommendations, April 2004, p. 5, https://reports.energy.gov/BlackoutFinal-Web.pdf.

The vast majority of the transmission system in the United States is an alternating current (AC) system. This is largely because the voltage of AC power can be stepped up and down with relative ease. A small portion of the system runs on high voltage direct current (DC) lines. This technology is very efficient but requires expensive converter stations to connect with the AC system.

The transmission grid was not built in conformance with a plan like the interstate highway system. The grid is a patchwork of systems originally built by individual utilities as isolated transmission islands to meet local needs. These small networks were unsystematically linked when utilities decided to jointly own power plants or to connect to neighboring companies to

---

[2] Lines rated at 345 kilovolts (kV) or 500 kV are referred to as extra high voltage (EHV) lines. Lines rated at 765 kV are referred to as ultra high voltage (UHV) lines.

[3] In addition to the 167,000 miles of high voltage transmission lines, the transmission system includes about another 300,000 miles of lower voltage transmission lines. Note that the division between the transmission and distribution systems is not clear-cut. Depending on the application, a 69kV line might be considered a transmission or distribution line. For more information see Douglas R. Hale, *Electricity Transmission in a Restructured Industry: Data Needs for Public Policy Analysis*, Energy Information Administration (EIA), DOE/EIA-0639, Washington, DC, December 2004, p. 16, http://www.eia.doe.gov/cneaf/electricity/page/transmission/DOE_EIA_0639.htm.

facilitate power sales.[4] The grid eventually evolved into three major "interconnections," Eastern, Western, and the Electric Reliability Council of Texas (ERCOT, which covers most but not all of the state) (**Figure 2**). Within each interconnection the AC grid must be precisely synchronized so that all generators rotate at 60 cycles per second (synchronization failure can cause damage to utility and consumer equipment, and cause blackouts). There are only eight low capacity links (called "DC ties") between the Eastern, Western, and ERCOT Interconnections.[5] In effect, the 48 contiguous states have three separate grids with limited connections.

Figure 2. United States Power System Interconnections

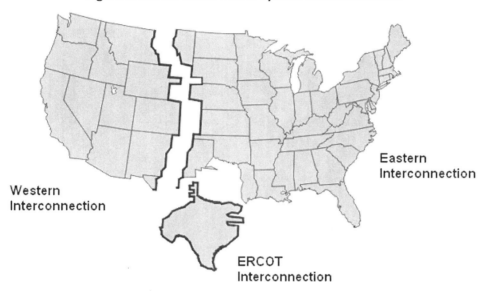

Source: adapted from a map located on the Energy Information Administration website at http://www.eia.doe.gov/cneaf/electricity/page/fact_sheets/transmission.html.

Notes: ERCOT = Electric Reliability Council of Texas. For the extensions of the interconnections into Canada and Mexico see **Figure 3**. Neither figure shows the Quebec Interconnection.

---

[4] As recently as 1962 the systems that now constitute the Eastern Interconnection were not fully connected (**Figure 2**). Securities and Exchange Commission, Prepared Direct Testimony of Paul B. Johnson on Behalf of the American Electric Power System, *In the Matter of American Electric Power Company, Inc.*: File No. 3-11616, December 7, 2004, pp. 9 and 11, http://www.sec.gov/divisions/investment/opur/filing/3-11616-120704aepex2.pdf.

[5] The direct current DC ties permit limited power transfers between the interconnections without synchronizing the systems. For example, a synchronization problem in the Eastern Interconnection cannot propagate across a DC tie into the Western Interconnection. ERCOT has two ties with the Eastern Interconnection and there are six ties between the Eastern and Western Interconnections. See http://www.wapa.gov/about/faqtrans.htm and Bill Bojorquez and Dejan J. Sobajic, "AC-DC Ties @ ERCOT," The 8th Electric Power Control Centers Workshop, Les Diablerets, Switzerland, June 6, 2005, http://www.epccworkshop.net/archive/2005/paper/pdf_monday/PanelSession/Sobajic_ERCOT.pdf. The typical capacity of these ties appears to be about 200 megawatts. Total generating capacity in the United States is about one million megawatts.

Within the three interconnections, the grid is operated by a total of about 130 balancing authorities.[6] These are usually the utilities that own transmission systems, but in some cases (such as ERCOT) a single authority supervises an entire regional grid. The balancing authorities operate control centers which monitor the grid and take actions to prevent failures like blackouts.

The transmission grid is owned by several hundred private and public entities. **Table 1** shows the miles of high voltage transmission line in the 48 contiguous states by region and type of owner. The table also shows the data expressed as ownership percentages (values in brackets).

Table 1. High Voltage Transmission by Owner and Region

Data in Miles [and Regional %] for the 48 Contiguous States for Transmission Lines of 230 kV and Higher

Owner Type	Northeast /Midwest	Southeast	Southwest	Upper Plains	West	U.S. Total
Federal	21 [0%]	2,768 [7%]	0 [0%]	2,541 [17%]	18,214 [27%]	23,544 [14%]
Other Public Power	964 [3%]	2,079 [5%]	731 [5%]	1,798 [12%]	5,525 [8%]	11,098 [7%]
Cooperative	0 [0%]	2,993 [8%]	387 [2%]	2,908 [20%]	4,496 [7%]	10,784 [6%]
Subtotal – All Public Power and Cooperatives	986 [3%]	7,840 [20%]	1,118 [7%]	7,247 [49%]	28,235 [42%]	45,426 [27%]
Independent Transmission Companies	4,640 [15%]	0 [0%]	351 [2%]	1,045 [7%]	0 [0%]	6,036 [4%]
Investor Owned Utilities	24,968 [81%]	31,412 [79%]	12,408 [80%]	5,402 [36%]	37,034 [56%]	111,223 [66%]
N/A	260 [1%]	264 [1%]	1,686 [11%]	1,148 [8%]	1,250 [2%]	4,609 [3%]
Total	30,853 [100%]	39,516 [100%]	15,563 [100%]	14,843 [100%]	66,519 [100%]	167,294 [100%]

Source: Data downloaded from Platts POWERmap, information on entity ownership type provided by the Energy Information Administration, and CRS estimates.

Notes: The Northeast/Midwest region is the combination of the RFC and NPCC NERC regions; the Southeast is the combination of SERC and FRCC; the Southwest is the combination of ERCOT and SPP; the Upper Plains is the MRO region; and the West is the WECC region. For a NERC regional map, see **Figure 3**. N/A signifies that ownership information is not available. Other Public Power includes municipal and state systems. kV = kilovolt. Detail may not add to totals due to independent rounding.

The table illustrates how ownership patterns vary greatly across the country. In the West and Upper Plains regions, public power owns more than 40% of the high voltage grid. In the other regions about 80% of the grid is owned by investor owned utilities.

Figure 3 (below) shows the eight North American Electric Reliability Corp. (NERC) regions. As discussed later in the report, NERC and its regions play important roles in maintaining the

---

[6] U.S. Department of Energy, *20% Wind Energy by 2030*, Washington, D.C., July 2008, p. 91. http://www1.eere.energy.gov/windandhydro/pdfs/41869.pdf. For a map that displays balancing authorities see the NERC website at http://www.nerc.com/fileUploads/File/AboutNERC/maps/NERC_Regions_BA.jpg.

reliability of the power system. Like the interconnections, some of these NERC regions extend into Canada or Mexico. However, this report is concerned only with the U.S. transmission grid.

Figure 3. NERC Reliability Regions
(North American Grid Interconnections Outside of Quebec Also Shown)

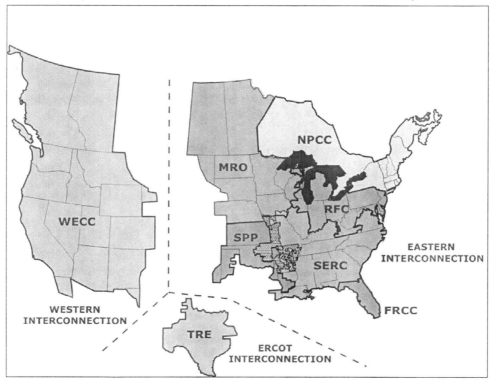

Source: North American Electric Reliability Corp. website at http://www.nerc.com/fileUploads/File/AboutNERC/maps/NERC_Interconnections_color.jpg.

Notes: NERC = North American Electric Reliability Corp. The NERC regional entities are ERCOT (Electric Reliability Council of Texas); FRCC (Florida Reliability Coordinating Council); MRO (Midwest Reliability Organization); NPCC (Northeast Power Coordinating Council); RFC (Reliability*First* Corp.); SERC (SERC Reliability Corp.); SPP (Southwest Power Pool); TRE (Texas Regional Entity, an independent division of ERCOT); and WECC (Western Electricity Coordinating Council). Quebec Interconnection is not shown.

Regulatory Framework

Electric power regulation is divided among federal, state, and regional authorities. The scope of federal authority is different for rates and reliability. The following discussion reviews:

- State regulation and self-governing public power.

- Federal regulation of the transmission system and the reliability of the bulk power system.

State Regulation and Self-Governing Public Power

State regulation of the electric power industry is usually centered in a public utility commission (PUC). The authority of the commissions is often limited to investor-owned utilities (IOU; i.e., private corporations, usually with publicly traded stock). The PUCs set retail rates, review utility operations, and, most importantly for the purposes of this report, issue siting approvals (permits) for new transmission lines.

Publicly-owned utilities (POUs) are owned by municipal, state, and federal governments. The term is also sometimes applied to customer-owned rural electric cooperatives. POUs are typically small and operate only distribution systems, but some have large transmission and generation systems, such as the Tennessee Valley Authority. POUs are self-regulated by their governing boards, are generally not subject to state or federal economic regulation, and make their own decisions on adding new generating capacity and building transmission lines.[7]

Federal Regulation of Electric Power Transmission and Power System Reliability

This part of the report first discusses federal regulation of transmission, and then federal regulation of the reliability of the power system.

Federal Transmission Regulation

Federal regulation of the power industry is exercised by the Federal Energy Regulatory Commission (FERC), an independent agency administratively housed within the Department of Energy. FERC regulates wholesale electricity rates,[8] approves transmission line projects, and sets transmission rates. However, FERC's authority is limited in important respects:

- For the most part FERC's rate-making and transmission authorization authority is limited to IOUs in the 48 contiguous states outside of ERCOT.[9]

- While FERC must approve transmission projects proposed by jurisdictional utilities and establishes rates, a project also needs a siting permit from every state the line will traverse.

---

[7] Rural cooperatives as a group are somewhat more subject to state and federal regulation than government-owned utilities. Close to half the state PUC's have authority over cooperatives, and a small number of cooperatives are also subject to FERC rate regulation.

[8] Wholesale electricity sales are transactions between a generator and a reseller of power, or between two resellers. An example is when one utility sells power to another. This is also referred to as sales for resale. A retail transaction is a sale to the final end user, as when a utility sells electricity to a homeowner.

[9] The entities which eventually formed ERCOT severed non-emergency connections with outside grids in August 1935. Their objective was to avoid falling under the ratemaking jurisdiction of the Federal Power Commission (FERC's predecessor) by maintaining a purely intrastate system. DC ties between ERCOT and the Eastern Interconnection were constructed in the 1980s, but these do not put ERCOT under FERC rate jurisdiction due to a specific exemption in the law (16 U.S.C. § 824k(k)). For additional information, see Richard D. Cudahy, "The Second Battle of the Alamo: The Midnight Connection," *Natural Resources and Environment*, Summer 1995; *West Texas Utilities Co. and Central Power and Light Co. v. Texas Electric Service Co. and Houston Lighting and Power Co.*, U.S. District Court of the Northern District of Texas, Dallas Division, Memorandum Opinion, January 30, 1979 (470 F. Supp. 798).

Critics have argued that multi-state permitting of transmission lines has delayed the construction of needed transmission lines. In the Energy Policy Act of 2005 (EPACT05), Congress gave FERC "backstop" siting authority.[10] This authority operates as follows:

- The Department of Energy (DOE) is to conduct a triennial study of transmission system congestion. Based on this study, DOE may designate as National Interest Electric Transmission Corridors (NIETC) areas with severe transmission congestion.

- A special permitting rule applies to transmission projects proposed for a NIETC. If a state has not acted on the permit application for a NIETC project within a year, the developer can bypass the state and bring its application to FERC for approval.

DOE completed its first congestion study in 2006 and in 2007 designated two NIETCs, one in southern California–Arizona, and a second covering a large part of the Northeast.[11] As of early 2009 no use had been made by transmission developers of the backstop process. Moreover, in February 2009, the Fourth Circuit Court of Appeals ruled that FERC had overstepped its authority in implementing the backstop process. FERC had interpreted the law to mean that the backstop process could be used if a state has not acted within a year *or if the state has affirmatively decided to reject the project*. However, the court ruled that a state's decision to reject a project could not be appealed to FERC.[12] On April 2, 2009, FERC asked the Fourth Circuit to reconsider its decision.

Another aspect of FERC regulation is its efforts to encourage competition in the electric power market. In 1996, FERC mandated "open access" to the transmission system.[13] Open access requires transmission owners to make available, at cost-based or market-based fees, available transmission capacity to any generator or power buyer that is or can be connected to the system. The objective is to prevent transmission owners from using their control of the power system from stifling competition. To further facilitate open access, in 1999 a FERC order[14] encouraged the creation of regional transmission organizations (RTO; see **Figure 4**). RTOs take over operation of the transmission network in a region or large state, although utilities continue to own their systems. RTO's ensure open access to the grid, coordinate transmission planning, and establish mechanisms to pay for new transmission lines.[15]

---

[10] 18 U.S.C. § 824p.

[11] The first congestion study, the NIETC designation report, maps of the current NIETCs, and information on the upcoming second congestion study are available on the website of DOE's Office of Electricity Delivery and Energy Reliability at http://www.oe.energy.gov/index.htm. The second congestion study is due in 2009 and could result in revisions to the NIETCs.

[12] *Piedmont Environmental Council v. FERC*, Case No. 07-1651 (4th Cir 2009), http://pacer.ca4.uscourts.gov/opinion.pdf/071651.P.pdf.

[13] See Orders 888 and 889 at http://www.ferc.gov/legal/maj-ord-reg.asp.

[14] See Order 2000 at http://www.ferc.gov/legal/maj-ord-reg.asp.

[15] RTOs also operate short-term markets for electricity sales and in some cases operate capacity markets which arrange for new power plants to be built. (Similar in function to RTOs are independent system operators (ISOs) and the terms are sometimes used interchangeably. However, the only ISOs to be qualified as RTOs under the terms of FERC's Order 2000 are ISO-New England, PJM, the Midwest ISO, and the SPP RTO.) As shown in **Figure 4**, RTOs and ISOs cover only part of the nation. Some regions that have been opposed to "restructuring" of the power market, such as the southeast and northwest, do not have RTOs. A full discussion of the competitive restructuring of the electric power market is beyond the scope of this report. For summaries of these developments, see The Electricity Advisory (continued...)

- Expanding the grid to reach areas where renewable electricity plants can be built.

- Resolving grid congestion and reliability problems.

The next two parts of the report discuss the following transmission planning issues:

- The objectives of the planning process.

- Planning authority.

Objectives of the Planning Process

This discussion looks at the objectives of federal transmission planning from three perspectives: renewable energy development; congestion relief and reliability; and alternatives to transmission.

Expansion for Renewable Energy

Currently the most important source for new renewable electricity generation is wind power,[21] but many of the best wind production areas are in thinly populated areas in the Midwest and northern plains that have limited access to transmission lines. The best region for solar development is the isolated desert southwest. Some planning process proposals, including S. 539, are explicitly focused on expanding the grid to serve these remote renewable resource areas. In the S. 539 process, the purpose of the plan is to develop transmission lines to serve "National Renewable Energy Zones," and 75% of the generating capacity connected to the lines must be renewable (e.g., wind and solar energy). Other objectives, such as congestion relief, are included in the legislation but are not the primary aims of the bill.

There are two basic concepts for expanding the transmission grid to reach remote renewable energy regions. One concept is to plan and construct a continent-spanning ultra-high voltage "overlay" system of AC or DC transmission lines that would be the electrical equivalent of the interstate highway system. This system of "transmission superhighways" would be designed to move large amounts of renewable electricity to customers across the country. No firm plans exist for such a system, but conceptual layouts have been proposed.[22] The second, less ambitious, concept relies on interconnection-wide or regional plans for identifying discrete transmission projects to connect renewable energy zones to load centers.[23]

Renewable energy-focused transmission planning could accelerate the development of renewable power. However, some critics argue that such a renewable-centric approach to transmission planning would produce costly facilities. This is because a transmission line built for peak renewable power output would be underutilized much of the time (since the output of wind and

---

[21] In 2007, 5,193 net summer megawatts (MW) of new wind generating capacity was installed in the United States, compared to 89 megawatts of solar generation. The largest source of new generating capacity in 2007 was gas-fired power plants (6,673 MW). Total additions were 13,845 MW. EIA, *Electric Power Annual 2007*, Table 2.6. For more information on wind power see CRS Report RL34546, *Wind Power in the United States: Technology, Economic, and Policy Issues*, by Jeffrey Logan and Stan Mark Kaplan.

[22] For example, see U.S. Department of Energy, *20% Wind Energy by 2030*, Washington, D.C., July 2008, p. 12, http://www1.eere.energy.gov/windandhydro/pdfs/41869.pdf.

[23] Examples include the Joint Coordinated System Plan conceptual design of lines for moving wind power to the industrial Midwest and Northeast (http://www.jcspstudy.org/), and the ITC Holdings' Green Power Express, a plan for tying wind power from the Great Plains into the Midwestern grid (http://www.thegreenpowerexpress.com/).

---

solar power vary with the weather and time-of-day). Critics also claim that a renewable-centric planning approach might not adequately meet congestion relief and reliability objectives.[24]

Expansion for Congestion Relief and Reliability

Transmission congestion occurs when use of a power line is restricted (for example, to prevent overloading and failure of the line). Utilities and RTOs can work around transmission congestion by using alternative transmission paths or by changing power plant operations, but these steps (which often involve running expensive power plants that would otherwise be less-utilized or idle) can be costly. Studies suggest that the annual costs of transmission congestion range from the hundreds of millions to billions of dollars.[25] The solution for congestion costs is not necessarily massive transmission construction. For example, DOE found that in the Eastern Interconnection "a relatively small portion of constrained transmission capacity causes the bulk of the congestion cost that is passed through to consumers. This means that a relatively small number of selective additions to transmission capacity could lead to major economic benefits for many consumers."[26]

Transmission system reliability is defined by NERC has having two aspects: whether a transmission system has enough capacity to continuously meet customer needs, and whether the system has the resiliency to withstand major failures, such as the loss of a key transmission line.[27] As with congestion relief, the solutions to reliability problems do not necessarily involve building new transmission lines. For example, sometimes reliability can be enhanced by building new or expanded substations or by installing certain types of specialized equipment that helps maintain system voltage levels.[28]

When new transmission lines are needed for congestion relief or to improve reliability they can be expensive, multi-year projects. An example of a large transmission project for reliability is the 210 mile Trans-Allegheny Interstate Line (TrAIL) from southwestern Pennsylvania through West Virginia to Northern Virginia. This $820 million project involves construction of 210 miles of new transmission lines and substations. According to NERC, the project is needed to "relieve anticipated overloads and voltage problems in the Washington, DC area, including anticipated

---

[24] According to the Large Public Power Council, "Such [renewable energy] requirements simply may not work when one considers the physics of the electric grid and the intermittent nature of renewable resources." U.S. Congress, Senate Committee on Energy and Natural Resources, *To Receive Testimony On Pending Legislation Regarding Electricity Transmission Lines*, Prepared Testimony of James A. Dickenson on Behalf of the Large Public Power Council, 111th Cong., 1st sess., March 12, 2009, p., 7, http://energy.senate.gov/public/index.cfm?FuseAction=Hearings.Testimony& Hearing_ID=b9e47ea9-c62b-23fc-33ff-30fda7b3a744&Witness_ID=ed6a79eb-6664-412c-b66b-7e87e4e55435.

[25] Bernard Lesieutre and Joseph Eto, Electricity Transmission Congestion Costs: A Review of Recent Reports, Lawrence Berkeley National Laboratory, p. 2, http://certs.lbl.gov/pdf/54049.pdf, and U.S. Department of Energy, *National Transmission Grid Study*, May 2002, pp. 16–18, http://www.pi.energy.gov/documents/TransmissionGrid.pdf.

[26] U.S. DOE, *National Electric Transmission Congestion Study*, August 2008, p. 28, http://www.pi.energy.gov/documents/TransmissionGrid.pdf. Emphasis in the original not shown. DOE also found that only 20% of the constrained transmission capacity accounted for 60% of national congestion costs (p. 29).

[27] It would be uneconomic to build a transmission system capable of withstanding any conceivable failure, so systems are designed to meet "credible contingencies."

[28] The specialized equipment referred to injects "reactive power" into the grid. Reactive power supply is a relatively obscure topic that plays a critical role in power system operations. For more information see Federal Energy Regulatory Commission, *Principles for Efficient and Reliable Reactive Power Supply and Consumption*, February 4, 2005, http://www.ferc.gov/eventcalendar/files/20050310144430-02-04-05-reactive-power.pdf.

Senate Energy Majority Draft allows "National High Priority Transmission Projects" identified through a federally-sanctioned planning process to go directly to FERC for approval.

A common element in these and some other proposals is that projects eligible for FERC permitting must be included in an interconnection-wide plan. Other projects would remain under state purview. Another approach is to simply give FERC permitting authority over all transmission projects, or at least all high voltage transmission projects.[44]

An alternative view is that the current permitting process is not broken and at most needs tweaking. The National Association of Regulatory Utility Commissioners (NARUC), an association of state PUCs, passed a resolution in March 2009 urging "Congress and the White House to move cautiously, if at all, in expanding federal jurisdiction over siting and planning of new transmission infrastructure."[45] According to the NARUC president, "Siting and planning transmission is one of the most difficult yet essential jobs of a State regulator, and no federal agency will have the resources or local knowledge on its own to balance all the considerations that must be taken into account."[46]

Where FERC does have siting authority, as with natural gas pipelines and LNG terminals, it has sometimes been intensely criticized by state officials and members of Congress who believe FERC has made poor decisions.[47] However, this may simply argue for giving an agency other than FERC any new federal transmission siting authority.

Transmission Permitting: Summary of Policy Issues

In summary, in considering how much additional transmission siting authority, if any, the federal government should assume, Congress may want to consider the following policy questions:

- *Should the grid be viewed from a national perspective?* The grid began as local systems regulated by states. Now that the system has evolved into three separate synchronized interconnections, each spanning (other than ERCOT) many states, a question is whether a state-by-state view of the grid or a national perspective is most appropriate. The question is made pressing by proposals to make more use of the grid for long distance power transactions, such as for renewable energy. The issue does not necessarily have a single answer; for example, a state perspective may be appropriate for "routine" projects, while a national perspective could be applied to high priority interstate projects (however "routine" and "high-priority" are defined).

- *Can transmission system reliability be separated from authority over new transmission construction?* In EPACT05 Congress put the reliability of the grid

---

[44] Esther Whieldon, "FERC Needs Eminent Domain to Site Transmission Lines, Kelliher Tells Senate," *Inside F.E.R.C.*, August 4, 2008.

[45] National Association of Regulatory Utility Commissioners, "States Reiterate Vital Role in Grid Expansion, List Principles as Congress Mulls Action," press release, March 11, 2009, http://www.naruc.org/News/default.cfm?pr=133.

[46] Ibid.

[47] An example are current disputes in Oregon over the permitting of an LNG terminal and a new gas pipeline (Ted Sickinger, "State Asks Court to Toss Bradwood Site's Approval," *The (Portland) Oregonian*, January 27, 2009). Also see Jason Fordney, "Connecticut Governor Blasts Idea of Giving FERC More Authority Over Power Line Siting," *Platts Electric Utility Week*, March 30, 2009, p. 6.

under federal jurisdiction. By extension, should the federal government have control over the permitting of transmission lines aimed at enhancing system reliability (which could mean almost any new line in an interconnected power system)? As discussed in more detail later in the report, failures at one point in a synchronized system can spread widely, and these failures (in a worst case, blackouts) do not respect state lines.

- *How important is it to accelerate the construction of new transmission lines?* One criticism of current regulation is that it takes many years to permit a project.[48] Expanding federal authority over permitting is viewed as a means of accelerating the process. The underlying assumption is that it is indeed important to build transmission lines faster. For example, if national priorities include quickly putting low carbon generating plants on line to reduce greenhouse gas emissions and to speed the introduction of electric vehicles, then a rapid permitting process may be critical. But with other assumptions about the shape of the future power market, acceleration of the permitting process may be less pressing, and therefore expanding federal permitting authority less important. (An example of such an alternative assumption would be more reliance on large nuclear or coal plants built near existing transmission networks, rather than many small wind plants in remote locations.)

- *Management of the permitting process.* Whether FERC or another agency is assigned a federal permitting role, it will need the resources to expeditiously process applications. Otherwise the whole point of giving more permitting power to the federal government would largely be obviated.

Transmission Financing and Cost Allocation

Background

Between 1977 and 1998, real dollar investment in the transmission system by investor-owned utilities generally declined, from about $4 billion annually to a trough of $2.1 billion annually by 1997 and 1998 (constant 2000 dollars; see **Figure 5**, below). Although spending picked up to $4.2 billion by 2004 (constant 2000 dollars),[49] Congress was still sufficiently concerned about

---

[48] According to DOE's Electricity Advisory Committee, "A 'poster child' example of this problem is American Electric Power's Jacksons Ferry, Wyoming, 765 kV transmission line. It required 16 years to complete, and nearly 14 of those years and $50 million was spent on siting activities." The Electricity Advisory Committee, *Keeping the Lights on in a New World,* U.S. Department of Energy, Washington, DC, January 2009, p. 49, http://www.oe.energy.gov/eac.htm.

[49] This data is only for investor owned utilities (see **Figure 5** for source notes). No source appears to capture all investment in the transmission system. For a discussion of some of the related data issues, see Energy Security Analysis, Inc., *Meeting U.S. Transmission Needs,* Edison Electric Institute, Washington, DC, July 2005, p. vi and footnote 3. The reasons for the decline in real dollar transmission investment are unclear. While insufficient return on investment is cited as one factor, another issue may have been excess capacity on the transmission system that reduced the need for transmission spending. Another factor may have been new patterns in building power plants closer to load centers. See Steve Huntoon and Alexandra Metzner, "The Myth of the Transmission Deficit," *Public Utilities Fortnightly,* November 1, 2003, p. 30. The lack of comprehensive data on the transmission system makes it difficult to sort out these issues. For a discussion of transmission system data needs see Douglas R. Hale, *Electricity Transmission in a Restructured Industry: Data Needs for Public Policy Analysis,* Energy Information Administration, DOE/EIA-0639, Washington, DC, December 2004, http://www.eia.doe.gov/cneaf/electricity/page/transmission/ (continued...)

---

transmission investment to include construction incentives in EPACT05 (in the form of more profitable rates for projects that met certain criteria).[50]

Figure 5. Transmission Investment by Investor-Owned Utilities
1977 – 2007, in Millions of Dollars

Source: nominal dollar from the Edison Electric Institute website at http://www.eei.org/whatwedo/DataAnalysis/IndustryData/Documents/Transmission-Investment-Expenditures.pdf and http://www.eei.org/whatwedo/DataAnalysis/IndustryData/Pages/default.aspx. Values were converted to constant dollars by CRS using the implicit deflator for gross domestic product.

Some critics claim that FERC has awarded incentives to projects that did not need special rates.[51] Nonetheless and for whatever reason, transmission investment has continued to grow since 2004, reaching a 30-year high of $6.5 billion (constant 2000 dollars) in 2007.

More growth in annual investment may be needed. Estimates of the cost of expanding the transmission grid to increase renewable power delivery and other goals run into the tens of billions of dollars. For example (all figures in nominal dollars):

- The estimated transmission cost of the Joint Coordinated System Plan to bring Great Plains wind power to the East Coast range from $49 to $80 billion.[52]

(...continued)

DOE_EIA_0639.htm.

[50] 16 U.S.C. § 824s. The implementing rule authorizes FERC to award incentive rates to new transmission projects where the project will "either ensure reliability or reduce the cost of delivered power by reducing transmission congestion" and "there is a nexus between the incentive sought and the investment being made." Federal Energy Regulatory Commission, Order No. 697, *Promoting Transmission Investment through Pricing Reform*, Final Rule, July 20, 2006, pp. 207 – 208, http://www.ferc.gov/legal/maj-ord-reg.asp.

[51] Esther Whieldon, "FERC Grants Incentive Rates For Two Major Grid Projects Proposed In New England," *Platts Inside F.E.R.C.*, November 24, 2008; Will Harrington, "Consumer Groups Expect New FERC Will Trim Transmission Incentives," EnergyWashington.com, January 28, 2009.

- A DOE study of expanding the use of wind power estimated transmission expansion costs of $60 billion by 2030.[53]

- A study of transmission funding requirements for all purposes for the period 2010 to 2030 estimated total costs of about $300 billion.[54]

There are two major transmission financing policy issues: early financing for new projects, and how to allocate the costs of interstate projects to customers. Each issue is discussed below.

Early Financing

The early funding or "chicken and egg" problem particularly applies to renewable power. Renewable power plant developers may have difficulty getting funding because the transmission to bring their power output to market is not in place, while the transmission projects cannot get loans because the generation that would justify construction of the new lines has not been built. This early funding issue is exacerbated by the typical development pattern for many renewable energy projects. The projects are built in phases over several years.[55] However, it is not economic to build a transmission line in phases; the line must be built at once for the maximum anticipated capacity even if the full load will not be developed until years after the line is first put into operation.

The FERC, RTOs, and the states have been developing regulatory solutions for the early funding problem, but there is no standard or widely used approach.[56] The Western Governors' Association

---

(...continued)

[52] Executive summary to the Joint Coordinated System Plan 2008, p. 6, http://www.jcspstudy.org/. Note that the cost of the transmission is modest compared to the estimated cost of the generation needed to meet demand and, in one scenario, renewable energy goals ($674 billion to $1,050 billion).

[53] U.S. Department of Energy, *20% Wind Energy by 2030*, Washington, D.C., July 2008, p. 98, http://www1.eere.energy.gov/windandhydro/pdfs/41869.pdf.

[54] Marc Chupka et al., *Transforming America's Power Industry: The Investment Challenge 2010 - 2030*, prepared by the Brattle Group for The Edison Foundation, Washington, DC, November 2008, p. 40, http://www.eei.org/ourissues/finance/Documents/Transforming_Americas_Power_Industry.pdf.

[55] Renewable generation is built incrementally for a number of reasons, including the nature of the technology (which lends itself to incremental development), the small size and limited financial resources of some developers, and the way in which state renewable electricity standards ratchet up goals over time. For further discussion of this and related issues see California Independent System Operator, *Petition for Declaratory Order*, before the Federal Energy Regulatory Commission, Docket EL07-33-000, January 25, 2007, http://www.caiso.com/1b71/1b71d1263dad0.pdf.

[56] One approach involves changes to FERC's open transmission policies, which have required merchant transmission developers to bid-out all of the capacity of a proposed line (via an "open season" auction) before beginning construction. Developers had been precluded from making pre-auction capacity sales to a lead group of generators as a means of jump-starting the project. FERC's new ruling (involving two 500 kV DC transmission lines that would bring wind power from Montana and Wyoming to the southwest) allows large shares of a merchant project's capacity to be pre-sold to "anchor" customers. See Federal Energy Regulatory Commission, *Order Authorizing Proposals and Granting Waivers*, February 19, 2009, Docket Nos. ER09-432-000 and ER09-433-000. In this case the commission allowed the developers to pre-sell 50% of the capacity of each line prior to auctioning the balance of the capacity. At the state level, the California Independent System Operator (CAISO, the transmission authority covering most of the state) has implemented a unique financing policy for utilities developing new high voltage transmission lines that access renewable resource areas. The policy allow utilities to allocate across the utility's customer base the development costs that are not recovered from the first wave of new renewable plants that connect to the project. As more generators are connected each picks up its pro rata share of the development costs, until the full capacity of the line is subscribed and all costs are being paid by generators. Transmission projects that quality for this arrangement must have pre-sold about 25% to 35% of planned capacity in order to demonstrate commercial viability. The total cost (continued...)

has proposed that the federal government and the federal power marketing administrations[57] step in with direct funding and other incentives that will allow transmission developers to "supersize" planned lines to meet potential future generation, not just the renewable power expected to be built in the near term.[58]

Note that although the early funding issue and cost allocation issue (discussed immediately below) are currently viewed as largely problems for renewable energy development, they could also apply to new coal plants with carbon capture and sequestration (CCS) equipment. This is because one option for siting coal plants with CCS is to place them in remote locations where captured CO_2 can be stored or used for enhanced oil recovery. In this scenario, a long-term build-out of new coal capacity may face transmission funding issues similar to that of renewable development in remote areas.

Cost Allocation

Perhaps the most contentious transmission financing issue is cost allocation for new interstate transmission lines – that is, deciding which customers pay how much of the cost of building and operating a new transmission line that crosses several states. DOE's Electricity Advisory Committee concluded that "cost allocation is the single largest impediment to any transmission development."[59] The committee also noted that "cost allocation disagreements can also impact transmission siting; therefore, resolution of these two issues must be linked."[60] This is an important point, and most current transmission proposals fold the cost allocation issue into the transmission planning process. For example, S. 539 and the Senate Energy Majority Draft both require the regional planning authorities to submit cost allocation proposals along with their

---

(...continued)

exposure of ratepayers is limited by a cap formula. FERC ruled that this proposal was just and reasonable, and not unduly discriminatory, in part because it "advances state, regional and federal initiatives to encourage the development of renewable generation" (Federal Energy Regulatory Commission, *Order Granting Petition for Declaratory Order*, Docket EL-033-000, April 19, 2007, pp. 1 – 2). This decision and related statements made by FERC commissioners can be retrieved through FERC docket search at http://elibrary.ferc.gov/idmws/docket_search.asp. CAISO's initial request for FERC approval is also available at this site or at http://www.caiso.com/1b71/1b71d1263dad0.pdf.

[57] For more information on the power marketing administrations see CRS Report RS22564, *Power Marketing Administrations: Background and Current Issues*, by Richard J. Campbell.

[58] Letter from Jon Huntsman, Jr., Chairman and Governor of Utah, and Brian Schweitzer, Vice-Chairman and Governor of Montana, Western Governors' Association, to The Honorable Nancy Pelosi, Harry Reid, John Boehner, and Mitch McConnell, January 27, 2009, http://www.westgov.org/wga/testim/transmission-for-renewables1-27-09.pdf; and "Western Governors Eye Senate Bills to Push New Transmission Policy," *EnergyWashington.com*, March 20, 2009.

[59] The Electricity Advisory Committee, *Keeping the Lights on in a New World*, U.S. Department of Energy, Washington, DC, January 2009, p.50, http://www.oe.energy.gov/eac.htm.

[60] Ibid. An example of a cost allocation dispute is arguments over a proposed 150 mile transmission line to connect wind power in Maine to other parts of New England. According to an article on the project, "States had disagreed about fair allotment of costs for the transmission project. Maine utilities pushed for all states to pitch in because they would gain economic benefits – access to Maine's renewable energy. However, Connecticut and Massachusetts disagreed and said the region should socialize costs only when transmission ensures reliability." In another instance, the Illinois utility commission characterized the PJM Interconnection's proposed allocation of regional transmission costs to a state utility as "not only unjust and unreasonable, but patently irrational." The commission has protested the allocation to FERC. See, respectively, Lisa Wood, "Maine Regulators Reject 345-MW Line to Connect Wind Power to New England," *Platts Electric Utility Week*, February 9, 2009 and Jason Fordney, "Exelon, Illinois Commission and DP&L Protest PJM Allocation of Upgrade Costs," *Platts Electric Utility Week*, February 9, 2009.

---

transmission plans. If cost allocation proposals are not submitted or are rejected by FERC, then FERC can order its own cost-allocation scheme.

Another suggestion is to simply allocate the costs of new projects that are part of an interconnection-wide plan to all customers in the interconnection (sometimes referred to as "socializing" costs). For example, every ratepayer in the Eastern Interconnection would help pay for a line from Maine to New Hampshire. The idea is that in a synchronized grid all ratepayers benefit to some extent from all transmission system enhancements. A related concept is that new transmission for renewable power yields environmental benefits to all ratepayers. And whether explicit or implicit, the notion is also that interconnection-wide cost sharing makes transmission projects more palatable by minimizing the rate impact on any one group of customers, and accelerates project approvals by substituting a simple cost allocation rule for lengthy rate hearings.[61]

A criticism of interconnection-wide cost allocation is that cost responsibility arguably becomes more diffuse and the incentives for cost discipline decline. Another criticism is that especially favorable funding for transmission could bias policymakers and investors away from other solutions to electric market problems, such as demand response or local renewable power.[62] Other cost allocation approaches are being explored across the country but no approach is standard or even widely used.[63]

Financing and Cost Allocation: Summary of Policy Issues

Transmission financing issues for Congress include:

- *Should the federal government help pay for new transmission lines?* Some proposals call for the federal government, possibly acting through the federal utilities, to help pay for new transmission lines, pay for expanding projects to meet future needs, or actually build new transmission. How far should the federal government go into financing the expansion of the transmission grid?

---

[61] For example, the Energy Future Coalition argues that "Just as local electric ratepayers currently fund local electricity infrastructure investments, broad based groups of ratepayers should cover the costs of national grid investments which provide broad-based national benefits. This will ensure all beneficiaries of the National Clean Energy Smart Grid support the cost of its development.... Cost allocation policies should be as simple as possible (e.g., allocating designated costs proportionately to all load in the interconnection) to avoid lengthy regulatory proceedings and provide greater predictability for developers and ratepayers." Energy Future Coalition, *The National Clean Energy Smart Grid: An Economic, Environmental, and National Security Imperative*, undated, p. 4, http://www.energyfuturecoalition.org/files/webfmuploads/Smart%20Grid%20Docs/EFC%205-page%20Vision%20Statement%20-%20FINAL.pdf.

[62] U.S. Congress, Senate Committee on Energy and Natural Resources, *To Receive Testimony On Pending Legislation Regarding Electricity Transmission Lines*, Prepared Testimony of James A. Dickenson on Behalf of the Large Public Power Council, 111th Cong., 1st sess., March 12, 2009, pp., 8 – 10, http://energy.senate.gov/public/index.cfm?FuseAction=Hearings.Testimony&Hearing_ID=b9e47ea9-c62b-23fc-33ff-30fda7b3a744&Witness_ID=ed6a79eb-6664-412c-b66b-7e87e4e55435.

[63] For example, the Southwest Power Pool (SPP) RTO has adopted a policy of developing cost allocation plans for packages of transmission projects that cover all the zones in the RTO. This way, all ratepayers benefit from the package even if individual projects in the package cover a small geographical area. The SPP approach has been approved and praised by FERC, but may only be applicable to RTOs like SPP that cover a relatively small area (see **Figure 4**). Alan Kovski, "FERC Approves SPP's Plan to Handle Grid Upgrade Costs on an Economic Basis," *Platts Global Power Report*, October 23, 2008.

- *Should the Congress establish a national cost allocation rule for new transmission projects?* In order to expedite transmission development, the federal government may need to implement standard, generally applicable cost allocation methodologies. An approach included in several proposals would require all ratepayers in an interconnection to pay for new projects anywhere in the interconnection. The notion is that in an interconnected system all customers benefit to some degree from enhancements to the grid, but a preferential cost allocation mechanism for transmission may bias investment away from other alternatives.

Transmission System Modernization and the Smart Grid

Background

Distinct from proposals for expanding the grid are proposals for modernizing the transmission system. Modernization proposals are often made under the rubric of the "smart grid," a term that encompasses technologies that range from advanced meters in homes to advanced software in transmission control centers. There is no standard definition of the smart grid.[64] For the purposes of this report, the smart grid can be viewed as a suite of technologies that give the grid the characteristics of a computer network, in which information and control flows between and is shared by individual customers and utility control centers. The technologies will allow customers and the utility to better manage electricity demand, and will include self-monitoring and automatic protection schemes to improve the reliability of the system.[65] Although grid technology has not been static over the years,[66] the smart grid concept would implement capabilities well beyond any existing electric power system.

The smart grid primarily involves the development of software and small-scale technology (e.g., smart meters for homes and businesses that would interface with grid controls) rather than construction of new transmission lines. However, full implementation of the smart grid also requires new electricity rate structures, especially for residential customers, and as discussed below, this and other aspects of the smart grid may prove contentious.

The following discussion is divided into three sections:

- A more detailed description of smart grid functions.

---

[64] DOE's Electricity Advisory Committee noted that "there are many working definitions of a Smart Grid." Electricity Advisory Committee, *Smart Grid: Enabler of the New Economy*, U.S. Department of Energy, Washington, DC, December 2008, p. 1.

[65] Other descriptions of the smart grid emphasize its environmental benefits through reducing fossil-fueled electric generation and air pollution emissions. See the comments of FERC Commissioners Moeller and Spitzer in Federal Energy Regulatory Commission, "FERC Accelerates Smart Grid Development with Proposed Policy, Action Plan," press release, March 19, 2009, http://www.ferc.gov/news/news-releases/2009/2009-1/03-19-09.asp.

[66] Scott Gawlicki, "Demonstrating the Smart Grid," *Public Utilities Fortnightly*, June 2008, p. 51; and Kenneth Martin and James Carroll, "Phasing in the Technology: Phasor Measurement Devices and Systems for Wide-Areas Monitoring," *IEEE Power and Energy*, September/October 2008.

- A summary of current federal support for the smart grid.

- Smart grid cost and rate issues.

Smart Grid Functions

Because the smart grid involves integrated operation of the power system from the home to the power plant, this discussion will go beyond the transmission system to cover the distribution network. Within this integrated system the smart grid has two scopes. *One scope is transmission monitoring and reliability*, and includes the following capabilities:

- Real-time monitoring of grid conditions;

- Improved automated diagnosis of grid disturbances, and better aids for the operators who must respond to grid problems;

- Automated responses to grid failures that will isolate disturbed zones and prevent or limit cascading blackouts that can spread over wide areas.

- "Plug and play" ability to connect new generating plants to the grid, reducing the need for time consuming interconnection studies and physical upgrades to the grid.

- Enhanced ability to manage large amounts of wind and solar power. Some (though not all) analysts believe deployment of the smart grid is essential to the large scale use of wind and solar energy.[67]

The second scope is consumer energy management. An essential part of this scope is the installation of smart meters (also referred to as advanced metering infrastructure or AMI). These meters and other technology would implement the following capabilities:

- At a minimum, the ability to signal homeowners and businesses that power is expensive and/or in tight supply. This can be done, for instance, via special indicators or displayed through web browsers or other personal computer software. The expectation is that the customer will respond by reducing its power demand.

- The next level of implementation would allow the utility to automatically reduce the customer's electricity consumption when power is expensive or scarce. This would be managed through links between the smart meter and the customer's equipment or appliances.

---

[67] This issue is different from constructing new power lines to reach renewable energy production zones. Because the output of wind and solar plants varies with the weather and time of day, integrating large amounts of these variable resources into the power system is challenging. The smart grid, with its theoretical ability to monitor and balance load, generation, and power storage across the whole electricity network – from the batteries in plug-in hybrid vehicles in a homeowner's garage to the dispatch of power plants – is sometimes viewed as the solution to these integration challenges. For example, see David Talbot, "Lifeline for Renewable Power," *Technology Review*, January/February 2009, http://www.technologyreview.com/printer_friendly_article.aspx?id=21747&channel=energy&section=. The article's summary states that "Without a radically expanded and smarter electrical grid, wind and solar will remain niche power sources." However, other studies do not see the smart grid as a prerequisite to large scale introduction of renewable power. For example, see U.S. Department of Energy, *20% Wind Energy by 2030*, Washington, D.C., July 2008, http://www1.eere.energy.gov/windandhydro/pdfs/41869.pdf (the term "smart grid" does not appear in the report).

Electric Power Transmission: Background and Policy Issues

- The smart grid system would automatically detect distribution line failures, identity the specific failed equipment, and help determine the optimal plan for dispatching repair crews to restore service. The smart grid would automatically attempt to isolate failures and prevent local blackouts from spreading.

- The smart grid would make it easier to install distributed generation, such as rooftop solar panels, and to implement "net metering," a ratemaking approach that allows operators of distributed generators to sell surplus power to utilities. The smart grid would also manage the connection of millions of plug-in hybrid electric vehicles into the power system.

The transmission and customer energy management scopes described above are integrated in the full smart grid concept. For example, if the transmission system becomes overloaded, the smart grid could respond at the distribution system level by automatically reducing customer demand.

Federal Support for the Smart Grid

The Energy Independence and Security Act of 2007 (EISA) articulated a national policy to modernize the power system with smart grid technology, and authorized research and development programs, funding for demonstration projects, and matching funds for investments in smart grid technologies. [68] These and related programs received $4.5 billion in funding in the 2009 stimulus bill. [69] In addition, the Emergency Economic Stabilization Act of 2008 shortens the depreciation period for smart meters and other smart grid equipment from 20 years to 10 years (which increases each year's depreciation tax deduction for the equipment). The value of this tax change to the power industry is reportedly $915 million over 10 years. [70]

EISA assigned to the National Institute of Standards and Technology (NIST), a unit of the Department of Commerce, the lead in developing interoperability standards for smart grid equipment. [71] This is a critical role, because it is essential that the smart grid technologies installed by one utility be able to communicate with those of another and with control centers. This work has been lagging. DOE, FERC, and NIST have reportedly begun interagency efforts to accelerate development of the standards, and NIST has created and filled a new National Coordinator on Smart Grid Interoperability to push the effort forward. [72]

---

[68] 42 U.S.C. § 17381, et seq.

[69] For additional information see CRS Report R40412, *Energy Provisions in the American Recovery and Reinvestment Act of 2009 (P.L. 111-5)*, coordinated by Fred Sissine.

[70] The Electric Advisory Committee, *Smart Grid: Enabler of the New Energy Economy*, U.S. Department of Energy, December 2008, p. 16, http://www.oe.energy.gov/eac.htm.

[71] 42 U.S.C. § 17385.

[72] "Stakeholders Look To Jump-Start Stalled Smart Grid Standards," EnergyWashington.com, January 20, 2009; John Siciliano, "Administration Pursuing Major Interagency Plan to Deploy Smart Grid," EnergyWashington.com, March 25, 2009. According to this article, during March 17, 2009 testimony before the House Science and Technology Committee, Secretary of Energy Chu expressed his displeasure with the lack of progress. The NIST smart grid site is located at http://www.nist.gov/smartgrid/. For a discussion of interoperability issues, see Federal Energy Regulatory Commission, *Proposed Policy Statement and Action Plan*, Smart Grid Policy, Docket PL09-4-000, March 19, 2009, http://www.ferc.gov/whats-new/comm-meet/2009/031909/E-22.pdf.

Pursuant to EISA, once NIST's work is sufficiently advanced FERC is to establish, through a rulemaking, national smart grid interoperability standards.[73] On March 19, 2009, FERC published for comment a proposed smart grid policy statement and action plan, intended "to articulate its policies and near-term priorities to help achieve the modernization of the Nation's electric transmission system, one aspect of which is 'Smart Grid' development."[74] According to FERC, the statement focuses on "Prioritizing the development of key standards for interoperability of Smart Grid devices and systems; [and] a proposed rate policy for the interim before the standards are developed."[75] Comments are due back to FERC in May 2009.

Smart Grid Cost and Rate Issues

Advocates believe the potential benefits from the smart grid are enormous. For example, the Electric Power Research Institute, a research arm of the power industry, estimated that implementation of the smart grid and related technologies could increase annual gross domestic product by 10% annually by 2020.[76] The Galvin Institute, a proponent of grid modernization, claims that among other benefits a modernized grid would "reduce the need for massive [electric power] infrastructure investments by between $46 and $117 billion over the next 20 years."[77]

Nonetheless, because the smart grid concept and technology are still evolving and there are no operational systems to evaluate, the benefits and costs are uncertain. According to Xcel Energy, which is developing a large smart grid demonstration in Boulder, Colorado:[78]

> Everybody says they have technology that can be applied to this project. How much really exists and how much of it still needs to be developed? Right now we think 60 percent of the data architecture is already there, while the other 40 percent will probably need tweaking. Then we will determine what is or isn't scalable [to larger installations].... As an industry we haven't really demonstrated the benefit of combining all these technologies. Until we do, there will be skepticism. That's the real value of this project.[79]

It does seem likely that costs of rolling out the smart grid will be high. Just installing the metering equipment is expensive. Pacific Gas and Electric, a large utility in California, plans to install 10.3 million smart meters by 2012 at a cost of $1.7 billion.[80] Estimates of installing smart meters

---

[73] 42 U.S.C. § 17385(d)

[74] Federal Energy Regulatory Commission, *Proposed Policy Statement and Action Plan*, Smart Grid Policy, Docket PL09-4-000, March 19, 2009, p. 1, http://www.ferc.gov/whats-new/comm-meet/2009/031909/E-22.pdf.

[75] Federal Energy Regulatory Commission, "Proposed Smart Grid Policy Statement and Action Plan," fact sheet, March 19, 2009, http://www.ferc.gov/news/news-releases/2009/2009-1/03-19-09-E-22-factsheet.pdf.

[76] Electric Power Research Institute, *Electricity Sector Framework For The Future, Volume I: Achieving The 21st Century Transformation*, August 6, 2003, p. 42, Table 5-1, http://positiveenergydirections.com/ESFF_volume1.pdf. The estimated gains through an improved grid are described in the report as achievable stretch goals (p. 41).

[77] See the Galvin Institute website at http://www.galvinpower.org/resources/galvin.php?id=27.

[78] For information on the Boulder project see Xcel Energy website at http://smartgridcity.xcelenergy.com/ and Stephanie Simon, "The More You Know ... " The Wall Street Journal, February 9, 2009. For information on other demonstration projects, see Scott Gawlicki, "Demonstrating the Smart Grid," Public Utilities Fortnightly, June 2008 and Peter Slevin and Steven Mufson, "Stimulus Dollars Energize Efforts To Smarten Up the Electric Power Grid," The Washington Post, March 10, 2009. For an example of an overseas project see Todd Woody, "IBM to Build World's First National Smart Utility Grid [in Malta]," Green Wombat Blog – Fortune on CNNMoney.com, February 4, 2009, http://greenwombat.blogs.fortune.cnn.com/.

[79] Scott Gawlicki, "Demonstrating the Smart Grid," *Public Utilities Fortnightly*, June 2008, pp. 56 – 57.

[80] Lisa Weinzimer, "PG&E's Advanced Meter Upgrade Would Cost Ratepayers $900 million," *Platts Electric Utility* (continued...)

nationwide are in the $40 billion to $50 billion range.[81] Some utilities are incurring costs to replace smart meters installed just a few years ago with newer models, indicating both the rapidity with which the technology is changing and the absence of firm standards.[82]

Some consumer advocacy groups have expressed concern that utilities and regulators are pressing ahead with smart grid investments, especially the installation of smart meters, without knowing whether the benefits will justify the costs. A claim from critics is that some utilities are enthusiastic about immediate spending on smart grid technology because once the investment is reflected in the company's rate base it will result in higher profits.[83]

Another consumer advocate concern relates to the change in utility rate structures that will likely accompany implementation of the smart grid. As discussed above, one function of the smart grid is to signal consumers when electricity is expensive or in short supply. The question is whether the consumer will act on this information by reducing power usage. In typical utility rate structures, consumers pay a rate for power that reflects annual average costs. The consumer's rate does not vary from day to day or hour to hour. But if the consumer's rates do not reflect real-time power costs, then the consumer has no immediate economic incentive to respond to utility price signals. For this reason, the smart grid concept is accompanied by new rate structures, such as "dynamic" pricing in which charges to consumers reflect actual market prices (or marginal production costs) for electricity. As put by the President of NARUC, "You can't have a smart grid and dumb rates. We have been used to – for over 100 years – rates that are the same all day, every day. That's not the way electricity is produced."[84]

Dynamic rates mean that the price of power would be much higher in the afternoon of a hot summer day when demand peaks and the most expensive generating plants are on-line, than in evening of the same day or on the weekend. With dynamic rates, consumers would have an incentive to respond to utility price signals by reducing demand by turning down the air conditioner or delaying the laundry. If the capability exists, the consumer might sign-up for direct utility control of appliances.

In theory, this demand response scenario has consumer benefits in the short-term (less use of expensive fuels and inefficient peaking power plants) and long-term (less need for new power plants to meet growth in peak load and reduced air emissions). However, in the view of critics these benefits are much more nebulous than the certainty that under dynamic rates consumers will

(...continued)

Week, December 17, 2007.

[81] Ahmad Faruqui and Sanem Sergici, Household Response to Dynamic Pricing: A Survey of the Experimental Evidence, The Brattle Group, January 10, 2009, p. 6, http://www.hks.harvard.edu/hepg/Papers/2009/The%20Power%20of%20Experimentation%20_01-11-09_.pdf, and Katie Fehrenbacher, "Even With Stimulus, Smart Grid Could Face Rough Year," Earth2Tech, February 6, 2009, http://earth2tech.com/2009/02/06/even-with-stimulus-smart-grid-could-face-rough-year/#more-22357 (citing comments by an analyst with the Edison Electric Institute).

[82] Lisa Weinzimer, "PG&E's Advanced Meter Upgrade Would Cost Ratepayers $900 million," *Platts Electric Utility Week*, December 17, 2007, and Tom Tiernan, "Utilities Sometimes In Middle As Enthusiasm, Wariness Circle Each Other In Smart Grid Push ," *Platts Electric Utility Week*, November 3, 2008.

[83] Under traditional regulation, which still applies throughout the country to distribution system rates, investor-owned utilities earn a return on their invested capital. This means that if they make a PUC-approved investment in smart meters (which, as discussed above, for a large utility can exceed a billion dollars), other things being equal the company's profits will increase proportionally to the size of the investment.

[84] Frederick Butler, President of NARUC, quoted in Daniel Vock, "Smart Grid's Growth Now Depends On States," Stateline.org, March 17, 2009, http://www.stateline.org/live/details/story?contentId=384804.

face higher power costs. The critics also argue that lower income people may not have the schedule flexibility to shift cooking and laundry to less expensive hours of the day; however, there is some evidence that lower income people will actually be more responsive to price signals than higher-income households.[85] Another argument is that the elderly or ill may face the choice of paying higher power bills or risking their health by turning down the air conditioning or electric heat.[86]

It is also unclear how much smart grid technology and cost needs to be incurred to get most of the available demand response benefits. For example, a dynamic pricing pilot program in Chicago used minimal technology (e.g., price notifications by phone) but still produced substantial reductions in peak demand.[87] Some studies suggest that the more sophisticated the technology used in a demand response pilot program the greater the savings,[88] but the optimal balance between technology cost and benefits is still unclear. Industrial customers will reportedly recommend adding a cost-benefit test for smart grid investments to FERC's final policy.[89]

Modernization and Smart Grid: Summary of Policy Issues

Congress has already put in place federal programs to help develop the smart grid. Continuing policy issues for Congress include:

- *Program oversight.* The American Recovery and Reinvestment Act provided funding for previously authorized smart grid programs, including one key effort – development of interoperability standards by the National Institute for Standards and Technology – that has been lagging. Congress may want to monitor how these programs progress.

- *Smart grid cost/benefit oversight.* The balance of costs and benefits that the smart grid will produce for customers has been hotly debated. Many smart grid investment decisions will be made by state utility commissions. However, other investments and rate decisions will involve transmission systems and RTOs under FERC jurisdiction, or will relate to bulk power system reliability standards that are under federal jurisdiction throughout the 48 contiguous states. (This federal role will be even larger if an interconnection-wide planning process under federal supervision is made into law, because these plans will inevitably have to deal with grid modernization.) These responsibilities create ample room for

---

[85] Summit Blue Consulting, *Evaluation of the 2006 Energy-Smart Pricing Plan*, CNT Energy, Boulder, CO, November 2007, p. 9, http://www.cntenergy.org/reports.php.

[86] For a discussion of smart grid consumer issues and responses, see Tom Tiernan, "Utilities Sometimes in the Middle as Enthusiasm, Wariness Circle Each Other in Smart Grid Push," *Platts Electric Utility Week*, November 3, 2008.

[87] Summit Blue Consulting, *Evaluation of the 2006 Energy-Smart Pricing Plan*, CNT Energy, Boulder, CO, November 2007, pp. 4 and 11, http://www.cntenergy.org/reports.php.

[88] Ahmad Faruqui and Sanem Sergici, Household Response to Dynamic Pricing: A Survey of the Experimental Evidence, The Brattle Group, January 10, 2009, pp. 43 (Table 31) and 46, http://www.hks.harvard.edu/hepg/Papers/2009/The%20Power%20of%20Experimentation%20_01-11-09_.pdf.

[89] "FERC 'Single Issue Rate Cases' Smart Grid Policy Draws Consumer Fire," EnergyWashington.com, March 23, 2009. According to this article, some industrial power users are also alarmed by an element of FERC's smart grid policy proposal that would allow utilities to request rate increases to recover smart grid costs in isolation from all other transmission expenses. The critics object that other transmission expenses might have decreased, but these costs would not be examined in a "single issue" smart grid rate case at FERC.

Congressional oversight of the actual costs, benefits, and performance of smart grid investments.

Transmission System Reliability

This section of the report will discuss the reliability of the transmission system from three perspectives:

- Problems in evaluating the current reliability condition of the grid;
- Modernization and reliability;
- Reliability and changes in the energy market.

Problems in Evaluating the Current Reliability Condition of the Grid

As discussed earlier, power system reliability has two dimensions: adequate capacity to consistently meet customer demands, and the ability to withstand disturbances such as failed transmission lines or power plants. It is currently impossible to judge the reliability of the national transmission system by either criteria because the data does not exist to make an assessment. According to the Energy Information Administration, "The Government does not have the [analytical tools] and data necessary to verify that existing and planned transmission capability is adequate to keep the lights on."[90]

This is not to say that transmission risks cannot be evaluated for specific parts of the transmission grid. These studies are performed routinely.[91] What is missing is uniform, nationwide data on the frequency and causes of transmission outages that can be used to determine whether the overall performance of the system is improving or deteriorating, and what factors are driving these changes.

A contrast can be drawn between the data available on generating plant reliability and operations versus that for the transmission system. For decades NERC has managed a highly detailed collection of data on the reliability of power plants, and other relevant data are available from EIA and the Environmental Protection Agency (EPA).[92] In contrast to the wealth of information

---

[90] Douglas R. Hale, *Electricity Transmission in a Restructured Industry: Data Needs for Public Policy Analysis*, Energy Information Administration, DOE/EIA-0639, Washington, DC, December 2004, p. 4, http://www.eia.doe.gov/cneaf/electricity/page/transmission/DOE_EIA_0639.htm. Also see U.S.-Canada Power System Outage Task Force, Final Report on the August 14, 2003 Blackout in the United States and Canada: Causes and Recommendations, April 2004, pp. 147 – 148 (Recommendation I.F.10), https://reports.energy.gov/BlackoutFinal-Web.pdf.

[91] For example, DOE was able to conclude that transmission congestion is a "serious threat to the reliability of electricity supply" to southern California. Department of Energy, "National Electric Transmission Congestion Report," 72 *Federal Register* 57016, October 5, 2007.

[92] Information on NERC's Generating Availability Data System (GADS) is available from the NERC website at http://www.nerc.com/page.php?cid=4|43. EIA collects data on power plant monthly operations and plant characteristics, available through the agency's website at http://www.eia.doe.gov/fuelelectric.html or by calling the National Energy Information Center at 202-586-8800. EPA collects power plant data as part of its air emissions compliance programs. For more information see the EPA website at http://camddataandmaps.epa.gov/gdm/ and http://www.epa.gov/cleanenergy/energy-resources/egrid/index.html.

---

on power plant operations, minimal data has been collected by government or industry on transmission system reliability. The most significant existing source is information on major transmission outages collected on DOE's Form OE-417, which is compiled by EIA and NERC.[93] A recent Carnegie Mellon University study of this data was able to conclude "that the frequency of large blackouts in the United States has not decreased over time," but could not determine why this is because of the lack of detailed information.[94]

This information gap leaves policy makers without a full understanding of transmission reliability risks or able to determine the best steps for improving reliability.[95] To help fill this gap, NERC has launched a new Transmission Availability Data System (TADS) to provide the data "needed to support decisions with respect to improving reliability and performance."[96] TADS reporting, which began in 2008, is mandatory for all high voltage transmission owners in the 48 contiguous states. NERC is still developing metrics to display and analyze the data in a meaningful way, and believes it may take up to five years before the data can be used to analyze trends."[97]

It may also take several years to judge whether TADS is collecting all the necessary data or if it needs to be revised or expanded.[98] Pursuant to EPACT05, NERC and FERC have been promulgating and enforcing new, mandatory, power system reliability standards.[99] Until a useful

---

[93] Information on the OE-417 form is available on the DOE website at http://www.eia.doe.gov/cneaf/electricity/page/forms.html. Major power system disruptions are listed in EIA's Electric Power Monthly, Appendix B (http://www.eia.doe.gov/cneaf/electricity/epm/epm_sum.html). EIA also collects transmission data on its EIA-411 form, but in the view of NERC this information is not useful for reliability analyses. Letter from David Nevius, Senior Vice President, North American Electric Reliability Corp., to OMB Desk Officer for DOE, Office of Management and Budget, "NERC Comments on EIA-411," October 24, 2007.

[94] Paul Hines, Jay Apt, and Sarosh Talukdar, "Large Blackouts in North America: Historical Trends and Policy Implications," Carnegie Mellon Electricity Industry Center, Working Paper CEIC-09-01, March 4, 2009, p. 28, http://wpweb2.tepper.cmu.edu/ceic/PDFS/CEIC_09_01_blt.pdf.

[95] In the absence of solid reliability measures, data intended for other purposes are sometimes used as indicators of transmission system reliability. An example is counts of transmission load relief (TLR) requests on a power system. TLRs are used in parts of the Eastern Interconnection to reallocate and sometimes curtail transmission service when power lines are congested. TLR requests have been growing, which is sometimes cited as an indicator of increasing stress on the transmission grid (for example see Eric Hirst, *U.S. Transmission Capacity: Present Status and Future Prospects*, Edison Electric Institute and U.S. Department of Energy, Washington, DC, June 2004, pp. 7-8. http://www.oe.energy.gov/DocumentsandMedia/transmission_capacity.pdf). However, TLRs are used for economic as well as reliability reasons, and part of the increase in TLRs is an artifact of procedural changes by the Southwest Power Pool. For more information see North American Electric Reliability Corp., *2008 Long-Term Reliability Assessment*, October 2008, pp. 58 – 61, http://www.nerc.com/files/LTRA2008.pdf and Steve Huntoon and Alexandra Metzner, "The Myth of the Transmission Deficit," *Public Utilities Fortnightly*, November 1, 2003, p. 31 (text box). Another example is calls on customers who have signed up for demand response programs to reduce load when power supplies are tight or transmission lines are overloaded. NERC regions may record use of demand response as reliability problem events even if it is a routine use of demand control tools; see North American Electric Reliability Corp., *2008 Long-Term Reliability Assessment*, October 2008, p. 57, http://www.nerc.com/files/LTRA2008.pdf.

[96] Transmission Availability Data System Task Force, *Transmission Availability Data System Revised Final Report*, North American Electric Reliability Corp., September 26, 2007, p. 1, http://www.nerc.com/filez/tadstf.html. Detailed information on TADS is available at this website and a brief summary is at http://www.nerc.com/page.php?cid=4|62.

[97] Ibid., p. 13. At this time the detailed TADS data will be proprietary to NERC and not released to EIA. Personal communication with Robert Schnapp, Energy Information Administration, March 26, 2009. As noted above, NERC is working on how to report the aggregated data.

[98] EIA has already suggested that the TADS data coverage may be incomplete. E-mail from Robert Schnapp, Energy Information Administration, to David Nevius, North American Electric Reliability Corp. "Phase II TADS Request for Comments," June 16, 2008.

[99] For information on the FERC and NERC reliability activities, see the FERC website at http://www.ferc.gov/industries/electric/indus-act/reliability.asp and the NERC website generally (http://www.nerc.com/).

data collection and analysis system are in place, it will be difficult to judge whether these standards and other actions are actually improving the reliability of the transmission system.

Reliability and Grid Modernization

The transmission grid is sometimes portrayed as a decrepit victim of underinvestment; one recent press report described the grid as "frayed" like grandmother's quilt.[100] There is, in fact, no clear evidence that the transmission grid is physically deteriorating. But this does not mean that the grid is universally well managed or is as up-to-date as it should be. The grid probably needs to be modernized to improve reliability.[101] This is not necessarily the same as installing the full smart grid discussed above. The smart grid is an ambitious concept for integrated operation of the power system. The full smart grid is not needed to use a subset of "intelligent" technologies to improve the reliability of the transmission system.

The need for modernization is illustrated by the causes of the August 14, 2003 northeastern blackout. The blackout, which interrupted service to 50 million people in the United States and Canada for up to a week, started with transmission line trips (automatic shutdowns) and resulting overloads on the FirstEnergy utility system in Ohio. The blackout was not the result of insufficient transmission capacity or deteriorated equipment. As identified by the joint United States – Canada investigating task force, the blackout was caused by factors such as the following:[102]

- FirstEnergy and the NERC reliability region within which it operated did not understand the strengths and weaknesses of the FE system. FirstEnergy consequently operated its system at dangerously low voltages.[103]

- FirstEnergy's system operators lacked the "situational awareness" that would have revealed the blackout risk as lines began to trip. The operators were blinded by monitoring and computer system breakdowns, combined with training and procedural deficiencies which led to those failures going undetected until it was too late.[104]

---

[100] Peter Slevin and Steven Mufson, "Stimulus Dollars Energize Efforts to Smarten Up the Electric Power Grid," *The Washington Post*, March 10, 2009.

[101] Beyond the scope of this report is the issue of "cybersecurity" (i.e., steps taken to prevent malicious acts that would compromise the electronic or physical security perimeter of a critical cyber asset. In this context, a "critical cyber asset" includes the electronic elements of facilities, systems, and equipment which, if destroyed, degraded, or otherwise rendered unavailable, would affect the reliability or operability of the electric power system). The importance of this issue was emphasized in April 2009 by press reports of apparently hostile penetrations of electric power industry computer systems (Siobhan Gorman, "Electricity Grid in U.S. Penetrated by Spies," *The Wall Street Journal*, April 8, 2009).

[102] The following points list some of the key factors that contributed to the collapse of the First Energy system and the consequent cascading blackout. For a full analysis of this complex event see U.S.-Canada Power System Outage Task Force, *Final Report on the August 14, 2003 Blackout in the United States and Canada: Causes and Recommendations*, April 2004, https://reports.energy.gov/BlackoutFinal-Web.pdf. Perhaps the best brief description of the causes of the blackout is the "Voltage Collapse" text box on page 81.

[103] Ibid., p. 33.

[104] "Transcripts of telephone conversations, released by the House Energy Committee, show bewilderment after the first control room computer went down. 'We have no clue,' one operator said. Another, speaking to a regional controller at MISO just before the blackout, said, 'We don't even know the status of some of the stuff around us.'" Ralph G Loretta and James E Anderson, "The Near Term Fix," *Public Utilities Fortnightly*, November 1, 2003, p. 34. (continued...)

- FirstEnergy did not adequately trim the trees under its transmission lines. As a result, three key transmission lines tripped when they sagged (as the lines are designed to do as they heat up with use) and came in contact with trees.[105]

- The Midwest Independent System Operator (MISO), the RTO that manages the grid in FirstEnergy's service area, did not have the real-time information necessary to assess the situation on FirstEnergy system and provide direction to the utility.[106]

Once the FirstEnergy system collapsed, overloads and power swings spread out across the Northeast, causing a cascading series of transmission line and power plant trips that left tens of millions of people without electricity. One reason the outage spread over such a wide area was because many power plants were equipped with unnecessarily sensitive automatic protection mechanisms that tripped the units prematurely.[107] The speed of the cascade allowed almost no time for manual intervention. The elapsed time from the start of the cascade (i.e., when failures began to radiate out from the collapsed FirstEnergy grid) to its full extent was about seven minutes.[108]

In summary, as discussed in the official blackout report and other analyses, the 2003 blackout was not caused by a utility having built too few transmission lines, or because power line towers and substations were falling apart. The blackout was apparently due to such factors as malfunctioning if not obsolete computer and monitoring systems, human errors that compounded the equipment failures, mis-calibrated automatic protection systems on power plants, and FirstEnergy's failure to adequately trim trees.

One part of a strategy for preventing repetitions of the 2003 blackout is to modernize the grid from a reliability standpoint. This will not always entail building more power lines. One analysis written shortly after the 2003 blackout concluded that "The common contributing factor to the recent blackout, based on investigations to date, is confusion-communication breakdowns both technical and human….[W]e maintain that much can be solved by updating technology and by changing procedures followed within the operating companies. This fix is cheaper and much more immediate than huge investment in new power lines."[109]

Modernization involves installing new technology into the existing system so that:

- Operators have accurate real time data on the status of the power network.

(…continued)

The blackout report notes that FirstEnergy had no automatic load-shedding schemes in place, and did not attempt to begin manual load-shedding. U.S.-Canada Power System Outage Task Force, Final Report on the August 14, 2003 Blackout in the United States and Canada: Causes and Recommendations, April 2004, p. 70, https://reports.energy.gov/BlackoutFinal-Web.pdf.

[105] U.S.-Canada Power System Outage Task Force, *Final Report on the August 14, 2003 Blackout in the United States and Canada: Causes and Recommendations*, April 2004, pp. 18 and 57, https://reports.energy.gov/BlackoutFinal-Web.pdf.

[106] Ibid., p. 19, 46-49, and 55.

[107] Ibid., p. 94.

[108] The "full cascade" started at 4:05:57 pm and reached its maximum extent by 4:13 pm. Ibid., pp. 77 and 82.

[109] Ralph G. Loretta and James E. Anderson, "The Near-Term Fix," *Public Utilities Fortnightly*, November 2003, p. 34.

- Operators also have advanced simulation tools to assist them in evaluating incipient problems and formulating responses.

- The grid can automatically respond to certain types of problems. This is sometimes referred to as the "self-healing" grid.

Some of these technologies are being implemented. An example is two new control centers installed by the Western Electricity Coordinating Council (WECC), the NERC reliability region covering the western states. According to WECC:

> These centers have a view of the entire Western Interconnection. They can see every tower, line, and transmission element over 100 kV. They will be able to see the entire Western bulk system, identify its status, and respond to outages.... they have the tools now to see and head off problems as they develop and they have the authority to contact grid operators and direct them to take certain actions to protect the interconnection as a whole.[110]

On the other hand, the control centers will not be able to remotely actuate equipment such as transmission line circuit breakers. As is typically the case, a crew will still need to be sent to manually reset the equipment, so the control system is still several steps away from automated, "self-healing" responses to grid problems.

In summary, depending on the case, building new transmission lines is not the only or best approach to enhancing power system reliability.[111] In some instances investments in new monitoring and control technology may be the better solution.

Reliability and Changes in the Energy Market

The transmission grid was built for a specific business and technical model: power plants would use transmission lines to move electricity to distribution networks for delivery to customers. The power plants were large "central station" facilities using fossil, nuclear, or hydroelectric energy sources, and were designed to run as-needed, when-needed. The power flow was one-way, from the power plant to the customer.

This model is already changing:

- *Variable Renewable Generation*: One factor is the introduction of large amounts of wind power onto the grid. Unlike conventional power plants, the output of wind plants varies with the weather. Power systems were not designed to handle this kind of power supply variability and uncertainty. Total wind capacity is now

---

[110] Daniel Guido, "WECC's Two New Reliability Centers Replace Three Operations; Interconnection Now Is One," *Platts Electric Utility Week*, January 12, 2009, pp. 24 – 25. Another example is installation across the grid of phasor measurement units (PMU), a technology that provides system operators with real time data on wide areas of the power system. This is a new technology which will reportedly take five or more years to reach its full potential for enhancing system reliability. Saikat Chakrabarti et al., "Measurements Get Together," *IEEE Power and Energy*, January/February 2009, pp. 42-43.

[111] The Carnegie Mellon study cited earlier observes that "While transmission investment can, but is not guaranteed to, have a positive impact on cascading failure risk and reliability, transmission construction alone is a costly, and potentially ineffective, solution to reliability problems." Paul Hines, Jay Apt, and Sarosh Talukdar, "Large Blackouts in North America: Historical Trends and Policy Implications," Carnegie Mellon Electricity Industry Center, Working Paper CEIC-09-01, March 4, 2009, p. 29, http://wpweb2.tepper.cmu.edu/ceic/PDFS/CEIC_09_01_blt.pdf.

large enough in some parts of the country, such as the ERCOT Interconnection (covering most of Texas), to be an important influence on how the power system is operated.

The variable output of wind plants can be dealt with in a variety of ways, including improved wind forecasting, adding electricity storage and/or quick start natural gas-fired peaking plants to the grid, and drawing wind power from a wide geographic area to smooth out local changes in wind speed. However, these capabilities will have to be added rapidly to the grid if, as some expect, the use of wind power grows quickly.

- *Demand Response*: Another factor is the increasing use of demand response programs, in which large commercial and industrial customers agree to interruptible power service in return for lower rates. For example, in the Florida and northeastern NERC reliability regions, significant parts of peak demand (respectively, 6% and 4%) can now be met by customers reducing output rather than by operating power plants.

  Demand response reverses the conventional power system operating model: instead of changing power plant output to match demand, demand is reduced to match the available supply of electricity. An issue is how much real time information and control (also referred to as "visibility") system operators will have over industrial and commercial facilities that have signed on to demand response programs. Another issue is whether industrial and commercial loads will become less willing to participate in demand response programs if cycling of their operations becomes routine rather than a rarity. These issue are clearly not insuperable, given the success to date with these programs, but they may have to be dealt with on a much larger scale in the future.[112]

- *Distributed Generation*: A third factor is the use of distributed generation (local power generation controlled by the customer), which can vary from rooftop solar units to large industrial cogeneration[113] facilities. A distributed generation facility will sometimes take power off the grid. Other times it will have excess power to sell to the utility, reversing the normal flow of electricity. Buying power from customers is inconsistent with standard utility technology, accounting, and rates. This is especially true when the generation is hooked up to the distribution system, which was designed to make final delivery of power to customers, not receive power from the customer.

  Distributed generation poses control and visibility issues similar to demand response. Wide use of distributed generation will also pose institutional issues. One is that generation connected to the distribution system (in contrast to the transmission system) is not covered by NERC reliability standards. Second,

---

[112] Not all demand response is directly controllable by the utility, which makes integration more difficult. For information on the various flavors of demand response and issues with grid integration see North American Electric Reliability Corp., *2008 Long-Term Reliability Assessment*, October 2008, pp. 41-43 and 270-271. http://www.nerc.com/files/LTRA2008.pdf.

[113] Cogeneration (also referred to as combined heat and power or CHP) is an integrated process to produce electricity and process heat for industrial or commercial use, such as space heating. Because the CHP plant makes use of the waste heat lost in a stand-alone power plant or steam plant, it is much more energy efficient than those types of facilities.

realizing the full potential of distributed generation may require the states to implement net metering laws that allow owners to sell surplus power back to the grid.

As with demand response, these issues are neither new or insuperable, although the scale may increase greatly. On the other hand, plug-in hybrid electric vehicles would pose a truly unique challenge, since their batteries would be a load on the power system at times and a source of stored electricity at other times. System operators would have to be able to decide on a daily or hourly basis how much they can rely on electricity storage scattered over thousands or millions of batteries, none of which are owned by the utility.

Integrating non-traditional resources into the grid will be a reliability challenge. This is not because these resources are new. For example, distributed generation in the form of industrial cogeneration has been increasingly common since Congress passed the Public Utility Regulatory Policies Act (P.L. 95-617) in 1978. The issue is integration of *much larger amounts* of these resources into a power system primarily designed around a different model. For example, NERC has concluded that "Demand response will become a critical resource for maintaining system reliability over the next ten years."[114] In 2008 NERC reported proposals to connect 145,000 MW of new wind capacity to the transmission grid by 2017, equivalent to about 14% of current total generating capacity in the United States.[115] Even if all of the proposed wind capacity is not built, many more wind plants will probably be connected to the grid. The most recent EIA long-term forecast, which assumes no changes to current laws, estimates that wind generation will increase by 300% by 2030.[116]

A characteristic that variable renewable generation, demand response, and distributed generation have in common is potentially less predictability (in respect to availability and level of service) than traditional resources. Improved real time monitoring, analysis, and control of the grid could help compensate for this issue. Another system-wide response may be to collapse the 130 balancing authorities that currently operate the transmission system into a smaller number that could call on a wider range of resources for managing electricity supply and demand.

Transmission Reliability: Summary of Policy Issues

In response to the 2003 northeastern blackout, Congress gave FERC authority over the reliability of the bulk power system in the 48 contiguous states. Continuing policy issues include:

- *Transmission system information gap.* There is currently no good source of data that measures the reliability of the transmission grid or allows trend analysis. NERC is developing a new process for collecting and analyzing transmission

[114] North American Electric Reliability Corp., *2008 Long-Term Reliability Assessment*, October 2008, p. 20, http://www.nerc.com/files/LTRA2008.pdf.

[115] Ibid., p. 12. In 2007, total wind capacity in the United States was 16,515 MW. In 2000 it was 2,400 MW. Total net summer capacity of all types in 2007 was 994,888 MW. Energy Information Administration, *Electric Power Annual 2007*, Table 2.2, http://www.eia.doe.gov/cneaf/electricity/epa/epa_sum.html, and Energy Information Administration, *Annual Energy Review 2007*, Table 8.11a, http://www.eia.doe.gov/emeu/aer/contents.html.

[116] Energy Information Administration, *Annual Energy Outlook 2009 Early Release*, December 17, 2008, slide 16. http://www.eia.doe.gov/oiaf/aeo/aeo2009_presentation.html (select presentation with data). Related materials are at http://www.eia.doe.gov/oiaf/aeo/index.html.

reliability data. The progress of this effort may be of interest to Congress, because without good data it will be difficult to judge whether FERC's new reliability standards and other actions are actually improving the reliability of the transmission system.

- *Modernization and reliability.* The implementation of modernized technology and management may be an alternative, or necessary supplement, to building new transmission lines to improve the reliability of the grid. In considering new spending and planning approaches for the transmission system, Congress may wish to ensure that the right balance is struck between modernization and new construction.

- *Reliability and the changing power market.* The power system is changing from a model based on central station power plants to a more diverse range of resources, including variable renewable power, demand response, and distributed generation. Congress may want to exercise oversight to ensure that FERC and NERC are developing reliability standards for a changing grid. Also, certain kinds of distributed generation are not covered by federal reliability authority, a situation Congress may want to revisit in the future.

Summary of Transmission Policy Issues

This concluding section summarizes policy issues of potential interest to Congress.

Federal Transmission Planning

S. 539 and other proposals call for a much larger federal role in transmission planning, and suggest that planning should be conducted on a larger geographic scope than in the past. Policy issues include:

- *What should be the objectives of the planning process?* For example, planning could be focused on renewable power development or on broader objectives, such as congestion relief and reliability enhancement.

- *What should be the scope of authority of the planning entities.* Federal transmission planning could be run by interconnection-wide centralized authorities (the top-down approach) or be conducted primarily at a regional level (the bottom-up approach), or as a hybrid.

- *What is the appropriate scope of the planning process?* Should the planning process extend beyond transmission planning narrowly defined to a include a broader array of solutions to power system issues, such as demand response, distributed power, or conventional power plant construction.

- *Could preferential treatment tied to the planning process distort transmission investment?* The planning proposals typically make available certain benefits, such as a federal permitting option, to projects included in the plan. These benefits could lead developers to add unnecessary features and costs to qualify proposals to meet plan criteria. Avoiding these distortions will require careful oversight or, arguably, limiting the benefits associated with the plan (for example, putting all new power lines or none, whether or not they are in the plan, under federal government permitting authority).

- *Is the scheme for managing and financing the planning process realistic?* An effective planning process will need realistic schedules and sufficient resources to timely develop and update transmission plans.

Permitting of Transmission Lines

Transmission line permitting is primarily under the control of the states. Current proposals would extend federal authority, perhaps by completing displacing the state role. Issues include:

- *Should the grid be viewed from a national perspective?* The grid evolved as local systems serving limited utility service areas. Now that the system has evolved into three separate synchronized interconnections, each spanning (other than ERCOT) many states. The question is whether a state-by-state or national view of the grid is most appropriate. The issue does not necessarily have a single answer; for example, a state perspective may be appropriate for "routine" projects, while a national perspective could be applied to "national interest" projects.

- *Can transmission system reliability be separated from authority over new transmission construction?* In EPACT05 Congress put the reliability of the grid under federal jurisdiction. By extension, should the federal government have control over the permitting of transmission lines aimed at enhancing system reliability (which could mean almost any new line in an interconnected power systems)?

- *How important is it to accelerate the construction of new transmission lines?* One criticism of the current regulatory regime is that it takes many years to move a transmission project through the permitting steps. Expanding federal authority over permitting is viewed as a means of accelerating the process. The question is how important is it to quickly build transmission lines to meet reliability, environmental, and other objectives.

- *Management of the permitting process.* If FERC or some other agency is assigned a federal permitting role, it will need the resources to expeditiously process applications. Otherwise the whole point of giving more permitting power to the federal government would largely be obviated.

Transmission Line Funding and Cost Allocation

Building new transmission lines could cost billions of dollars. Even more contentious than how to fund these projects is the question of how the costs of interstate transmission lines should be allocated to utility customers. Issues include:

- *Should the federal government help pay for new transmission lines?* Some proposals call for the federal government, possibly acting through the federal utilities, to help pay for new transmission lines, pay for expanding projects to meet future needs, or actually build new transmission. How far should the federal government go into financing the expansion of the transmission grid?

- *Should the Congress establish a national cost allocation rule for new transmission projects?* An approach included in several proposals would require all ratepayers in an interconnection to pay for new projects anywhere in the interconnection. The notion is that in a interconnected grid all customers benefit to some degree from enhancements to the system, but a preferential cost allocation mechanism for transmission may bias investment away from other alternatives.

Transmission Modernization and the Smart Grid

The smart grid is a concept for modernizing the grid with information technology and intelligent features. Congress has already established and funded programs for encouraging development of the smart grid. Policy issues include:

- *Program oversight.* The American Recovery and Reinvestment Act provided funding for previously authorized smart grid programs, including one key effort – development of interoperability standards by the National Institute for Standards and Technology – that has been lagging. Congress may want to monitor how these programs progress.

- *Smart grid cost/benefit oversight.* The balance of costs and benefits that the smart grid will produce for customers has been hotly debated. Many smart grid investment decisions will be made by state utility commissions. However, other investments and rate decisions will be under FERC jurisdiction, so there is ample room for Congressional oversight of the actual costs, benefits, and performance of smart grid investments.

Transmission System Reliability

In response to the 2003 northeastern blackout, Congress gave FERC authority over the reliability of the bulk power system in the 48 contiguous states. Continuing policy issues include:

- *Transmission system information gap.* There is no good source of data that measures the reliability of the transmission grid or allows trend analysis. NERC is developing a new process for collecting and analyzing transmission reliability data. The progress of this effort may be of interest to Congress, because without good data it will be difficult to judge whether FERC's new reliability standards and other actions are actually improving the reliability of the transmission system.

- *Modernization and reliability.* The implementation of modernized technology and management may be an alternative, or necessary supplement, to building new transmission lines to improve the reliability of the grid. In considering new spending and planning approaches for the transmission system, Congress may wish to ensure that the right balance is struck between modernization and new construction.

- *Reliability and the changing power market.* The power system is changing from a model based on central station power plants to a more diverse range of resources, including variable renewable power, demand response, and distributed generation. Congress may wish to exercise oversight to ensure that FERC and

NERC are developing reliability standards for a changing grid. Also, certain kinds of distributed generation are not covered by federal reliability authority, a situation Congress may want to revisit in the future.

Author Contact Information

Stan Mark Kaplan
Specialist in Energy and Environmental Policy
skaplan@crs.loc.gov, 7-9529

Congressional Research Service

Electric Transmission: Approaches for Energizing a Sagging Industry

Amy Abel
Section Research Manager

January 30, 2008

Congressional Research Service

7-5700

www.crs.gov

RL33875

CRS Report for Congress ———————————————

Prepared for Members and Committees of Congress

43

Summary

The electric utility industry is inherently capital-intensive. At the same time, the industry must operate under a changing and sometimes unpredictable regulatory system at both the federal and state level. The transmission system was developed to fit the regulatory framework established in the 1920 Federal Power Act—utilities served local customers in a monopoly service territory. The transmission system was not designed to handle large power transfers between utilities and regions. Enactment of the Energy Policy Act of 1992 (P.L. 102-486) created tension between the regulatory environment and the existing transmission system: The competitive generation market encouraged wholesale, interstate power transfers across a system that was designed to protect local reliability, not bulk power transfers.

The blackout of 2003 in the Northeast, Midwest, and Canada highlighted the need for infrastructure improvements and greater standardization of operating rules. The Energy Policy Act of 2005 (P.L. 109-58) set in place government activities intended to relieve congestion on the transmission system. The law creates an electric reliability organization that is to enforce mandatory reliability standards for the bulk-power system. In addition, processes are established to streamline the siting of transmission facilities. Many observers predict that until the electric power industry reaches a new equilibrium with more regulatory certainty, investment in transmission infrastructure and technology will continue to be inadequate.

This report discusses factors that have contributed to the lack of new transmission capacity and some of the resulting issues, including

- background on the evolution of the regulatory structure, including the creation of an electric reliability organization (ERO);

- issues associated with operating a congested transmission system;

- security of the physical assets;

- siting of transmission lines;

- cost implications of burying power lines;

- pricing of new transmission projects; and

- funding of these projects.

In addition, this report reviews approaches being taken to address the lack of investment in transmission infrastructure and transmission congestion.

Contents

Figures

Tables

Contacts

Introduction

The electric utility industry is inherently capital-intensive. At the same time, the industry must operate under a changing and sometimes unpredictable regulatory system at both the federal and state level. Inconsistent rules and authorities can result in inefficient operation of the interstate transmission system. The electric transmission system has been affected by a combination of factors that has resulted in insufficient investment in the physical infrastructure.

This report discusses factors that have contributed to the lack of new transmission capacity and some of the resulting issues, including

- background on the evolution of the regulatory structure, including the creation of an electric reliability organization (ERO);
- issues associated with operating a congested transmission system;
- security of the physical assets;
- siting of transmission lines;
- cost implications of burying power lines;
- pricing of new transmission projects; and
- funding of these projects.

In addition, this report reviews approaches being taken to address the lack of investment in transmission infrastructure and transmission congestion.

The transmission system was developed to fit the regulatory framework established in the 1920 Federal Power Act[1]—utilities served local customers in a monopoly service territory. The transmission system was not designed to handle large power transfers between utilities and regions. Enactment of the Energy Policy Act of 1992 (EPACT92)[2] created tension between the regulatory environment and the existing transmission system. EPACT92 effectively deregulated wholesale generation by creating a class of generators that were able to locate beyond a typical service territory with open access to the existing transmission system. The resulting competitive market encouraged wholesale, interstate power transfers across a system that was designed to protect local reliability, not bulk power transfers.

The blackout of August 2003 in the Northeast, Midwest, and Canada highlighted the need for infrastructure and operating improvements. However, a conflict exists between the apparent goal of increasing competition in the generation sector and assuring adequate transmission capacity and management of the system to move the power. Additions to generating capacity are occurring at a more rapid pace than transmission additions. The traditional vertically integrated utility no longer dominates the industry structure.[3] In addition, demand for electric power continues to

---

[1] 16 U.S.C. 791a et seq.

[2] P.L. 102-486.

[3] Seventeen states and the District of Columbia are implementing retail choice for electricity. According to the Energy Information Administration, in 1996, 10% of generating capacity was owned by non-utility generators. By 2005, 43% of net summer generating capacity was owned by non-utility generators. See http://www.eia.doe.gov/cneaf/electricity/epa/epat2p3.html.

increase. Unresolved regulatory issues that have emerged after 1992 have resulted in considerable uncertainty in the financial community. As a result of all of these factors, investment in the transmission system has not kept pace with demand for transmission capacity.

The Energy Policy Act of 2005 (EPACT05) addresses electric reliability and infrastructure investment.[4] In part, Title XII creates an electric reliability organization (ERO) that is to enforce mandatory reliability standards for the bulk-power system. These standards are necessary for reliable operation of the grid. The Federal Energy Regulatory Commission (FERC) will be reviewing the ERO's proposed reliability standards before granting its approval.[5] Under this title, the ERO could impose penalties on a user, owner, or operator of the bulk-power system that violates any FERC-approved reliability standard. FERC approved the North American Electric Reliability Corporation (NERC) as the ERO.[6] NERC is a nonprofit corporation whose membership is composed of the eight regional reliability councils.[7]

Title XII also addresses transmission infrastructure issues. As required by EPACT05, the Department of Energy issued the first *National Electric Transmission Congestion Study* in August 2006.[8] Additional studies are required every three years. The study identified two areas of critical congestion: Southern California and the eastern coastal area from metropolitan New York to Northern Virginia. This congestion study included detailed information on the transmission congestion in the western United States (**Figure 1**) but did not provide comparable detail on congestion in the eastern United States.[9] In determining whether to designate national interest electric transmission corridors, DOE is required to identify transmission congestion that adversely affects consumers. However, EPACT05 does not define "congestion that adversely affects consumers," nor does it require empirical analysis of the specific adverse effects of transmission congestion.

As a result of EPACT05, DOE designated two national interest electric transmission corridors on October 2, 2007: The Mid-Atlantic Area National Interest Electric Transmission Corridor and the Southwest Area National Interest Electric Transmission Corridor based on the 2006 congestion study (**Figures 2** and **3**).[10] This designation allows FERC, under certain circumstances to

---

[4] P.L. 109-58.

[5] FERC Docket No. RM06-22-000. Approval status is available at http://www.ferc.gov/industries/electric/indus-act/ reliability/standards.asp.

[6] *Order Certifying North American Electric Reliability Corporation As the Electric Reliability Organization in Ordering Compliance Filing.* 116 FERC, 61,062. Docket No. RR06-1-000. Issued July 20, 2006.

[7] The regional reliability councils are Electric Reliability Council of Texas, Inc. (ERCOT); Florida Reliability Coordinating Council (FRCC); Midwest Reliability Organization (MRO); Northeast Power Coordinating Council (NPCC); Reliability First Corporation (RFC); Southeastern Reliability Council (SERC); Southwest Power Pool, Inc. (SPP); and Western Electricity Coordinating Council (WECC).

[8] U.S. Department of Energy, National Electric Transmission Congestion Study, August 2006, available at http://www.oe.energy.gov/DocumentsandMedia/Congestion_Study_2006-9MB.pdf.

[9] **Figure 1** shows how many hours in a year the transmission system was loaded at or above 75% of Operating Transfer Capability (OTC) in the Western Electricity Coordinating Council (WECC) region. The most heavily loaded lines include Bridger West, which delivers power from the Bridger, Montana coal-fired plants to loads in Utah and Oregon; Southwest of Four Corners-to-Cholla-to-Pinnacle Peak in Arizona, which is designed to deliver power from baseload plants to load; western Colorado to Utah; Wyoming to Colorado; and southern New Mexico to El Paso. The Department of Energy will not be publishing a detailed map of eastern congestion until the release of the next congestion study (due August 8, 2009). (E-mail communication with Agrawal Poonum, Manager, Markets and Technical Integration, Office of Electricity Delivery and Energy Reliability, Department of Energy, January 19, 2007.)

[10] Department Of Energy Docket Numbers 2007-OE-01 and 2007-OE-02. Issued October 2, 2007. *Federal Register,* (continued...)

---

authorize "the construction or modification of electric transmission facilities."[11] A permit holder would still need to obtain rights-of-way from property owners. If the permit holder is not able to successfully negotiate with each affected property owner, then FERC would entitle a permit holder to acquire the rights-of-way by exercising the right of eminent domain.[12] In its designation, DOE stated that:

> A National Corridor designation is not a determination that transmission must, or even should, be built. Whether a particular transmission project, some other transmission project, or a non-transmission project is an appropriate solution to a congestion or constraint problem identified by a National Corridor designation is a matter that market participants, applicable regional planning entities, State authorities, and potentially FERC will consider and decide before any project is built. A National Corridor designation itself does not preempt State authority or any State actions, including action to approve or order the implementation of non-transmission solutions to congestion and constraint problems.... FERC is committed to considering non-transmission alternatives, as appropriate, during its permit application review process.[13]

(...continued)

Vol. 72, No. 193, October 5, 2007. p. 56992-57028.

[11] Federal Power Act §216(b), 16 U.S.C. 824p(b).

[12] Federal Power Act §216(e)(1), 16 U.S.C. 824p (f)(2).

[13] Department Of Energy Docket Numbers 2007-OE-01 and 2007-OE-02. October p. 7-8.

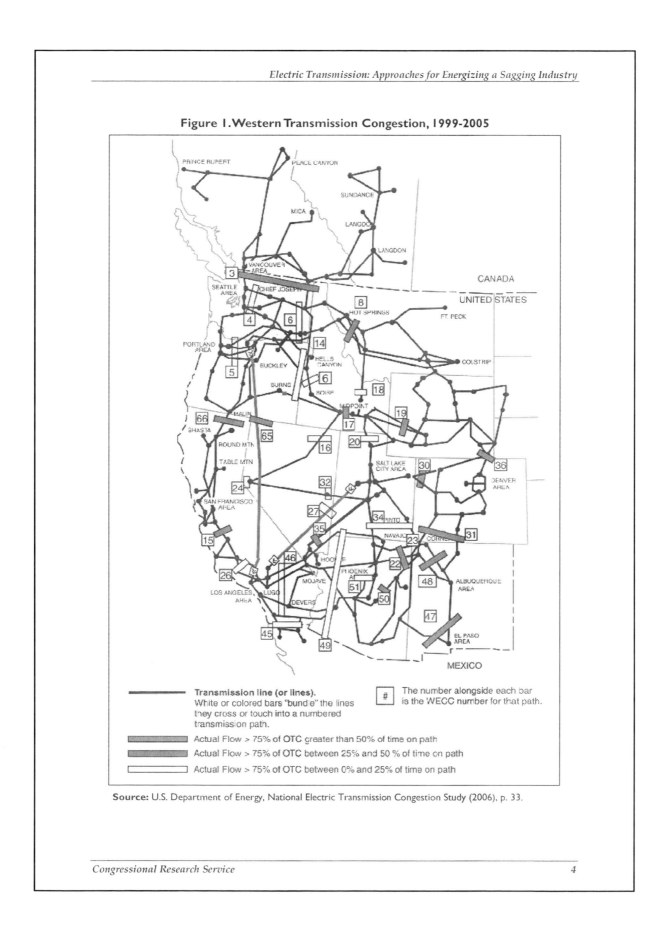

Figure 2. Mid-Atlantic Area National Interest Electric Transmission Corridor

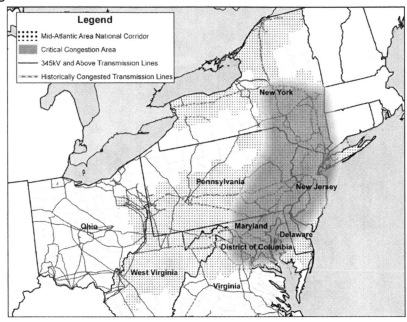

Source: U.S. Department of Energy, October 2007.

Figure 3. Southwest Area National Interest Electric Transmission Corridor

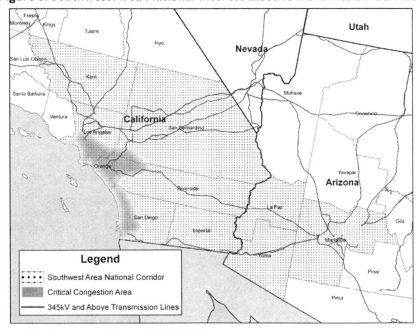

Source: U.S. Department of Energy, October 2007.

Transmission pricing was also addressed in EPACT05 (§1241) to encourage investment in transmission. FERC issued Order 679 on this issue, *Promoting Transmission Investment through Pricing Reform*, on July 20, 2006.[14] The order identifies specific incentives that FERC will allow, but the burden remains on an applicant to justify the incentives.

Historical Context

There are three components to electric power delivery: generation, transmission, and distribution. Transmission, by its nature, is generally considered an interstate transaction, whereas distribution is considered intrastate. State public utility commissions regulate the siting of all transmission and distribution lines within each state's borders, as well as distribution charges and retail electric rates. In states that have not restructured, the system operates as it has since enactment of the Federal Power Act, with retail consumers paying one price that includes transmission, distribution, and generation. This is referred to as a *bundled transaction*. In states that have restructured, consumers are billed for separate transmission, distribution, and generation charges. This is referred to as *unbundled electricity service*. FERC regulates all transmission, including unbundled retail transactions.[15]

Generators of electricity need to move their power to their ultimate customers through the transmission system. The current system allows for power transfers within, but not between, three major regions of the United States: the area west of the Rockies (Western Interconnection), Texas, and the Eastern Interconnection. Transmission lines and distribution lines are categorized by their voltage rating. Transmission lines are typically rated 230 kilovolts (kV) and higher (765 kV is the highest installed). Subtransmission systems are 69 kV to 138 kV, and distribution systems are rated less than 69 kV.[16] Existing transmission infrastructure was designed to accommodate the old system of central station power plants with nearby customers. Since enactment of the Energy Policy Act of 1992, there has been an increase in interstate bulk power transfers, a purpose for which the existing system was not designed.

The Energy Policy Act of 1992 (EPACT92) created a new category of wholesale electric generators called Exempt Wholesale Generators (EWGs) that are not considered utilities.[17] EWGs, also referred to as *merchant generators*, were intended to create a competitive wholesale

---

[14] Federal Energy Regulatory Commission Final Rule, Order Number 679, *Promoting Transmission Investment through Pricing Reform*, July 20, 2006, Docket Number RM06-4-000.

[15] On October 3, 2001, the U.S. Supreme Court heard arguments in a case (*New York et al. v. Federal Energy Regulatory Commission*) that challenged FERC's authority to regulate transmission for retail sales if a utility unbundles transmission from other retail charges. In states that have opened their generation market to competition, unbundling occurs when customers are charged separately for generation, transmission, and distribution. Nine states, led by New York, filed suit, arguing that the Federal Power Act gives FERC jurisdiction over wholesale sales and interstate transmission and leaves all retail issues up to the state utility commissions. Enron in an amicus brief argued that FERC clearly has jurisdiction over all transmission and FERC is obligated to prevent transmission owners from discriminating against those wishing to use the transmission lines. On March 4, 2002, the U.S. Supreme Court ruled in favor of FERC and held that FERC has jurisdiction over transmission, including unbundled retail transactions.

[16] Transmission lines generally carry bulk-power transfers between utilities and move power to load centers. Distribution lines move power to ultimate customers. Subtransmission is sometimes considered transmission and other times considered distribution for regulatory purposes.

[17] Exempt Wholesale Generators may sell electricity only at wholesale. EWGs may be located anywhere, including foreign countries. Before enactment of EPACT05, utility generators were limited by the Public Utility Holding Company Act of 1935 (PUHCA) to operate within one state.

electric generation sector. In addition, EPACT92 provided a means for these non-utility generators to have access to the transmission system. As a result of EPACT92, FERC issued a policy statement on transmission pricing policy:

> Greater pricing flexibility is appropriate in light of the significant competitive changes occurring in wholesale generation markets, and in light of our expanded wheeling authority under the Energy Policy Act of 1992 (EPACT92)[footnote omitted]. These recent events underscore the importance of ensuring that our transmission pricing policies promote economic efficiency, fairly compensate utilities for providing transmission services, reflect a reasonable allocation of transmission costs among transmission users, and maintain the reliability of the transmission grid. The Commission also recognizes that advances in computer modeling techniques have made possible certain transmission pricing methods that once would have been impractical.[18]

In May 1994, FERC established general guidelines for comparable access to the transmission system.[19] In April 1996, FERC clarified its open-access transmission tariff policy with Orders 888 and 889, making it easier for merchant generators to gain access to the transmission grid and requiring utilities to "functionally unbundle" their operations. In practice, this means that a utility's generation and transmission operations must be conducted separately, without the sharing of resources, books, and records. Some states that have opened their retail markets to competition, including California, have required utilities to divest of either transmission and distribution or of generation. In these states, most utilities have divested generation assets and maintained their transmission and distribution business. By July 9, 1996, all utilities that own or control transmission had filed a single open-access tariff with FERC that provides transmission service to eligible wholesale customers at comparable terms to the service that the utilities provide themselves. Some merchant generators asserted that they continued to be discriminated against by incumbent transmission utilities and were denied access to the system.

Orders 888 and 889 established a pro forma open-access transmission tariff (pro forma OATT). Many argued that this pro forma OATT allowed for opportunities for the exercise of undue discrimination by transmission owners. On May 18, 2006, FERC issued a Notice of Proposed Rulemaking (NOPR), *Preventing Undue Discrimination in Preference in Transmission Service*, to remedy some of the deficiencies in the pro forma OATT.[20] According to FERC, the major reforms in the NOPR include

- greater consistency and transparency in the Available Transfer Capability (ATC) calculation;[21]

- open, coordinated, and transparent planning;

- reform of energy imbalance penalties;

---

[18] Inquiry Concerning the Commission's Pricing Policy for Transmission Services Provided by Public Utilities Under the Federal Power Act, policy statement, October 26, 1994, Docket No. RM 93-19-000, 18 CFR 2, 59 FR 55031. *Wheeling* is defined as the movement of electricity from one system to another over transmission facilities of interconnecting systems.

[19] 67 FERC 61,168.

[20] FERC Docket Numbers RM05-25-000 and RM05-17-000.

[21] ATC is the transfer capability remaining on a transmission provider's transmission system that is available for further commercial activity over and above already committed uses.

- clarification of tariff ambiguities; and
- increased transparency and customer access to information.

On February 16, 2007, FERC issued Order 890, *Preventing Undue Discrimination Preference in Transmission Service*.[22] The final Order reflected much of what was addressed in the NOPR, including calculations of available transfer capability, coordination of the transmission planning process, establishing a requirement for conditional firm long-term point-point service contracts, reforming energy and generator imbalance charges, and increasing the transparency of the *existing pro forma* OATT.

Current Issues

Physical Limitations

Three types of constraints limit the transfer capability within the transmission system: thermal constraints, voltage constraints, and system operating constraints. Thermal constraints limit the capability of a transmission line or transformer to carry power because the resistance created by the movement of electrons causes heat to be produced. Overheating can lead to two possible problems: The transmission line loses strength, which can reduce the expected life of the line, and the transmission line expands and sags between the supporting towers. This presents safety issues as the lines approach the ground, as well as reliability concerns. If a transmission line comes in contact with the ground, trees, or other objects, the transmission line will trip off-line and not be able to carry power.

Voltage can be likened to the pressure inside the transmission system. Constraints on the maximum voltage levels are set by the design of the transmission line. If voltage levels exceed the maximum, short-circuits, radio interference, and noise may occur. Low voltages are also a problem and can cause customers' equipment to malfunction and can damage motors.

System operating constraints refer to reliability and security. Maintaining synchronization among generators on the system and preventing the collapse of voltages are major aspects of the role for transmission operators.[23] North American Electric Reliability Council guidelines require utilities to be able to handle any single outage through redundancy in the system. When practical, NERC recommends the ability to handle multiple outages within a system. Reducing the constraints on the system through technology improvements is one way to increase the transfer capability over existing lines.[24]

The regulatory regime has shifted the operations of the electric utility industry, creating larger and more frequent bulk power transfers across a transmission system designed largely for local intrastate service. However, investment and infrastructure have not kept up with increases in the

---

[22] FERC Order 890. Docket Numbers RM05-17-000 and RM05-25-000. *Preventing Undue Discrimination Preference in Transmission Service*. Issued February 16, 2007.

[23] Within each interconnection, all generators rotate in unison at a speed that produces a consistent frequency of 60 cycles per second.

[24] See Energy Information Administration, *Upgrading the Transmission Capacity for Wholesale Electric Power Trade*, available at http://www.eia.doe.gov/cneaf/pubs_html/feat_trans_capacity/w_sale.html.

bulk power transfers and electricity demand. Between 1978 and 1998, electricity demand had been growing at an average rate of 2.8% per year.[25] Transmission capacity expressed in relation to electricity demand increased by 3.5% per year between 1978 and 1982 and then declined by 1.2% per year between 1982 and 1998.[26] Actual annual transmission investment had declined from nearly $5 billion in 1975 to about $2.25 billion in 1998.[27] Reversing this trend, between 1999 and 2005, transmission investment increased at a 12% annual rate.[28] However, during the same period, total circuit miles of 230 kV and above transmission lines owned and operated by investor-owned utilities increased by 0.8% annually.[29] This long period of insufficient transmission investment has led to transmission lines that are congested in several regions of United States.

Similarly, as is shown in **Figure 4**, investment in generation capacity has not kept pace with electricity demand growth. This has led to lower capacity margins for electric utilities.[30] According to NERC, available capacity margins, which include only committed resources, are projected to drop below minimum regional target levels in several regions of the United States and Canada in two to three years.[31] The ERO has proposed enforceable standards for cyber- and physical security and reliability of the electric system that are intended to ensure optimum operation of the transmission system. FERC will be reviewing NERC's proposed standards before granting approval.[32]

---

[25] Energy Information Administration, Annual Energy Review, Electricity Overview, 1949-2005. Available at http://www.eia.doe.gov/emeu/aer/elect.html.

[26] Hirst, Eric, *Expanding U.S. Transmission Capacity* (August 2000), p. 5. Available at http://www.eei.org/industry_issues/energy_infrastructure/transmission/hirst2.pdf.Normalized transmission capacity is calculated using megawatt-miles of transmission per megawatts of summer peak demand.

[27] Real $2003. Edison Electric Institute, *EEI Survey of Transmission Investment: Historical and Planned Capital Expenditures (1999-2008)*, Washington (May 2005), p. 3.

[28] Edison Electric Institute, *EEI Statistical Yearbook/2005 Data* (2006), p. 107.

[29] NERC 2006 Long-Term Reliability Assessment and NERC Reliability Assessment 2000-2009. Available at http://www.nerc.com/~filez/rasreports.html.

[30] *Capacity margin* is the amount of unused available capability of an electric power system at peak load as a percentage of capacity resources. This gives an indication of the ability of the system to meet demand. A narrow capacity margin indicates a risk that supply will be interrupted because of a shortage of generation. On the other hand, excessive day-ahead capacity margin could add to the cost of electricity.

[31] North American Electric Reliability Council, *2006 Long-Term Reliability Assessment* (October 2006), p. 6.

[32] FERC Docket No. RM06-22-000.

Figure 4. Real Private Fixed Investment in Electrical Power Generation, and Electricity Consumption, Generation, and Real Prices

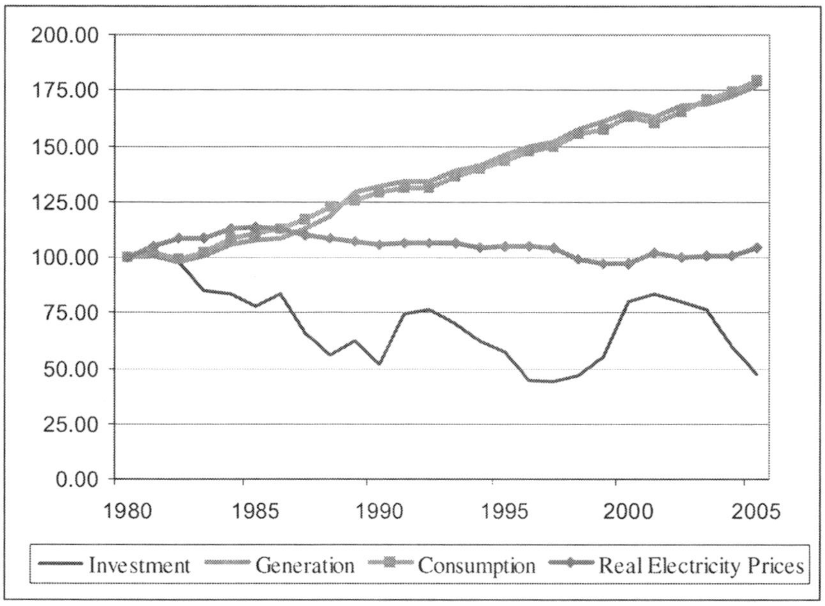

Source: Kliesen, Kevin L. , "Electricity: The Next Energy Jolt?" The Regional Economist, The Federal Reserve Bank of St. Louis, October 2006, p. 6.

Note: Index, 1980 = 100.

Congestion

DOE has identified areas in the Eastern Interconnection and Western Interconnection as congested. Problems with congestion on the transmission system are not new. In 1987, CRS noted that bulk power transmission lines in many parts of the country were already operating at or near capacity and the chief capacity-related barrier to bulk-power transfers (wheeling) was that the transmission system was not built for bulk-power transfers.[33] According to NERC, the number of requests to use the transmission system that were denied because of congestion (transmission line relief, TLRs) rose from 305 in 1998 to 1,494 in 2002. By 2005, there were 2,397 TLRs, dropping slightly in 2006 to 1,901 TLRs.[34] Over the next 10 years, the line-miles of high-voltage transmission are expected to increase 6%, in contrast to a 20% expected increase in generation demand and capacity.[35] If this projection is accurate, further pressure on reliability could occur in several regions.[36]

---

[33] See CRS Report 87-289, *Wheeling in the Electric Utility Industry*, by Alvin Kaufman et al. (out of print; available from the author of this report).

[34] NERC data on Transmission Loading Relief (TLR) requests are available at http://www.nerc.com/pub/sys/all_updl/oc/scs/logs/trends.htm.

[35] Department of Energy, *National Transmission Grid Study*, May 2002.

[36] See CRS Report RL31469, *Electric Utility Restructuring: Maintaining Bulk Power System Reliability*, by Amy Abel, Larry Parker, and Steven Stitt.

Security

Another issue surrounding the reliability of the electric system involves security. The system operates with built-in redundancies to minimize the risk of outages resulting from myriad causes, including weather, equipment failure, and terrorist activity. In general, physical attacks could target transformers, transmission towers, substations, control centers, power plants (including nuclear reactors or dams), and fuel-delivery systems. NERC reported that in 2006, of the 57 events that resulted in a system disturbance (outage), only one (1.75%) was caused by physical attack. Between 2001 and 2006, 11.21% of the system disturbances were caused by "system protection and controls" issues, which include physical attacks and cyber-attacks.[37]

High-voltage transformers are a critical and vulnerable part of the nation's electric power network. High-voltage (HV) units make up less than 3% of transformers in U.S. power stations, but they carry 60% to 70% of the nation's electricity.[38] Power grid planners generally anticipate the possible loss of a single HV transformer substation and are prepared to reroute power flows as necessary to maintain regional electric service. Loss of multiple HV transformers simultaneously could cause extended regional outages.

Utilities generally do not maintain a stockpile of transformers to replace more than a small percentage of their operating units. Large transformers generally cost $2 million to $3 million, are custom made, require long lead times to build, and are bulky and difficult to move around.[39] NERC maintains a transformer information database, a 15-year-old program used primarily for weather-related outages of large transformers. In response to the growing need for a stockpile of large transformers, the Edison Electric Institute (EEI) has begun a FERC-approved spare transformer sharing program, which is to be used solely to deal with terrorist activity or deliberate damage to utility substations.[40] At the time of the FERC approval in September 2006, 43 entities had joined the EEI program, representing more than 60% of the FERC-jurisdictional bulk-power transmission system. Unlike the NERC program, which does not charge utilities to participate, EEI charges a $10,000 sign-up fee to join the transformer sharing program, as well as annual dues of about $7,500. Some utilities need to obtain state approval before joining the EEI program.

NERC also operates the Electricity Sector Information Sharing and Analysis Center (ESISAC), as required by Presidential Decision Directive (PDD) 63 on critical infrastructure protection. The ESISAC is a voluntary means for utilities to share security-related information. In turn, the ESISAC is tasked with providing "timely, reliable and actionable warnings of threats and impending attacks on our critical infrastructures."[41]

---

[37] A physical attack involves human caused damage to transmission infrastructure that generally results in a major outage. Past incidents have included removing stabilizing bolts from transmission lines and shooting transformers. NERC Reported System Events available at http://www.nerc.net/dashboard/.

[38] Loomis, William M., consulting engineer for Strategic Partners-Technical Systems, "Super-Grade Transformer and Defense: Risk of Destruction and Defense Strategies," presentation to NERC Critical Infrastructure Working Group, Lake Buena Vista, Florida. (December 10-11, 2001).

[39] Stan Johnson, NERC Manager of Situation Awareness and Infrastructure Security, as quoted in *Electric Utility Week*, "More Utilities Sign up to Share Transformers, Information As Cost-Consciousness Grows"(January 8, 2007), p. 9.

[40] FERC approved the Spare Transformer Equipment Program (STEP) on September 21, 2006. FERC Docket Nos. EC06-140-000 and EL06-86-000.

[41] The ESISAC website is available at http://www.esisac.com.

Siting

One reason additional transmission lines have not been built in recent years is the problems encountered when siting them. Siting and building transmission lines have been very difficult because of citizen opposition, as well as inconsistent siting requirements among states. Even though the transmission of electricity is considered interstate commerce, the siting of transmission lines has been the responsibility of the states. In addition, several federal agencies play various roles in the siting process, primarily with regard to environmental impacts.

Since the blackout of 2003, FERC commissioners have supported federal siting backstop authority to help transmission companies overcome some of the siting obstacles,[42] although such support has been controversial. The electric industry is in favor of giving FERC siting authority.[43] States are generally opposed to federal backstop authority.[44]

EPACT05 established that the Secretary of Energy is required to certify congestion on the transmission lines, but EPACT05 does not specifically define congestion. The first congestion study was completed in August 2006.[45] As a result of this study, the Secretary may designate "any geographic area experiencing electric energy transmission capacity constraints or congestion that adversely affects consumers as a national interest electric transmission corridor." FERC may issue permits for construction of transmission lines to transmission owners within a national interest transmission corridor if FERC finds that a state does not have the authority to approve the siting of the facilities or that a state commission that has authority to approve the siting of facilities has withheld its approval for more than one year. Permit holders are able to petition in U.S. district court to acquire rights-of-way for the construction of transmission lines through the exercise of the right of eminent domain.[46]

On April 26, 2007, DOE issued two draft National Interest Electric Transmission Corridor designations (draft report), one stretching from the mid-Atlantic region through New York, and the other in Southern California.[47] EPACT05 created a new Federal Power Act §216(a)(2) which states that:

> After considering alternatives and recommendations from interested parties (including an opportunity for comment from affected States), the Secretary shall issue a report, based on the study, which may designate any geographic area experiencing electric energy transmission capacity constraints or congestion that adversely affects consumers as a national interest electric transmission corridor.

DOE broadly defines the term "constraints that adversely affects consumers" to include "not only constraints that cause persistent congestion, but also constraints that hinder the development or delivery of a generation source that is in the public interest ... which allows for a National

---

[42] Statement of Nora Mead Brownell, *FERC Reverses Position, Will Now Take Federal Backstop Authority*, at http://www.Energywashington.com, September 2, 2003.

[43] Edison Electric Institute, *Federal Siting Authority: Key to Expanding Electricity Infrastructure*, available at http://www.eei.org/industry_issues/energy_infrastructure/transmission/federalsiting.pdf.

[44] Statement of National Governors Association.

[45] U.S. Department of Energy, *National Electric Transmission Congestion Study*, August 2006, available at http://www.oe.energy.gov/DocumentsandMedia/Congestion_Study_2006-9MB.pdf.

[46] P.L. 109-58, § 1221.

[47] DOE Draft Report available at http://nietc.anl.gov/.

Corridor designation when there is a constraint that adversely affects consumers even though there is no present congestion...."[48] DOE did not consider alternatives to new transmission, such as conservation or efficiency improvements in existing lines, in identifying National Corridors. DOE received comments from many stakeholders that argued that DOE should conduct a cost-benefit analysis on transmission versus non-transmission solutions for congestion before designating a National Corridor.[49] DOE argues that nothing in § 216 of the Federal Power Act requires that the DOE demonstrate that transmission is the best or most cost-effective solution to a congestion problem.[50] In its final order, DOE concluded that:

> Consideration of non-transmission solutions to the congestion problems facing the Mid-Atlantic Critical Congestion Area is neither required nor necessary as a precondition to designating the Mid-Atlantic Area National Corridor. FPA section 2 16(a)(2) calls for the Secretary to consider "alternatives and recommendations from interested parties" before making a National Corridor designation. The statute, however, does not specify what the term "alternatives" refers to. Numerous commenters would have us to interpret the phrase to mean alternative solutions to congestion or constraint problems, which would then necessitate a comparison of non-transmission solutions against transmission solutions. Nothing in the language of FPA section 216 requires or suggest such an interpretation.[51]

Another controversial aspect of the National Corridor designation is their geographic boundaries. EPACT05 did not define the term "corridor," and DOE has concluded "that, while there may be circumstances where a project-based approach would be appropriate, in general the Department (DOE) will use a source-and-sink approach to define National Corridor boundaries."[52] In this process, the sink (the congested or constrained load area) and the source (an area of *potential* supply) are identified and the two areas are connected as a National Corridor. DOE does not intend to designate specific projects as National Corridors, but rather large areas with specific geographic boundaries. The final order used county borders to determine the corridor boundaries.

In advance of DOE designating national interest electric transmission corridors, on November 16, 2006, FERC issued its Final Rule on Regulations for Filing Applications for Permits to Site Interstate Electric Transmission Facilities.[53] In part, EPACT05 (§ 1221) allows FERC to issue construction permits for transmission facilities within a national interest electric transmission corridor if a state has "withheld approval for more than one year." The term *withheld approval* was not unanimously interpreted by the commissioners. The final rule adopted that *withheld approval* could mean a state failing to act on the siting permit application or denying the application. Commissioner Kelly in her dissent found this interpretation to be a preemption of state rights:

---

[48] Ibid. p. 20.

[49] For examples of these comments, see those submitted by the National Association of Regulatory Utility Commissioners (NARUC), New York Public Service Commission, New Jersey Board of Public Utilities, Electric Power Supply Association, Northern Indiana Public Service Co., Old Dominion Electric Cooperative, Piedmont Environmental Council, and the Wilderness Society. Available at http://nietc.anl.gov/.

[50] Ibid. p. 24.

[51] Department of Energy Order on the Designation of Mid-Atlantic Area National Interest Electric Transmission Corridor [Docket Number 2007-OE-01], and the Southwest Area National Interest Electric Transmission Corridor [Docket Number 2007-OE-02]. October 2, 2007. p. 69.

[52] Draft Report. p. 34.

[53] FERC Order Number 689 (RM06-12). 117 FERC 61,202. 18 CFR Parts 50 and 380. Issued November 16, 2006.

> The authority to lawfully deny a permit is critically important to the states for ensuring that the interests of local communities and their citizens are protected. What the Commission does today is a significant inroad into traditional state transmission siting authority. It gives states two options: either issue a permit, or we'll do it for them. Obviously this is no choice. This is preemption.[54]

Defending the final rule before the media, Chairman Kelliher stressed that FERC must assume that the words Congress chooses to use in passing a law are "deliberate and done with care. If Congress meant only failure to act, why didn't they just say that."[55]

Several bills have been introduced in the 110th Congress that would repeal or reform § 216 of the Federal Power Act. H.R. 809 would repeal §216 of the Federal Power Act. H.R. 810 would amend § 216(e) of the Federal Power Act by repealing the section that allows a permit holder to acquire the-right-of-way for transmission facilities through the exercise of eminent domain in a federal district court and replaces it with acquisition of right-of-ways in accordance with state laws. H.R. 829 would place additional requirements before an area could be considered a National Corridor. This bill would require DOE to perform an analysis of alternatives to new transmission construction to alleviate congestion. The bill would prohibit National Corridors from including parks or historic battlefield sites that are designated as scenic, natural, cultural, or historic resources under federal or state law. H.R. 829 also amends the Federal Power Act to limit FERC's ability to issue construction permits to instances when a state's denial of transmission siting is found to have been arbitrary or capricious or the state unreasonably withheld or delayed a siting decision beyond one year. H.R. 1945 (§ 403) would repeal § 216 of the Federal Power Act, which provides for federal siting authority for transmission lines. H.R. 2337 (§103) would require the completion of a study on the need for energy corridors on public lands before a corridor could be designated on such land.

Alternatives to New Rights-of-Way

Capacity of the existing transmission system can be increased without siting new lines. In addition, new generation can be sited closer to demand, reducing the need to use the transmission system. Additional transmission lines could be added to existing rights-of-way or, in some cases, existing towers could be restrung with higher capacity lines. However, in some cases, reliability levels would increase with the redundancy of new transmission lines sited on new rights-of-way; storms and other events that may cause physical damage to one area may not affect transmission lines in another part of a state or region.

Many transmission systems could increase the capacity of the transmission system with technology improvements. While many new technologies would require significant capital investment, one study by the New York Independent System Operator concluded that relatively inexpensive equipment upgrades could significantly increase the line ratings and could reduce congestion.[56] The study indicated that a significant number of transmission lines operate below their thermal limits because of equipment limitations at substations. By remediating those

---

[54] Kelly, Commissioner, *Dissenting in part*. Docket No. RM06-12-000 (November 16, 2006), p. 3.

[55] Foster Electric Report, *FERC's Electric Transmission Siting Rule Sparked Strong Dissent by Kelly over Jurisdictional Concerns* (November 22, 2006), p. 2.

[56] New York Independent System Operator, *Investigation of Potential Low Cost Transmission Upgrades Within the New York State Bulk Power System*, Interim Report (April 19, 2001).

---

limitations with relatively inexpensive equipment (e.g., disconnect switches, bus connectors, relays), according to the New York study, operation at thermal capacities could be reached with little or no risk of service interruption.

Other technological improvements to increase transmission capacity and allow the transmission system to be operated more efficiently include upgrading transformers, retrofitting electromechanical devices with digital devices to allow operation of the system closer to thermal limits, and restringing existing towers with aluminum conductor composite core cable. These would require significant capital investment.

Burying Power Lines

Many reasons have been given for burying power lines, including reduced maintenance, less susceptibility to weather damage, fewer traffic accidents involving poles, improved aesthetics, and increased property values. The primary reason against burying power lines is the high cost. Design and installation of underground systems is more complex and expensive, and takes longer than for overhead systems. In addition, the cost and time involved to modify or repair an existing system is also reportedly higher.[57]

The overwhelming damage to the electricity transmission and distribution system in the wake of Hurricanes Katrina and Rita has increased interest in replacing overhead lines with underground cable. However, studies suggest that both overhead and underground lines have their vulnerabilities, and there are considerable cost differences in constructing and maintaining them. A review of several studies has found that overhead lines are more susceptible to storm and other damage, but the sites requiring repairs can be identified more quickly and repaired faster. Underground lines have above-ground transformers that are subject to immediate storm damage. Although underground distribution is generally more reliable during storms, corrosion from water infiltration can cause outages in the days and weeks after severe storms. The uprooting of trees can damage underground lines directly. Underground lines can be more expensive and take longer to repair. Replacing overhead lines with underground cable is also expensive. Analysis by the Florida Public Service Commission (FPSC) has found that replacing overhead *transmission lines* in Florida with underground lines over a 10-year period might require a rate increase of nearly 50% spread over all kilowatt hours. Converting overhead *distribution lines* to underground over the same period could boost rates by more than 80%.[58]

The majority of existing transmission and distribution lines are overhead, but in the 10 years between 1993 and 2002, capital expenditures for new power lines were almost equally divided between underground (49%) and overhead (51%) lines.[59] According to the Edison Electric

---

[57] North Carolina Public Utilities Commission Staff, *The Feasibility of Placing Electric Distribution Facilities Underground* (November 2003).

[58] Florida Public Service Commission, *Preliminary Analysis of Placing Investor-Owned Electric Utility Transmission and Distribution Facilities Underground in Florida* (March 2005), available at http://www.psc.state.fl.us/publications/pdf/electricgas/Underground_Wiring.pdf.

[59] FERC Form 1 Data 1993-2002, as compiled by Edison Electric Institute, *Out of Sight, Out of Mind? A Study on the Costs and Benefits of Undergrounding Overhead Power Lines*, available at http://www.eei.org/industry_issues/energy_infrastructure/distribution/UndergroundReport.pdf.

Institute (EEI), new underground distribution costs average $1 million per mile, or $29,854 per customer, compared with $73,666 per mile, or $2,199 per customer, for existing overhead lines.[60]

Undergrounding Transmission

During storms, large steel transmission towers generally withstand high winds and rain. Because of their height, they are also less susceptible to damage from falling trees. In the United States, there are 200,000 miles of transmission lines, only 5,000 of which are underground cable. Transmission is placed underground typically to address a localized constraint, such as an airport, river crossing, or a central business district. Several factors are considered in evaluating whether burying transmission lines is feasible. As already noted, underground transmission has been found to be less susceptible to damage, but any damage is more difficult and time-consuming to locate and repair, according to an Australian study.[61] While overhead transmission lines generally take a few hours to two days to repair, EEI reports that average outage durations for underground transmission ranges from five days to nine months, depending on the technology used.[62]

There are several reasons why laying underground cable is significantly more expensive than overhead transmission. According to FPSC, an underground transmission cable needs to be about 10 times more massive than an overhead cable to transmit the same amount of power, with the cable cost being about 10 times greater than overhead cable.[63] Trenches need to have either concrete conduits or metal pipes for both safety and operational reasons. Because of the weight and thickness of underground cable, splices need to be made every 900 to 3,500 feet. At the site of each splice, an underground vault needs to be constructed for maintenance access. Above ground, a right-of-way of at least 20 to 50 feet must be completely cleared. Some studies have found that electromagnetic fields (EMFs) are stronger immediately above underground transmission than immediately below overhead transmission. However the EMF fields diminish more quickly with distance for buried transmission than for overhead transmission.[64]

FPSC completed a comprehensive analysis of burying transmission and distribution facilities and updated the analysis in 2005.[65] The study calculated a cost to remove and replace existing 138 kV overhead transmission facilities for investor-owned, municipally owned, and rural electric cooperatives. The cost calculations included

---

[60] Ibid.

[61] Australian Department of Communications, Information Technology and the Arts, *Putting Cables Underground* (1998).

[62] Average outage durations for High-pressure Fluid Filled Pipe, 8-12 days; Extruded Dielectric, 5-9 days; High-pressure Fluid Filled Pipe, 2-9 months. See http://www.eei.org/meetings/nonav_meeting_files/nonav_2003-03-30-km/WiseSiting.ppt.

[63] Florida Public Service Commission. *Preliminary Analysis of Placing Investor-Owned Electric Utility Transmission and Distribution Facilities Underground in Florida* (March 2005).

[64] Wise, K., *Going Underground: A Growing Reality for Transmission Line Routing?* (April 2003), presentation at Edison Electric Institute Natural Resources Workshop, Burns & McDonnell, available at http://www.eei.org/meetings/nonav_meeting_files/nonav_2003-03-30-km/WiseSiting.ppt.

[65] Florida Public Service Commission, *Report on Cost-Effectiveness of Underground Electric Distribution Facilities*, vols. 1-4 (December 1991), and Florida Public Service Commission, *Preliminary Analysis of Placing Investor-Owned Electric Utility Transmission and Distribution Facilities Underground in Florida* (March 2005).

- planning and permitting,
- labor to remove existing facilities,
- new underground transmission facilities,
- labor to install the new underground facilities,
- trucks and other equipment to remove and install facilities,
- credits for existing overhead facilities that could be employed in the future, and
- disposal of facilities that could not be employed in the future.

The most recent study estimated that in 2003 dollars, the cost per mile to place transmission underground was $3.6 million or a total of $51.8 billion for all investor-owned utility transmission assets in Florida. The FPSC further calculated that converting overhead transmission facilities to underground would increase rates 49.7% for customers of investor-owned utilities (IOUs) over a 10-year period (**Table 1**).

Table 1. Revenue Requirements for IOUs To Convert Florida's Existing Transmission Facilities to Underground, and Rate Impact Over 10-Year Period

(in 2003 dollars)

Rate Impact	
Estimated cost of conversion	$51.8 billion
Estimated cost adjusted for inflation over 10 years	$57.9 billion
Levelized annual revenue requirement	$6.5 billion
Percentage rate impact (spread over all kilowatt-hours)	49.7%
Assumptions	
Weighted rate of return	12.04%
Property tax rate	1.86%
Operation and maintenance savings[a]	(0.7%)
Inflation rate	2.44%

Source: Florida Public Service Commission.

a. Federal Energy Regulatory Commission (FERC) Form 1 data do not separately identify transmission operation and maintenance (O&M) for overhead and buried transmission. FPSC used distribution O&M savings from the 2003 FERC Form 1 for its calculations.

Both the 1991 FPSC study and the 1998 Australian study included a cost savings to utilities due to fewer automobile collisions with utility poles. Both studies considered lost wages, medical expenses, insurance administration costs, property damage, and loss of life. According to FPSC, utilities would avoid approximately $117 million (2003 dollars) annually of accident-related costs.[66] Neither study considered that some communities would plant trees on old rights-of-way,

---

[66] FPSC 1991. Volume II. For consistency, CRS used the same GDP deflator index ratio of 1.299 as was used in the 1995 FPSC report to index the 1990 findings to 2003 dollars.

and a collision with a well-established tree could cause injury and death, though in this case, a utility would not likely be liable for costs associated with the accident.

Another benefit of burying power lines is a reduction of electrocutions from sagging or downed power lines. In addition, workers would be less likely to inadvertently make contact with a buried distribution line. The FPSC study calculated an annual avoided cost from contact accidents of $243,000 (2003 dollars) if all power lines were buried.[67]

Pricing

Some transmission-owning utilities argue that the current pricing mechanism for transmission discourages investment. FERC regulates all transmission, including unbundled retail transactions. Under the Federal Power Act (FPA), FERC is required to set "just and reasonable" rates for wholesale transactions.[68] FERC has traditionally determined rates by using an embedded cost method that includes recovery of capital costs, operating expenses, improvements, accumulated depreciation, and a rate of return. Traditionally, transmission owners have been compensated for use of their lines based on a contract path for the movement of electricity, generally the shortest path between the generator and its customer. However, electricity rarely follows a contract path and instead follows the path based on least impedance.[69] Transmission lines often carry electricity that has been contracted to move on a different path. As more bulk power transfers are occurring on the transmission system, transmission owners not belonging to RTOs (regional transmission organizations) are not always being compensated for use of their lines, because a contract path rarely follows the actual flow. This creates a disincentive for transmission owners to increase capacity.[70]

Under Order 2000,[71] FERC stated its interest in incentive ratemaking and, in particular, performance-based ratemaking. Those in favor of incentive ratemaking, including the electric utility industry, argue that incentives are needed (1) to encourage participation in regional transmission organizations (RTOs),[72] (2) to compensate for perceived increases in financial risk because of participation in a regional transmission organization, and (3) to facilitate efficient expansion of the transmission system.

FERC has used a "license plate" rate for transmission: a single rate based on customer location. As FERC is encouraging formation of large regional transmission organizations, FERC may move toward a uniform access charge, sometimes called *postage stamp rates*. With a postage stamp rate, users pay one charge for moving electricity anywhere within the regional transmission organization.

---

[67] Ibid.

[68] 16 U.S.C. 824(d)(a).

[69] *Impedance* is a measure of the resistive and reactive attributes of a component in an alternating-current circuit.

[70] National Economic Research Associates, *Transmission Pricing Arrangements and Their Influence on New Investments*, World Bank Institute (July 6, 2000).

[71] 89 FERC 61,285.

[72] A *regional transmission organization* is an independent organization that does not own the transmission lines but operates a regional transmission system on a non-discriminatory basis. For additional discussion on RTOs see, CRS Report RL32728, *Electric Utility Regulatory Reform: Issues for the 109th Congress*, by Amy Abel.

Postage stamp rates eliminate so-called rate *pancaking*, or a series of accumulated transmission charges as the electricity passes through adjacent transmission systems, and increases the pool of available generation. On the other hand, by moving to postage stamp rates, customers in low-cost transmission areas may see a rate increase, and high-cost transmission providers in the same area may not recover embedded costs, because costs are determined on a regional basis.

In early 2003, FERC began to consider raising the rate of return as a way to reflect the regulatory uncertainty in the industry and encourage transmission investment.[73] The proposal would give a 1% return-on-equity-incentive for *new* transmission projects operating under an RTO. Transfer of transmission assets to an RTO would also result in an incentive return on equity of between 0.5% and 2%. This could raise return on equity to approximately 14% for some transmission projects. Increases in the return on equity would increase consumers' electric bills. However, in 2000, the cost of transmission accounted for less than 10% of the final delivered cost of electricity.[74] While the industry is in favor of increasing the return on equity as a way of providing an incentive to invest, consumer groups are opposed to such proposals because of the potential to increase consumer rates.[75]

As required by § 1241 of EPACT05, FERC issued its Final Rule on transmission pricing on July 20, 2006.[76] Although the order identifies specific incentives that FERC will allow, the burden remains on an applicant to justify the incentives by showing that the new transmission capacity will reduce the cost of delivered power by reducing transmission congestion or will ensure reliability. The applicant will also have to show that the rate is just, reasonable, and not unduly discriminatory or preferential.[77]

Although the order identifies specific incentives that FERC will allow, the burden remains on an applicant to justify the incentives. Several consumer groups argue that the Final Rule is too permissive in offering rate incentives. Under the Final Rule, FERC requires that applicants pass a "nexus test," meaning that the requested incentives match the demonstrable risks and challenges faced by the applicant undertaking the project. The final rule applies the "nexus test" to each incentive, rather than to the package of incentives as a whole. The American Public Power Association (APPA) and the National Rural Electric Cooperative Association (NRECA) argue that this approach fails to protect consumers where an applicant seeks incentives that both reduce the risk of the project and offer an enhanced return on equity for increased risk. In response to comments on the Final Order, FERC issued an Order on Rehearing and determined that the nexus

---

[73] Federal Energy Regulatory Commission, *Proposed Pricing Policy for Efficient Operation and Expansion of the Transmission Grid*, Docket No. PL03-1-000 (January 15, 2003).

[74] Energy Information Administration, *Electric Sales and Revenue 2000*.

[75] Testimony of Gerald Norlander for the National Association of State Utility Consumer Advocates before the House Committee on Energy and Commerce, March 14, 2003, available at http://energycommerce.house.gov/108/Hearings/03132003hearing818/hearing.htm.

[76] Federal Energy Regulatory Commission Final Rule, Order Number 679, *Promoting Transmission Investment through Pricing Reform* (July 20, 2006), Docket Number RM06-4-000.

[77] The final rule authorizes FERC to approve the following incentive-based rate treatments: a rate of return on equity sufficient to attract new investment in transmission facilities; allowance of 100% of prudently incurred Construction Work in Progress (CWIP) in the rate base; recovery of prudently incurred pre-commercial operations costs; accelerated depreciation used for rate recovery; recovery of 100% of prudently incurred costs of transmission facilities that are canceled or abandoned due to factors beyond the control of the public utility; and deferred cost recovery.

---

requirement no longer will be applied separately to each incentive but that the total package of incentives must match the demonstrable risks or challenges.[78]

In addition, the National Association of Regulatory Utility Commissioners (NARUC), APPA, NRECA, Transmission Dependent Utility Systems (TDU Systems), and the Transmission Access Policy Study Group (TAPS) argued that under the Final Rule, FERC erred in rebuttably presuming that certain review processes such as state siting approvals and regional planning processes would satisfy the requirement that a transmission project ensure reliability or reduce congestion. Under FERC's Order on Rehearing, FERC will require that each applicant explain whether any process being relied upon for a rebuttable presumption includes a determination that the project is necessary to ensure reliability or reduce congestion.[79]

Since Order 689 was issued, projects have received transmission rate incentives, including American Electric Power (AEP) Service Corp. received approval from FERC for incentive rates for a new 765 kV, 550-mile transmission line that is expected to extend from West Virginia to New Jersey; Allegheny Energy Inc. (Allegheny) was granted rate incentives on a proposed 500 kilovolt transmission line within the PJM region; Duquesne Light Co.'s (Duquesne) petition for incentive rates was conditionally approved for several projects in Western Pennsylvania; and Commonwealth Edison Company (ComEd) was granted incentive rates for Phase II of the West Loop Project and Chicago. FERC has approved incentives for the AEP and Allegheny projects that include a return on investment (ROE) "at the high end of the zone of reasonableness, with the zone of reasonableness to be determined in a future proceeding," recovery of construction work in progress (CWIP) costs, the ability to expense and recover pre-construction and pre-operating costs, and accelerated depreciation.[80] FERC conditionally granted Duquesne's ROE request of up to one and one-half percentage points above a base-level ROE, recovery of CWIP costs, recovery of prudently incurred pre-commercial operations costs, and prudently incurred costs of the project in the event the project is cancelled due to factors beyond Duquesne's control.[81] ComEd was granted a one percentage point adder to their ROE and recovery of CWIP.

Regulatory Uncertainty

For many years, transmission owners and investors expressed concern that the regulatory uncertainty for electric utilities is inhibiting both new investment in and construction of transmission facilities. For example, repeal of the Public Utility Holding Company Act of 1935 (PUHCA) had been debated since 1996. Without clarification on whether PUHCA would be repealed, utilities stated that they were reluctant to invest in infrastructure. It was argued that repeal of PUHCA could significantly expand the ability of utilities to diversify their investment options.[82] EPACT05 repealed PUHCA, and FERC and state regulatory bodies are given access to

---

[78] Federal Energy Regulatory Commission Final Rule, Order on Rehearing, Order Number 679-A, *Promoting Transmission Investment through Pricing Reform* (December 22, 2006), Docket Number RM06-4-001, p. 21.

[79] Ibid., p. 4.

[80] 116 FERC 61,059, Docket Number EL06-50-000, p. 15, available at http://www.ferc.gov/whats-new/comm-meet/072006/E-15.pdf.

[81] FERC, Docket No. EL06-109-000, *et al.* (February 6, 2007), available at http://www.ferc.gov/EventCalendar/Files/20070206185852-EL06-109-000.pdf.

[82] For discussion of PUHCA repeal issues, see CRS Report RL32728, *Electric Utility Regulatory Reform: Issues for the 109th Congress*, by Amy Abel.

utility books and records. Removing this uncertainty could encourage additional investment in the transmission system.

In addition, FERC has been moving toward requiring participation in regional transmission organizations to create a more seamless transmission system. A fully operational regional transmission organization would operate the entire transmission system in a region and be able to replace multiple control centers with a single control center.[83] This type of control can increase efficiencies in the operation of the transmission system. RTO participants are required to adhere to certain operational guidelines, but these are not currently enforceable in court. Uncertainty over the form of an RTO, its operational characteristics, and the transmission rates for a specific region have apparently made utilities wary of investing in transmission. FERC has granted RTO status to several entities and conditionally approved others. If RTOs are able to operate successfully and develop a track record, some regulatory uncertainty will diminish.

On July 31, 2002, FERC issued a Notice of Proposed Rulemaking (NOPR) on standard market design (SMD).[84] This NOPR was highly controversial. FERC's stated goal of SMD requirements in conjunction with a standardized transmission service was to create "seamless" wholesale power markets that allow sellers to transact easily across transmission grid boundaries. The proposed rulemaking would have created a new tariff under which each transmission owner would be required to turn over operation of its transmission system to an unaffiliated independent transmission provider (ITP). The ITP, which could have been an RTO, would have provided service to all customers and would have run energy markets. Under the NOPR, congestion would have been managed with locational marginal pricing. FERC withdrew its SMD proposal shortly before passage of EPACT05.

---

[83] PJM operates with a single control center.
[84] FERC, Docket No. RM01-12-000.

Figure 5. Congested Lines in the Eastern Interconnection

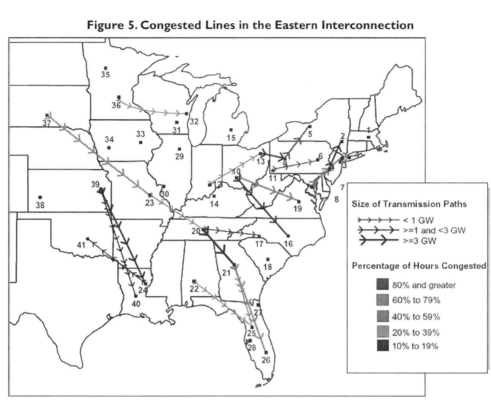

Source: U.S. Department of Energy, National Transmission Grid Study, May 2002.

Investment

Some contend that obtaining funding is the major impediment to transmission expansion.[85]
Utilities have traditionally raised capital from three sources: equity investors, internal cash flow,
and bondholders. Before 1978, utility stocks were seen as safe investments for investors. The
Three Mile Island nuclear accident and other cost overruns of nuclear facilities made utility
investment less attractive. Following enactment of the Energy Policy Act of 1992, many found
investing in non-traditional utilities (Enron, Mirant, etc.) to once again be an attractive option.
Following the California energy crises and the bankruptcy of several energy-related companies,
investors once again withdrew from heavily investing in utility stock. Between 2000 and 2002,
utility bonds had been unattractive to investors, according to Standard & Poor's.[86] Since then,
many utilities have had their bond ratings reduced. In 2002, there were 182 bond rating
downgrades of utility holding and operating companies and only 15 upgrades. A majority of
electric utilities (62%) had a bond rating of BBB or below while the number of those rated A- or
better fell from 51% to 38% in one year. Also, according to Standard & Poor's, debt and preferred

---

[85] Roseman, E., and Paul De Martini, *In Search of Transmission Capitalists*, Public Utilities Fortnightly
(April 1, 2003).

[86] Standard & Poor's, *U.S. Power Industry Experiences Precipitous Credit Decline in 2002; Negative Slope Likely to
Continue* (January15, 2003).

securities financing activity fell from $86 billion in 2001 to $74 billion in 2002. In addition, internal investment declined. The lack of investment options for utilities for transmission improvements had significantly slowed transmission capacity additions.

According to Standard & Poor's, the power sector had begun to experience an upward trend in bond ratings. During the first half of 2006, the U.S. power sector showed positive ratings gains, with company bond ratings being upgraded more often than downgraded. By the end of 2006, the investor-owned utility sector ratings leveled off with an equal number of upgrades and downgrades for the year.[87] This starkly contrasts with 2004 and 2005, when rating downgrades outpaced upgrades by about three to two.[88]

Conclusion

For the transmission system to operate efficiently and reliably, many observers argue that the tensions between economic, regulatory, and technology issues must be balanced. Currently, the transmission industry is widely viewed as being in a state of disequilibrium with significant regulatory and economic uncertainty. In addition, regional differences complicate regulatory solutions. A large component of regulatory uncertainty originates with a piece-meal approach to electric utility restructuring on both the federal and state level. In 1991, CRS stated that

> comprehensive regulatory reform of the electric power industry is neither desirable nor practical without a clearer vision of what form the industry should take. Too many uncertainties leave the future nature of the electric power industry such that a major overhaul of regulation would involve significant risks to the present stability of available and reliable electric power with little guarantee of improved service or lower costs.[89]

The Energy Policy Act of 1992 introduced competition to wholesale electric transactions without provisions for a comprehensive plan to address reliability issues and the development of efficient wholesale markets. In addition, many states have passed legislation or issued regulatory orders to introduce retail competition, each with its own set of rules for utilities to follow. Provisions in EPACT05 are intended to address many issues left outstanding by EPACT92. Although regulatory certainty may improve as a result of EPACT05, a clearer vision of the role of competition in the electric power industry, and additional investment in transmission infrastructure and technology, continues to be necessary to support an adequate and reliable transmission system.

Author Contact Information

Amy Abel
Section Research Manager
aabel@crs.loc.gov, 7-7239

---

[87] Standard & Poor's, *Pace of US Utility Raiding Activity Moderated in 2006* (January 23, 2007).

[88] Standard & Poor's, *Industry Report Card: U.S. Utility Second-Quarter Upgrade Surge Is Strongest in Years* (July 10, 2006).

[89] *Electricity: A New Regulatory Order?* Report prepared by the Congressional Research Service for the Committee on Energy and Commerce, U.S. House of Representatives, Committee Print 102-F (June 1991).

Congressional
Research
Service

Energy Independence and Security Act of 2007: A Summary of Major Provisions

Fred Sissine, Coordinator
Specialist in Energy Policy

February 22, 2008

Congressional Research Service

7-5700

www.crs.gov

RL34294

CRS Report for Congress ————————————————————
Prepared for Members and Committees of Congress

Summary

The Energy Independence and Security Act (P.L. 110-140, H.R. 6) is an omnibus energy policy law that consists mainly of provisions designed to increase energy efficiency and the availability of renewable energy. This report describes the key provisions of the enacted law, summarizes the legislative action on H.R. 6, and provides a summary of the provisions under each of the titles in the law.

The highlights of key provisions enacted into law are as follows:

- *Corporate Average Fuel Economy (CAFE).* The law sets a target of 35 miles per gallon for the combined fleet of cars and light trucks by model year 2020.

- *Renewable Fuels Standard (RFS).* The law sets a modified standard that starts at 9.0 billion gallons in 2008 and rises to 36 billion gallons by 2022.

- *Energy Efficiency Equipment Standards.* The adopted bill includes a variety of new standards for lighting and for residential and commercial appliance equipment. The equipment includes residential refrigerators, freezers, refrigerator-freezers, metal halide lamps, and commercial walk-in coolers and freezers.

- *Repeal of Oil and Gas Tax Incentives.* The enacted law includes repeal of two tax subsidies in order to offset the estimated cost to implement the CAFE provision.

The two most controversial provisions of H.R. 6 that were not included in the enacted law were the proposed Renewable Energy Portfolio Standard (RPS) and most of the proposed tax provisions, which included repeal of tax subsidies for oil and gas and new incentives for energy efficiency and renewable energy.

Congressional Research Service

Contents

Energy Independence and Security Act of 2007: A Summary of Major Provisions

Contacts

Congressional Research Service

Introduction

The Energy Independence and Security Act (P.L. 110-140, H.R. 6) is an omnibus energy policy law that consists mainly of provisions designed to increase energy efficiency and the availability of renewable energy. This report describes the key provisions of the enacted law, summarizes the legislative action on H.R. 6, and provides a summary of the provisions under each of the titles in the law.

Many analysts in the CRS Resources, Science, and Industry Division contributed to this report; their names and contact information are located on the back of the summary page.

Key Provisions

Provisions Included

The three key provisions enacted in P.L. 110-140 are the Corporate Average Fuel Economy (CAFE) Standards, the Renewable Fuel Standard (RFS), and the appliance/lighting efficiency standards.

Corporate Average Fuel Economy (CAFE) Standards

The law sets a target of 35 miles per gallon for the combined fleet of cars and light trucks by model year 2020. Also, a fuel economy program is established for medium- and heavy-duty trucks, and a separate fuel economy standard is created for work trucks. (For more details on issues related to the CAFE provision, see CRS Report RL33982, *Corporate Average Fuel Economy (CAFE): A Comparison of Selected Legislation in the 110th Congress*, by Brent D. Yacobucci and Robert Bamberger, and CRS Report R40166, *Automobile and Light Truck Fuel Economy: The CAFE Standards*, by Brent D. Yacobucci and Robert Bamberger.)

Renewable Fuel Standard (RFS)

The law sets a modified standard that starts at 9.0 billion gallons of renewable fuel in 2008 and rises to 36 billion gallons by 2022. Of the latter total, 21 billion gallons is required to be obtained from cellulosic ethanol and other advanced biofuels. (For more details on issues related to the RFS provision, see CRS Report RL34265, *Selected Issues Related to an Expansion of the Renewable Fuel Standard (RFS)*, by Brent D. Yacobucci and Tom Capehart)

Appliance and Lighting Efficiency Standards

Energy efficiency standards are set for broad categories of incandescent lamps (light bulbs), incandescent reflector lamps, and fluorescent lamps. A required target is set for lighting efficiency, and energy efficiency labeling is required for consumer electronic products. Also, efficiency standards are set by law for external power supplies, residential clothes washers, dishwashers, dehumidifiers, refrigerators, refrigerator-freezers, freezers, electric motors, residential boilers, commercial walk-in coolers, and commercial walk-in freezers. Further, DOE is directed to set standards by rulemaking for furnace fans and battery chargers.

Provisions Excluded

Two controversial provisions of H.R. 6 that were not included in the enacted law were the proposed Renewable Energy Portfolio Standard (RPS) and the proposed repeal of tax subsidies for oil and gas.

Renewable Energy Portfolio Standard (RPS)

Under an RPS, retail electricity suppliers (electric utilities) must provide a minimum amount of electricity from renewable energy resources or purchase tradable credits that represent an equivalent amount of renewable energy production. The minimum requirement is often set as a percentage share of a supplier's total retail electricity sales. The second degree amendment to H.R. 6 passed by the House on December 6, 2007, proposed a national RPS target that aimed to reach 15% of total electricity sales by 2020. Up to 4 percentage points of the 15% target could be met with energy efficiency measures. This provision was stripped out by the Senate and was not included in the final version of the bill. (For more details on issues related to the RPS provision, see CRS Report RL34116, *Renewable Energy Portfolio Standard (RPS): Background and Debate Over a National Requirement*, by Fred Sissine.)

Energy Tax Subsidies

The House-passed second degree amendment to H.R. 6 contained provisions that would have repealed about $22 billion of oil and gas subsidies that were designed to offset the cost of supporting a variety of energy efficiency and renewable energy tax incentives. These proposed incentives would have included a four-year extension of the renewable energy electricity production tax credit. Most of those provisions were stripped out by the Senate and were not included in the final bill. Enough tax revenue offsets were included to cover the estimated cost of the CAFE provision. (For more details about the proposed renewable energy incentives, see CRS Report RL34162, *Renewable Energy: Background and Issues for the 110th Congress*, by Fred Sissine. For more details about the proposed repeal of oil and gas subsidies, see CRS Report RL33578, *Energy Tax Policy: History and Current Issues*, by Salvatore Lazzari.)

Brief Legislative History of H.R. 6

House Passes H.R. 6

On January 18, 2007, the House passed the 14-page CLEAN Energy Act (H.R. 6) by a vote of 264-163.[1] The bill was crafted as part of the House Leadership's "Hundred Hours Legislation." It was designed only to establish a reserve to collect funds from repealed oil and gas subsidies that could be used to support new incentives for energy efficiency and renewable energy. The goal of the reserve was to reduce foreign oil dependence and to "serve other purposes." The actual uses of the reserve would have been determined at a later date by further legislation. The issues

---

[1] That version of H.R. 6 is described in CRS Report RS22571, *The Strategic Energy Efficiency and Renewables Reserve in the CLEAN Energy Act of 2007 (H.R. 6)*, by Fred Sissine.

involving the reserve were further addressed by the budget resolution process.[2] At that point, H.R. 6 was not an omnibus energy policy bill.

Senate Amends H.R. 6

On June 21, 2007, the Senate adopted an amendment in the nature of a substitute to H.R. 6 (S.Amdt. 1502).[3] This action transformed H.R. 6 into a 500-page omnibus energy policy bill, with a primary focus on energy efficiency and renewable energy. The new title was the Renewable Fuels, Consumer Protection, and Energy Efficiency Act of 2007. The Senate substitute was derived primarily from S. 1419 (of the same title), which, in turn, was composed from four major bills: the Energy Savings Act (S. 1321),[4] the Public Buildings Cost Reduction Act (S. 992), the Ten-in-Ten Fuel Economy Act (S. 357), and the Energy Diplomacy and Security Act (S. 193). Floor action to amend the substitute to attach tax provisions (S.Amdt. 1704) and a renewable energy electricity portfolio standard (S.Amdt. 1537) failed.[5] The key provisions of the Senate-passed substitute to H.R. 6 were appliance efficiency standards, an increase of the renewable fuel standard (RFS) to 36 billion gallons by 2022, and an increase of the combined corporate average fuel economy (CAFE) standards to 35 miles per gallon (mpg) by 2020.

House Approves H.R. 3221

On August 4, 2007, the House passed the omnibus energy policy bill, H.R. 3221, which had two divisions and 13 titles.[6] Division A contained provisions of the New Direction for Energy Independence, National Security, and Consumer Protection Act, which had nine titles composed from several bills.[7] An adopted floor amendment (H.Amdt. 748) added a 15% renewable energy portfolio standard (RPS).[8] Division B, the Renewable Energy and Energy Conservation Tax Act of 2007, contained the provisions of the House-approved version of H.R. 2776.[9] It added four tax titles to H.R. 3221, which included a four-year extension of the renewable electricity production tax credit (PTC) and several other tax incentives for energy efficiency and renewable energy.

---

[2] The budget resolution bills were H.Con.Res. 99 and S.Con.Res. 21.

[3] A summary of the Senate-passed substitute to H.R. 6 is presented in the appendices to CRS Report RL34135, *Omnibus Energy Efficiency and Renewable Energy Legislation: A Side-by-Side Comparison of Major Provisions in House-Passed H.R. 3221 with Senate-Passed H.R. 6*, coordinated by Fred Sissine.

[4] S. 1321 was, in turn, derived from several other Senate bills. Additional details about legislation that was incorporated into H.R. 6 are available in CRS Report RL33831, *Energy Efficiency and Renewable Energy Legislation in the 110th Congress*, by Fred Sissine, Mark Gurevitz, and Lynn J. Cunningham.

[5] For more details about efforts to include an RPS in H.R. 6, see CRS Report RL34116, *Renewable Energy Portfolio Standard (RPS): Background and Debate Over a National Requirement*, by Fred Sissine.

[6] A summary of House-passed H.R. 3221, and comparison with the Senate-passed substitute to H.R. 6, are presented in CRS Report RL34135, *Omnibus Energy Efficiency and Renewable Energy Legislation: A Side-by-Side Comparison of Major Provisions in House-Passed H.R. 3221 with Senate-Passed H.R. 6*, coordinated by Fred Sissine.

[7] The bills were H.R. 364, H.R. 2304, H.R. 2313, H.R. 2337, H.R. 2389, H.R. 2420, H.R. 2635, H.R. 2701, H.R. 2773, H.R. 2774, H.R. 2847, and a draft bill by the Committee on Energy and Commerce.

[8] The proposed RPS was nearly identical to the 15% RPS that had been proposed in S.Amdt. 1537, except that it would have allowed up to 4 percentage points of the 15% requirement to be met with energy efficiency measures.

[9] The rule (H.Res. 615) that brought H.R. 3221 to the floor directed that upon engrossment of H.R. 3221, the text of H.R. 2776 would be added to the end of H.R. 3221 as Division B.

Informal House-Senate Negotiations

Because the House omnibus bill (H.R. 3221) and the Senate omnibus bill (H.R. 6) had different bill numbers, the bills could not be taken directly to conference committee.[10] However, after the House completed action on H.R. 3221, informal bipartisan negotiations over the omnibus energy bills began between the House and Senate. Key issues included CAFE, the renewable fuel standard, the RPS provision in H.R. 3221, and a proposed repeal of certain oil and natural gas subsidies to offset costs for new energy efficiency and renewable energy tax incentives.[11] In November 2007, EIA issued a report on the possible impacts of the proposals in H.R. 3221 to establish an RPS and repeal selected oil and gas subsidies.[12]

On December 1, 2007, the Ranking Member of the Senate Committee on Energy and Natural Resources stated that the House Leadership's intent to include an RPS led him to cease negotiations.[13] Further, on December 3, 2007, the White House announced, by letter to the Speaker of the House, that it planned to veto the negotiated bill—if it included an RPS, repeal of oil and gas tax subsidies, and certain other provisions.[14] Subsequently, the Office of the Speaker issued a letter of response to the White House letter, which stated that the bill addressed the Administration's concerns.[15]

House Amends Senate Amendment to H.R. 6

On December 6, 2007, the House passed (235-181) its second-degree amendments to the Senate-passed amendments to H.R. 6. This "second version" of a House omnibus energy bill was derived primarily by trimming and modifying H.R. 3221 and adding major new provisions on CAFE and RFS. The House-passed bill included a proposed increase of the CAFE standard to 35 miles per gallon by 2020 and an increase of the renewable fuel standard to 36 billion gallons per year by 2022. The House bill also included a proposed 15% renewable electricity portfolio standard and $21 billion of new tax incentives for energy efficiency and renewable energy measures.[16] The bill proposed to offset the new tax incentives with revenue generated by a repeal of about $21 billion in tax subsidies for oil and natural gas.

---

[10] For an explanation of how the alternative to the conference process works, see CRS Report 98-696, *Resolving Legislative Differences in Congress: Conference Committees and Amendments Between the Houses*, by Elizabeth Rybicki.

[11] More details about the House and Senate bills developed up to that point are available in CRS Report RL34135, *Omnibus Energy Efficiency and Renewable Energy Legislation: A Side-by-Side Comparison of Major Provisions in House-Passed H.R. 3221 with Senate-Passed H.R. 6*, coordinated by Fred Sissine.

[12] EIA. Oil and Natural Gas Market Supply and Renewable Portfolio Standard Impacts of Selected Provisions of H.R. 3221. November 2007. 11 p. http://www.eia.doe.gov/oiaf/servicerpt/bmy/pdf/bmy.pdf.

[13] The statement is available on the Committee's website, at http://energy.senate.gov/public/ index.cfm?FuseAction=PressReleases.Detail&PressRelease_id=235405&Month=12&Year=2007.

[14] The White House. *Letter to House Speaker Nancy Pelosi from Allan B. Hubbard*. December 3, 2007. 2 p.; also, see UPI. *U.S. Energy Chief: Energy Bill Concerns*. December 4, 2007. http://www.upi.com/International_Security/ Emerging_Threats/Analysis/2007/12/04/us_energy_chief_energy_bill_concerns/3426/.

[15] Office of the Speaker of the House. *Pelosi to Hubbard: Bipartisan Energy Legislation Addresses White House Concerns*. December 5, 2007. http://www6.lexisnexis.com/publisher/ EndUser?Action=UserDisplayFullDocument&orgId=574&topicId=25148&docId=l:711920770&start=9.

[16] As with H.R. 3221, the proposed RPS would have allowed up to 4 percentage points of the 15% to be met with energy efficiency measures.

Subsequently, the White House released a Statement of Administration Policy on the House-passed bill. It threatened to veto the bill because it:

> ... raises taxes in a way that will increase energy costs facing consumers. It would also impose a national renewable electricity standard that would ignore the specific energy and economic needs of individual states.[17]

On the latter point, the *Statement* found a potential for higher electricity costs, noted regional differences in renewable energy resource availability, and called for leaving such standards to states' discretion. Regarding CAFE, it noted that H.R. 6 did not provide a clear role for the Environmental Protection Agency in regulating fuel economy. Regarding RFS, it objected to prescriptions that would employ greenhouse gas content as a criterion for fuel eligibility. Also, a strong objection was expressed toward the proposed repeal of oil and gas tax deductions for manufacturers. Further concerns were cited about tax credit bonds for renewables, Davis-Bacon prevailing wage requirements, federal building efficiency provisions, and appliance efficiency standards.

Senate Removes RPS and Most Tax Provisions of H.R. 6

On December 6, 2007, the House-passed version of H.R. 6—with provisions for an RPS and for the repeal of oil and gas subsidies—failed in a Senate cloture vote (52-43). A new Senate substitute amendment (S.Amdt. 3841) was prepared by stripping out the RPS and modifying the package of tax provisions somewhat.

In the morning of December 13, 2007 a cloture vote on the Senate substitute failed (59-40). The Senate then prepared another substitute amendment (S.Amdt. 3850) by stripping out the tax incentives for energy efficiency and renewable energy and removing the provisions that would have repealed subsidies for oil and natural gas producers.

In the evening of December 13, 2007, the Senate approved (86-8) that substitute amendment to the House-passed version of H.R. 6. The Senate substitute was nearly identical to the House-passed bill, except that the RPS provision and most tax provisions had been taken out. The resultant bill was subsequently approved by the House (314-100) and signed into law as P.L. 110-140.

Title I: Energy Security Through Improved Vehicle Fuel Economy

Subtitle A, Increased Corporate Average Fuel Economy

This subtitle requires an increase in CAFE standards and a restructuring of the fuel economy program. A single CAFE standard of 35 miles per gallon (mpg) by MY2020 is established, and the distinction between the passenger car and light truck fleet is preserved. The new standards

---

[17] Office of Management and Budget. *Statement of Administration Policy on H.R. 6, the Energy Independence and Security Act of 2007.* December 6, 2007. 3 p. http://www.whitehouse.gov/omb/legislative/sap/110-1/hr6sap-h_2.pdf.

---

will be based on vehicle attributes and expressed in the form of a mathematical function. Interim standards will be set, beginning with MY2011. Manufacturers will be required to come within 92% of the standard for a given model year. However, manufacturers can earn credits for exceeding the standards in one vehicle class that can be applied to boost, within limitations, the CAFE of a different vehicle class that is falling short of compliance. Additionally, credits may be sold and bought between manufacturers. CAFE credits for the manufacture of flexible-fueled vehicles (FFV) are retained but phased out by MY2020. Civil penalties assessed for non-compliance will be deposited to the general fund of the U.S. Treasury to support future rulemaking and to provide grants to manufacturers for research and development, and retooling in support of increasing the fuel efficiency of their fleets. The law requires the development of standards for "work trucks" and commercial medium- and heavy-duty on-highway vehicles. (For additional information, see CRS Report RL33413, *Automobile and Light Truck Fuel Economy: The CAFE Standards*, by Brent D. Yacobucci and Robert Bamberger.)

Subtitle B, Improved Vehicle Technology

This subtitle establishes a loan guarantee program for advanced battery development, grant programs for plug-in hybrid vehicles, incentives for purchasing heavy-duty hybrid vehicles for fleets, and credits for various electric vehicles.

Subtitle C, Federal Vehicle Fleets

Federal agencies are prohibited from acquiring any light-duty motor vehicle or medium-duty passenger vehicle that is not "a low greenhouse gas emitting vehicle" as defined in this subtitle. Alternatively, the agency may demonstrate that it has adopted cost-effective policies to reduce its petroleum consumption sufficiently to achieve a comparable reduction in greenhouse gas emissions. By 2015, federal agencies are required to achieve at least a 20% reduction in annual petroleum consumption and a 10% increase in annual alternative fuel consumption. These increases are to be calculated from a 2005 baseline. Interim milestones will be established and agencies will report annually on their progress. The regulations governing this program are required to be issued not later than 18 months after enactment.

Title II: Energy Security Through Increased Production of Biofuels

Subtitle A, Renewable Fuel Standard

This subtitle extends and increases the renewable fuel standard (RFS) set by P.L. 109-58 (§1501). The RFS requires minimum annual levels of renewable fuel in U.S. transportation fuel. The previous standard was 5.4 billion gallons for 2008, rising to 7.5 billion by 2012. The new standard starts at 9.0 billion gallons in 2008 and rises to 36 billion gallons in 2022. Starting in 2016, all of the increase in the RFS target must be met with advanced biofuels, defined as cellulosic ethanol and other biofuels derived from feedstock other than corn starch—with explicit carve-outs for cellulosic biofuels and biomass-based diesel. The EPA Administrator is given authority to temporarily waive part of the biofuels mandate, if it were determined that a significant renewable feedstock disruption or other market circumstance might occur. Renewable fuels produced from new biorefineries will be required to reduce by at least 20% the life cycle greenhouse gas (GHG) emissions relative to life cycle emissions from gasoline and diesel. Fuels produced from biorefineries that displace more than 80% of the fossil-derived processing fuels used to operate a biofuel production facility will qualify for cash awards. Several studies are required on the impacts of an RFS expansion on various sectors of the economy. (For more details on issues related to the RFS proposal, see CRS Report RL34265, *Selected Issues Related to an Expansion of the Renewable Fuel Standard (RFS)*, by Brent D. Yacobucci and Tom Capehart.)

Subtitle B, Biofuels Research and Development (R&D)

This subtitle promotes research on the expansion of the use of biodiesel and biogas as motor fuels. Grants are authorized for R&D and commercial applications of cellulosic biofuels technologies and for the conversion of existing corn-based ethanol plants to produce cellulosic biofuels. The Secretary of Energy is required to report to Congress on the feasibility of algae as a feedstock for biofuels production. The subtitle also promotes university-based R&D on biofuels.

Subtitle C, Biofuels Infrastructure

This subtitle aims to improve information about federal biofuels research programs, focus research on infrastructure and biorefineries, study potential impacts of increased biofuels use, and increase authorized funding for DOE biofuels research. A funding authorization of $25 million is established to provide grants for biofuels research, development, and demonstration (RD&D) and commercial applications in states that have low rates of ethanol production. A university-based program is authorized to provide grants of up to $2 million for R&D on renewable energy technologies. Priority is given to universities in low-income and rural communities with proximity to trees dying of disease or insect infestation.

DOE is directed to create a grant program to help establish or convert infrastructure to use renewable fuels, including E85 (85% ethanol). The Energy Policy Act of 2005 (EPACT, P.L. 109-58) authorization for grants to support cellulosic ethanol production is increased. A grant program is authorized to support production of flexible-fueled vehicles. Studies are also required on the market penetration of flexible-fueled vehicles, the feasibility of constructing dedicated ethanol pipelines, the feasibility of using greater percentages of ethanol in fuel blends, and the adequacy of railroad transportation for delivery of ethanol fuel.

Subtitle D, Environmental Safeguards

Previously, under the Clean Air Act (§211[f]), no new fuels or fuel additives could be introduced into commerce unless granted a waiver by the Environmental Protection Agency (EPA). If EPA did not act within 180 days of receiving a waiver request, the waiver was treated as granted.[18] Section 251 tightens the waiver provision. It amends the Clean Air Act to prohibit the introduction of new fuels or fuel additives unless EPA explicitly grants a waiver. After receiving a waiver request, EPA will now have 270 days to take final action.

---

[18] See 42 U.S.C. 7545(f).

Title III: Energy Savings Through Improved Standards for Appliances and Lighting

Subtitle A, Appliance Energy Efficiency

This title sets, by statute, new efficiency standards for external power supplies, residential clothes washers, dishwashers, dehumidifiers, refrigerators, refrigerator-freezers, freezers, electric motors, and residential boilers. DOE is allowed to establish regional variations in standards for heating and air conditioning equipment. DOE is required to complete a rulemaking process for furnace fans by 2013. Federal agencies are directed to purchase devices that limit standby power use. DOE is directed to issue a final rule that sets efficiency standards for battery chargers. Certain energy efficiency measures for walk-in coolers and walk-in freezers are set by law. Also, several procedural changes are now in place to expedite the DOE rulemaking process.

Subtitle B, Lighting Energy Efficiency

Section 321 sets an energy efficiency standard for general service incandescent lamps, provides for consumer education and lamp labeling, and requires market assessments and a consumer awareness program. Section 322 sets energy efficiency standards for incandescent reflector lamps and fluorescent lamps. For federal buildings, Section 323 sets energy efficiency requirements for GSA-leased space and for use of energy efficient lighting fixtures and bulbs in those leased spaces. Section 324 sets energy efficiency standards for metal halide lamp fixtures designed to be operated with lamps rated between 150 watts and 500 watts. Section 325 directs the Consumer Product Safety Commission to set energy efficiency labeling requirements for consumer electronic products.

Title IV: Energy Savings in Buildings and Industry

Subtitle A, Residential Building Efficiency

Section 411 increases the funding authorization for DOE's Weatherization Program, providing $3.75 billion over five years. Under Section 412, DOE is directed to conduct a study of the renewable energy system rebate program described in §206(c) of the Energy Policy Act of 2005. The study aims to determine the minimum funding the program would need to be viable. Further, DOE is directed to propose an implementation plan. Section 413 requires DOE to establish energy efficiency standards for manufactured housing.

Subtitle B, High-Performance Commercial Buildings

This subtitle encourages the development of more energy-efficient "green" commercial buildings. Section 421 creates an Office of Commercial High Performance Green Buildings at DOE. Section 422 establishes a zero-energy commercial buildings initiative. A national goal is set to achieve zero-net-energy use for new commercial buildings built after 2025. A further goal is to retrofit all pre-2025 buildings to zero-net-energy use by 2050. Section 423 requires that DOE establish a national clearinghouse for information and public outreach about high-performance green buildings.

Subtitle C, High-Performance Federal Buildings

Section 431 requires that total energy use in federal buildings, relative to the 2005 level, be reduced 30% by 2015. Section 432 directs that federal energy managers conduct a comprehensive energy and water evaluation for each facility at least once every four years. For new federal buildings and major renovations, Section 433 requires that fossil-fuel energy use—relative to the 2003 level—be reduced 55% by 2010 and be eliminated (100% reduction) by 2030. Section 434 requires that each federal agency ensure that major replacements of installed equipment (such as heating and cooling systems), or renovation or expansion of existing space, employ the most energy efficient designs, systems, equipment, and controls that are life-cycle cost effective. Section 435 prohibits federal agencies from leasing buildings that have not earned an EPA Energy Star label. Section 436 requires GSA to establish an Office of Federal High-Performance Green Buildings to coordinate green building information and activities within GSA and with other federal agencies. The Office must also develop standards for federal facilities, establish green practices, review budget and life-cycle costing issues, and promote demonstration of innovative technologies. Section 437 directs the Government Accountability Office (GAO) to audit the implementation of activities required under this subtitle. The audit must cover budget, life-cycle costing, contracting, best practices, and agency coordination. Section 438 requires federal facility development projects with a footprint exceeding 5,000 square feet to use site planning, design, construction, and maintenance strategies to control storm water runoff. Section 439 directs GSA to review the current use of, and design a strategy for increased use of, cost-effective lighting, ground source heat pumps, and other technologies in GSA facilities. Section 440 authorizes $4 million per year over five years to support work under sections 434-439 and 482. For the purpose of conducting life-cycle cost calculations, Section 441 increases the time period from 25 years, in prior law, to 40 years.

Subtitle D, Industrial Energy Efficiency

Section 451 directs DOE to conduct research on, develop, and demonstrate new processes, technologies, and operating practices and techniques to significantly improve the energy efficiency of equipment and processes used by energy-intensive industries. Section 452 directs EPA to establish a recoverable waste energy inventory program. This program must include an ongoing survey of all major industrial and large commercial combustion sources in the United States. EPA is required to identify the potential for economically feasible waste energy recovery, create a grant program to support waste energy recovery, and strengthen "clean energy centers" that analyze waste energy recovery. Section 453 directs DOE to initiate a voluntary national information program for widely used data centers and data center equipment for which there is significant potential for energy savings. DOE is also tasked with helping to devise strategies to improve energy efficiency at these data centers.

Subtitle E, Healthy High-Performance Schools

Section 461 creates a grant program for *Healthy High-Performance Schools* that aims to encourage states, local governments, and school systems to build green schools. EPA, in consultation with the Department of Education, is allowed to provide grants to state agencies to provide technical assistance and help with the development of state plans for school building design. Also, EPA is directed to develop model voluntary guidelines for school site selection. In addition to other environmental aspects, the grants and guidelines must have a focus on energy efficiency, natural daylighting, and other energy-related features. Section 462 directs EPA to lead a detailed study of how sustainable building features, such as energy efficiency, affect multiple perceived indoor environmental quality stressors on students in K-12 schools.

Subtitle F, Institutional Entities

Section 471 creates a program of grants and loans to support energy efficiency and energy sustainability projects at public institutions.

Subtitle G, Public and Assisted Housing

Section 481 directs the Department of Housing and Urban Development (HUD) to update energy efficiency standards for all public and assisted housing.

Subtitle H, General Provisions

Section 491 calls for the DOE Office of Commercial High Performance Buildings and the GSA Office of Federal High Performance Buildings to jointly develop guidelines for demonstration projects. In accordance with the guidelines, one federal project must be undertaken annually over a five-year period, supported by a $10 million funding authorization. Also, a total of four projects are to be undertaken at different universities over the five-year period, supported by an additional $10 million funding authorization. Section 492 calls for these two offices to undertake a joint survey of research on green buildings, coordinate efforts to develop a research plan, and identify potential benefits of green buildings for security, natural disasters, and emergency needs of the federal government. Section 493 requires EPA to create a program of competitive grants to local governments for green building demonstration projects. Section 494 directs the Office of Commercial High Performance Buildings and the Office of Federal High Performance Buildings to jointly appoint a Green Building Advisory Committee with representatives from a variety of backgrounds, including federal agencies, state and local governments, building industry experts, security advisors, and environmental health experts. Section 495 calls for DOE to create an advisory committee on energy efficiency finance to help lower costs and increase investment for energy efficiency technologies.

Title V: Energy Savings in Government and Public Institutions

Subtitle A, United States Capitol Complex

Section 501 allows the Architect of the Capitol (AOC) to perform a feasibility study regarding construction of a photovoltaic roof for the Rayburn House Office Building. Under Section 502, the AOC is allowed to construct a fuel tank and pumping system for E85 (85% ethanol) fuel at or within close proximity to the Capitol Grounds Fuel Station. Section 503 requires the AOC, to the maximum extent practicable, to include energy efficiency measures, climate change mitigation measures, and other appropriate environmental measures in the Capitol Complex Master Plan. Under Section 504, the AOC is directed to operate the steam boilers and chiller plant at the Capitol Power Plant in the most energy efficient manner possible to minimize carbon emissions and operating costs. Further, Section 505 requires the AOC to install technologies for the capture and storage or use of carbon dioxide emitted from coal combustion in the Capitol Power Plant.

Subtitle B, Energy Savings Performance Contracting

Section 511 eliminates the advance reporting requirement for Energy Savings Performance Contracts (ESPCs) that have a cancellation ceiling exceeding $10 million. Section 512 increases ESPC funding flexibility by allowing a combination of appropriated funds and private financing. Section 513 restricts federal agencies from limiting the duration of ESPCs to less than 25 years or limiting the total amount of obligations. Further, this section permits the criteria for savings verification to satisfy the requirement for energy audits. Also, it directs federal agencies to modify existing ESPCs to conform with the requirements of this subtitle. Section 514 permanently authorizes ESPCs.

Section 515 extends the definition of energy savings reduction to include increased use of an existing energy source by cogeneration or heat recovery, use of excess electrical or thermal energy generated from onsite renewable sources or cogeneration, and increased energy-efficient use of water resources. Section 516 permits agencies to retain the full amount of energy and water cost savings obtained from utility incentive programs. Section 517 authorizes $750,000 per year over five years for a program to train contract officers in negotiating ESPCs. Section 518 directs the Department of Defense (DOD) and DOE to study the potential use of ESPCs in nonbuilding applications, which is defined to include vehicles and federally owned equipment to generate electricity or transport water.

Subtitle C, Energy Efficiency in Federal Agencies

Under Section 521, GSA is directed to use up to $30 million—subject to appropriation—from FY2007 and prior years' unobligated balances of the Federal Buildings Fund to support the installation of a solar photovoltaic system for the DOE headquarters building in the District of Columbia. Section 522 prohibits, except under certain circumstances, the purchase of incandescent light bulbs for use in Coast Guard office buildings. Section 523 requires 30% of the hot water demand in new federal buildings (and major renovations) to be met with solar hot water equipment, provided it is life-cycle cost-effective. Section 524 encourages federal agencies to minimize standby energy use in purchases of energy-using equipment. Section 525 requires federal procurement to focus on use of Energy Star and Federal Energy Management Program (FEMP)-designated products. Section 526 prohibits federal agencies from procuring synfuel unless its life cycle GHG emissions are less than those for conventional petroleum sources. Section 527 directs each federal agency subject to any requirements under this title to issue an annual report that describes the status of initiatives to improve energy efficiency, reduce energy costs, and reduce GHG emissions. Section 528 requires the Office of Management and Budget (OMB) to submit an annual report to Congress that summarizes the information reported under Section 527, evaluates overall progress toward the goals of Section 527, and recommends additional actions needed to meet those goals. Section 529 directs the Federal Energy Regulatory Commission (FERC) to conduct a national assessment of demand response, including an estimate of nationwide demand response out to a 10-year horizon. Further, FERC is required to prepare a

National Action Plan on Demand Response, with cooperation from industry. Annual funding of $10 million per year is authorized over three years.

Subtitle D, Energy Efficiency of Public Institutions

Section 531 increases annual funding authorizations for DOE's state energy programs. Under Section 532, electric and natural gas utilities are required to make energy efficiency a priority resource and to integrate energy efficiency into resource plans and planning processes. Further, the utilities are directed to modify their rates to align their incentives with the delivery of cost-effective energy efficiency and promote energy efficiency investments. Utilities are encouraged to consider several policy options for achieving those goals.

Subtitle E, Energy Efficiency and Conservation Block Grants

This subtitle establishes an energy efficiency block grant program. Section 541 provides definitions of program elements. Section 542 directs DOE to establish an energy efficiency and conservation block grant program to help reduce energy use and emissions at the local and regional level. Section 543 establishes allocation percentages for grants provided under this subtitle. Section 544 enumerates the allowed purposes for the use of funds provided under this subtitle, which includes strategic planning, consultant services, and energy audits. Section 545 provides eligibility requirements for grants under this program, including payment of prevailing wage rates, submission of a strategic plan, and sharing of information. Section 546 sets criteria for minimum allocations of competitive grant funding. Section 547 specifies that DOE may review and evaluate the performance of grant recipients and withhold funds from those it deems have failed to achieve compliance. To support the grant program, Section 548 authorizes $2 billion annually over five years. Additional funding is authorized to cover administrative costs of the program. Section 548 stresses that funding will supplement, not replace, funding provided by DOE under the Weatherization and State Energy programs.

Title VI: Accelerated Research and Development

Subtitle A, Solar Energy

Section 602 aims to improve the cost and effectiveness of thermal energy storage technologies that could improve the operation of concentrating solar power electric generating plants. Section 603 calls for improved integration of concentrating solar power into regional electricity transmission systems.

Subtitle B, Geothermal Energy

DOE is directed to support programs of R&D, demonstration, and commercial application to expand the use of geothermal energy. Section 613 directs DOE to support programs that (1) develop advanced prospecting tools to locate and develop hidden geothermal resources, and (2) demonstrate advanced exploratory drilling technologies and techniques with industry partners. Section 614 directs DOE to support programs to develop components and systems necessary to develop, produce, monitor, and model the performance of geothermal reservoirs used to produce geothermal energy. In addition, Section 614 directs DOE to support programs that mitigate or prevent environmental damage from geothermal energy development.

Section 615 directs DOE to support enhanced geothermal system development, whereby geothermal reservoir systems are engineered (as opposed to naturally occurring systems) by creating fractures and permeable conduits via reservoir stimulation. DOE would support R&D programs for enhanced geothermal system technologies and for reservoir stimulation and support demonstration projects at a minimum of four sites.

DOE is directed to establish a program of R&D, demonstration, and commercial application for geothermal energy production from oil and gas fields and from geopressured resources.[19] Section 616 directs DOE to implement a grant program for at least three demonstration projects that use geothermal techniques to extract energy from marginal, unproductive, and productive oil and gas fields. Also, DOE is directed to establish a grant program for the recovery of energy from geopressured resources.

Section 618 directs DOE to establish a Center for Geothermal Technology Transfer, via a grant to an institution of higher learning or consortium thereof, that would serve as an information clearinghouse for the geothermal industry, make data available to the public, and coordinate R&D efforts among national and international partners. Section 619 would rename DOE's GeoPowering the West program as "GeoPowering America" and expand its geothermal technology transfer activities to cover the entire United States. Section 620 would award a grant on a competitive basis to an institution of higher education to establish a geothermal-powered energy generation facility on the institution's campus.

Section 624 directs DOE to support international geothermal energy development through collaborative efforts to promote geothermal R&D and deployment of geothermal technologies. Section 625 directs DOE to make grants to eligible entities from "high-cost regions"[20] of the United States for a feasibility study, demonstration, and commercial application of technologies related to geothermal energy.

Subtitle B authorizes $90 million annually for geothermal activities, of which $10 million is designated for activities under Section 616. An additional $5 million is authorized annually for the Intermountain West Geothermal Consortium, and $5 million is authorized annually for Section 624. All of the foregoing authorizations are in effect from 2008 to 2012.

Subtitle C, Marine and Hydrokinetic Renewable Energy Technologies

DOE is directed to create an R&D program focused on technology that produces electricity from waves, tides, currents, and ocean thermal differences (§633). A report to Congress is required. Further, DOE is instructed to award grants to institutions of higher education (or consortia

---

[19] Geopressured resources are geothermal deposits of hot water or steam found in sedimentary rocks under higher than normal pressures and that are saturated with oil and gas.

[20] These "high cost" regions are defined as places where the average cost of retail power exceeds 150% of the national average.

thereof) to establish National Marine Renewable Energy Research, Development, and Demonstration Centers (§634).

Subtitle D, Energy Storage for Transportation and Electric Power

The U.S. Energy Storage Competitiveness Act of 2007 directs DOE to conduct a cost-shared RD&D program to support the ability of the nation to remain globally competitive in energy storage systems for electric drive vehicles, stationary applications, and electricity transmission and distribution. An Energy Storage Advisory Council will be created, with responsibility for preparing a five-year research plan. Also, through competitive bids, DOE will establish four energy storage research centers managed by the Office of Science. DOE is required to conduct energy storage demonstration projects. Also, DOE is to investigate secondary applications of energy storage equipment and to examine technologies and processes for final recycling and disposal of energy storage equipment. After five years of program operation, the law will require a review of the program by the National Academy of Sciences. A total authorization of nearly $3 billion is provided over a 10-year period.

Subtitle E, Miscellaneous Provisions

Section 651 directs DOE to establish an RD&D program to determine ways in which the weight of motor vehicles could be reduced to improve fuel efficiency without compromising passenger safety. This will focus on the development of new materials and on reducing the cost of lightweight materials. An $80 million authorization is provided over a five-year period.

Section 652 directs DOE to report on the state of technology development for "advanced" insulation with an R-value greater than R35 per inch. The report is to include an estimate of potential cost savings by applying such insulation to covered refrigeration units. If sufficient cost savings are projected, DOE will then be directed to conduct a cost-shared demonstration program to show actual cost savings. An $8 million funding authorization is provided for that program.

Section 653 changes the sulfur dioxide (SO_2) criterion for clean coal power plants from a percentage basis (99% of SO_2 removed) to a weight-by-energy basis (no more than 0.04 pounds of SO_2 per million Btu).

Section 654 on the "H-Prize" directs DOE to conduct a competitive program to award cash prizes to advance R&D, demonstration, and commercial application of hydrogen energy technologies. Prizes can be a mix of federal appropriations and funds provided by an entity that DOE chooses to administer the program. The program sunsets in 2018. Prize categories include technology advancements in hydrogen production, storage, distribution, and use; prototypes of hydrogen vehicles and products; and technologies that "transform" distribution or production. DOE is required to report to Congress annually, identifying award recipients, technologies developed, and specific actions undertaken to commercialize the technologies. More than $1 billion is authorized over a 10-year period.

Section 655 directs DOE to create the "Bright Tomorrow" lighting prizes for solid state (LED) lighting developments that achieve targeted levels of energy efficiency and other traits. Two specific categories are a solid state replacement for a 60-watt incandescent light and a replacement for the PAR Type 38 halogen light. Also, a prize is established for a "twenty-first century lamp" that achieves certain output, efficiency, and color targets. After the awards are made, DOE is required to develop guidelines for federal agency purchases of the incandescent and halogen replacements, with the goal of complete replacement within five years.

Section 656 directs DOE to establish a cost-shared Renewable Energy Innovation Manufacturing Partnership Program to make awards to support RD&D on advanced manufacturing processes, materials, and infrastructure for renewable energy technologies. Further goals are to increase domestic renewable energy production and better coordinate federal, state, and private resources through partnerships. Solar, wind, biomass, geothermal, energy storage, and fuel cell systems are eligible forms of equipment.

Title VII: Carbon Capture and Sequestration

Subtitle A, Carbon Capture and Sequestration Research, Development, and Demonstration

DOE's program for carbon capture and sequestration R&D is expanded and will include large-scale demonstration projects. DOE is directed to engage the National Academy of Sciences (NAS) to conduct a review of the program. DOE is directed to work with the NAS to develop interdisciplinary graduate degree programs with emphasis on geologic sequestration science. A university-based R&D grant program will be established to study carbon capture and sequestration using various types of coal. EPA is directed to assess potential impacts of carbon sequestration on public health and safety and the environment. Further, injection and sequestration activities under this subtitle are subject to the requirements of the Safe Drinking Water Act.

Subtitle B, Carbon Capture and Sequestration Assessment and Framework

Section 711 directs the Department of the Interior (DOI) to develop a methodology for an assessment of the national potential for geologic storage of carbon dioxide. Following publication of the methodology, DOI will be required to complete an assessment of national capacity for carbon dioxide storage in accordance with the methodology.

Section 712 directs DOI to develop a methodology for an assessment of the total capacity of ecosystems to sequester carbon and the ability of ecosystems to reduce emissions of carbon dioxide, methane, and nitrous oxides in ecosystems through management practices. Following publication of the methodology, DOI will be required to complete a national assessment of the quantity of carbon stored in and released from ecosystems, and the annual flux of carbon dioxide, methane, and nitrous oxides in and out of ecosystems.

Section 713 calls for DOI to maintain records, and an inventory, of the quantity of carbon dioxide stored within federal mineral leaseholds.

Section 714 directs DOI to submit a report on a recommended regulatory framework for managing geologic carbon sequestration on public lands. The report must include an assessment of options to ensure that the United States receives fair market value for the use of public land, the proposed procedures for public review and comment, procedures for protecting natural and

---

cultural resources of the public land overlying the geologic sequestration sites, a description of the status of liability issues related to the storage of carbon dioxide in public land, identification of legal and regulatory issues for cases where the United States owns title to the mineral resources but not the overlying land, identification of issues related to carbon dioxide pipeline rights-of-way, and recommendations for additional legislation that may be required for adequate public land management and leasing to accommodate geologic sequestration of carbon dioxide and pipeline rights-of-way.

Title VIII: Improved Management of Energy Policy

Subtitle A, Management Improvements

Section 801 directs DOE to conduct a 10-year national media campaign to educate consumers to save energy and reduce oil use. Competitive bidding is required for contracting the media services. A funding authorization of $5 million per year is provided for five years. An annual report to Congress is required.

Section 802 authorizes the Federal Coordinator for the Alaska Natural Gas Transportation Projects to appoint and terminate personnel and to pay appointed and temporary personnel up to a maximum of the level III rate of the Executive Schedule. The Federal Coordinator is granted authority to establish various payment requirements and to use funds raised without further appropriation. This authority does not affect the authority of the Secretary of the Interior.

Section 803 creates a 50% matching grant program for constructing small renewable energy projects. Alaskan small hydroelectric power projects must have an electrical generation capacity less than 15 megawatts to qualify. Eligible applicants include local governments, utilities, and Indian tribes. Such sums as necessary are authorized for the program.

Section 804 requires the Energy Information Administration (EIA) to monitor planned petroleum refinery outages and report to the Secretary of Energy when such outages are affecting the price or availability of petroleum products. The Secretary will then be required to share data with refinery operators and encourage reductions in out-of-service refinery capacity.

Section 805 requires the Administrator of the Energy Information Administration (EIA) to develop a five-year plan for enhancing the scope, quality, and timeliness of the agency's data collection efforts. In addition, it requires closer coordination by EIA with state energy officials and with the Federal Energy Regulatory Commission. The section addresses state-level data in several respects and requires the Administrator to submit to Congress within a year an assessment of state level energy data needs. EIA is directed to revisit certain data series that had been terminated due to budget constraints and to identify data gaps that may have resulted from those terminations. To implement this section, $10 million is authorized for 2008, and additional sums are authorized through 2012.

Section 806 expresses the sense of Congress that there is a national goal to use renewable energy resources from agricultural, forestry, and working lands of the nation to provide at least 25% of the nation's energy use by 2025.

Section 807 directs the Department of the Interior's U.S. Geological Survey to conduct a comprehensive assessment of geothermal energy resources in the United States and report the findings of that assessment to Congress.

Subtitle B, Prohibitions on Market Manipulation and False Information

This subtitle prohibits crude oil and petroleum product wholesalers from using any technique to manipulate the market or provide false information. The law directs the Federal Trade Commission to treat such action as an unfair or deceptive practice, subject to civil penalties of not more than $1 million per incident.

Title IX: International Energy Programs

This title authorizes assistance to promote clean and efficient energy technologies in foreign countries, and it establishes an International Clean Energy Foundation.

Subtitle A, Assistance to Promote Clean and Efficient Energy Technologies in Foreign Countries

The U.S. Agency for International Development (USAID) is directed to report to Congress on efforts to support policies for clean and efficient energy technologies. The Department of Commerce is directed to increase efforts to export such technologies and report to Congress on the results. Other U.S. agencies with export promotion responsibilities are required to increase efforts to support these technologies. Also, a multi-agency Task Force on International Cooperation for Clean and Efficient Energy Technologies is created to support the implementation of clean energy markets in key developing countries.

Section 917 creates a U.S.-Israel Energy Cooperation partnership to support research, development, and deployment (RD&D) of energy efficiency and renewable energy measures.

Subtitle B, International Clean Energy Foundation

The Foundation is established with the long-term goal of reducing GHG emissions. It is directed to use the funds authorized by this subtitle to make grants to promote projects outside of the United States that serve as models of how to reduce emissions. An annual report to Congress is required.

Subtitle C, Miscellaneous Provisions

Section 931 calls for the Secretary of State to ensure that energy security is integrated into the core mission of the Department of State. Energy advisors are required at key embassies, and the Department is required to report to Congress every two years on its energy-related activities. Section 932 adds the Secretary of Energy to the National Security Council. Section 933 calls for the President to submit to Congress a comprehensive annual report that describes a national energy security strategy for the nation.

Section 934 implements the Convention on Supplementary Compensation for Nuclear Damage that was opened for signature in 1997. The convention has since been signed by the United States and 12 other countries but has not yet entered into force. Each party to the convention will be required to establish a compensation system within its borders for nuclear damages to the public. In the United States, this obligation will be fulfilled by the existing Price-Anderson Act (§170 of the Atomic Energy Act of 1954). The convention will also establish a second tier of damage compensation to be paid by all parties. Section 934 requires the U.S. contribution to the second tier to be paid by suppliers of nuclear equipment and services, under a formula to be developed by DOE. Supporters of the convention contend that it will help U.S. exporters of nuclear technology by establishing a predictable international liability system.

Section 935 has the stated purpose of improving national energy security by promoting anti-corruption initiatives in oil and natural gas rich countries and of improving global energy security by promoting programs such as the Extractive Industries Transparency Initiative (EITI) that aim to increase transparency and accountability into extractive resource payments. The sense of Congress is expressed that global energy security should be furthered by encouraging further participation in EITI by eligible countries and companies and by promoting the effectiveness of the EITI program by ensuring that a robust and candid review mechanism is put in place. The Secretary of State is required to report to Congress on progress made in promoting transparency in extractive industries resource payments. An authorization of $3 million is provided to support U.S. contributions to the Multi-Donor Trust Fund of EITI.

Title X: Green Jobs

This title authorizes up to $125 million in funding to establish national and state job training programs, administered by the Department of Labor, to help address job shortages that are impairing growth in green industries, such as energy efficient buildings and construction, renewable electric power, energy efficient vehicles, and biofuels development.

Title XI: Energy Transportation and Infrastructure

Subtitle A, Department of Transportation (DOT)

An Office of Climate Change and Environment is established at DOT to plan, coordinate, and implement strategies to reduce transportation-related energy use, mitigate the effects of climate change, and address the impact of climate change on transportation systems and infrastructure.

Subtitle B, Railroads

This subtitle directs DOT, in coordination with EPA, to establish and conduct a pilot grant program to assist railroad carriers in purchasing hybrid locomotives, including hybrid switch locomotives, in order to demonstrate the extent to which such locomotives increase fuel economy, reduce emissions, and lower costs of operation. Also, DOT is directed to create a program of capital grants for the rehabilitation, preservation, or improvement of railroad track (including roadbed, bridges, and related track structures) of class II and class III railroads.

Subtitle C, Marine Transportation

Short sea transportation is defined as commercial waterborne transportation that originates at a port in the United States and ends at another port in the United States or at a port in Canada located in the Great Lakes Saint Lawrence Seaway System. The same definition applies for the case where origination and end points are reversed.[21] This subtitle directs DOT to establish a short sea transportation program and designate short sea transportation projects to be conducted under the program to mitigate landside congestion. Short sea shipping activities are made eligible for support from DOT's capital construction fund. A report to Congress on the short sea transportation program is required.

Subtitle D, Highways

Section 1131 increases the federal share for congestion mitigation and air quality (CMAQ) projects up to 100% of project or program cost. Under Section 1132, DOT is directed to redistribute within each state any unobligated balances of the Highway Trust Fund that are rescinded in FY2008 or FY2009. Section 1133 expresses a sense of Congress that, in constructing new roadways or rehabilitating existing facilities, state and local governments should employ policies designed to accommodate all users, including motorists, pedestrians, cyclists, transit riders, and people of all ages and abilities.

Title XII: Small Business Energy Programs

Loans, grants, and debentures are established to help small businesses develop, invest in, and purchase energy efficient buildings, fixtures, equipment, and technology. Section 1201 empowers the Small Business Administration (SBA) to make "express" loans for certain energy efficiency and renewable energy projects. Section 1202 creates a two-year pilot loan program for purchasing energy efficient technologies under Section 7(a) of the Small Business Act at half the cost that would have otherwise been required. After the pilot program terminates, GAO is required to prepare a report to Congress that describes its energy-saving impact. Section 1203 creates small business energy efficiency, sustainability, and telecommuting programs. Reports to Congress are required for each of those programs. Section 1204 raises the Small Business Investment Act (SBIA) loan ceilings for certain energy efficiency and renewable energy projects undertaken by small businesses. Section 1205 enables qualified small business investment companies to issue energy-saving debentures. Section 1206 expands certain SBIA provisions to include investments in energy- saving small businesses. Section 1207 creates a Renewable Fuel Capital Investment (RFCI) pilot program that taps into venture capital to help small firms develop renewable energy sources and new technologies. A funding authorization of $30 million is provided for RFCI over two years. Section 1208 requires SBA to study the RFCI program and issue a report to Congress on its findings.

---

[21] This is the definition offered in Section 1122. Also, see the definition provided in *Short Sea Shipping: Practices, Opportunities, and Challenges*, by Gary A. Lombardo, http://www.insourceaudit.com/WhitePapers/Short_Sea_Shipping.asp.

Title XIII: Smart Grid

Section 1301 establishes a federal policy to modernize the electric utility transmission and distribution system to maintain reliability and infrastructure protection. The term "Smart Grid" refers to a distribution system that allows for flow of information from a customer's meter in two directions: both inside the house to thermostats, appliances, and other devices, and from the house back to the utility.[22] Smart Grid is defined to include a variety of operational and energy measures—including smart meters, smart appliances, renewable energy resources, and energy efficiency resources. Section 1302 calls for DOE to report to Congress on the deployment of Smart Grid technologies and any barriers to deployment. Section 1303 directs DOE to establish a Smart Grid Advisory Committee and a Smart Grid Task Force to assist with implementation. Section 1304 directs DOE to conduct Smart Grid RD&D and to develop measurement strategies to assess energy savings and other aspects of implementation. Section 1305 directs the National Institute of Standards and Technology to establish protocols and standards to increase the flexibility of use for Smart Grid equipment and systems. Section 1306 directs DOE to create a program that reimburses 20% of qualifying Smart Grid investments. Section 1307 directs states to encourage utilities to employ Smart Grid technology and allows utilities to recover Smart Grid investments through rates. Section 1308 requires DOE to prepare a report to Congress on the effect of private wire laws on the development of combined heat and power facilities. Section 1309 directs DOE to report to Congress on the potential impacts of Smart Grid deployment on the security of electricity infrastructure and operating capability. (For additional information, see CRS Report RL34288, *Smart Grid Provisions in H.R. 6, 110th Congress*, by Amy Abel.)

Title XIV: Pool and Spa Safety

Section 1401 identifies this title as the "Virginia Graeme Baker Pool and Spa Safety Act." Section 1402 finds that proper use of barriers or fencing could substantially reduce the number of childhood residential swimming and pool drownings. Section 1403 provides several definitions employed throughout this subtitle. Section 1404 sets an industry standard (ASME/ANSI A112.19.8) as a national performance standard for swimming pool and spa drain cover equipment. Section 1405 establishes a grant program and requires that at least 50% of the funding be used to assist states in hiring and training enforcement personnel to implement and enforce standards. The remaining funds must be used to educate pool construction and installation companies, pool owners and operators, and pool service companies, about the standard. Also, funding of $2 million per year over two years is authorized for the federal Consumer Product Safety Commission to implement the grant program. Section 1406 specifies minimum state law requirements to qualify for a grant under Section 1405. The requirement includes the enclosure of all outdoor residential pools and spas, installation of devices to prevent entrapment by pool or spa drains, and notification to pool owners about entrapment protection standards. Also, in setting minimum state law requirements, the Commission is directed to consider current or revised national standards for barrier and entrapment equipment, and to ensure that the requirements are consistent with the Commission's existing publications on pool safety guidelines. Section 1407

---

[22] The Smart Grid could allow appliances to be turned off during periods of high electrical demand and cost and give customers real-time information on constantly changing electric rates. The goal is to use advanced, information-based technologies to increase power grid efficiency, reliability, and flexibility, and reduce the rate at which additional electric utility infrastructure needs to be built.

directs the Commission to conduct a public education program on methods to prevent drowning and entrapment in swimming pools. A funding authorization of $5 million per year is provided over five years. Section 1408 directs the Commission to submit a report to Congress that evaluates the implementation of the state grant program.

Title XV: Revenue Provisions

Section 1500 specifies that, unless expressed otherwise, all tax provisions in this act refer to provisions of the Internal Revenue Code of 1986.

Section 1501 extends Federal Unemployment Tax Act (FUTA) taxes for one year. FUTA imposes a 6.2% gross tax rate on the first $7,000 paid annually by covered employers to each employee. In 1976, Congress passed a temporary surtax of 0.2% of taxable wages to be added to the permanent FUTA tax rate. The temporary surtax was subsequently extended through 2007. The President's FY2008 Budget had proposed extending the FUTA surtax. The Treasury Department stated that "extending the surtax will support the continued solvency of the federal unemployment trust funds and maintain the ability of the unemployment system to adjust to any economic downturns." This section enacts the President's proposal for one year, 2008. This provision is estimated to raise $1.446 billion over 10 calendar years.

Under Section 1502, the geological and geophysical costs of a major integrated oil company will be amortized (deducted proportionally) over a seven-year period instead of the current five-year period. (A major integrated oil company is defined as one with an average world production of at least 500,000 barrels per day, with 2005 gross receipts exceeding $1 billion, and which has at least a 15% interest in refinery operations.)

Title XVI: Effective Date

Section 1601 specifies that this act and the amendments it makes will take effect one day after enactment.

Author Contact Information

Fred Sissine, Coordinator
Specialist in Energy Policy
fsissine@crs.loc.gov, 7-7039

Energy Independence and Security Act of 2007: A Summary of Major Provisions

CRS Key Policy Staff

Area of Expertise	Name	Telephone
Agriculture-Based Energy	Randy Schnepf	7-4277
	Tom Capehart	7-2425
Biofuels	Brent Yacobucci	7-9662
Carbon Storage	Peter Folger	7-1517
Energy Prices	Robert Pirog	7-6847
Energy Taxes	Salvatore Lazzari	7-7825
Energy-Saving Performance Contracts	Anthony Andrews	7-6843
Fuel Economy Standards	Robert Bamberger	7-7240
International Energy	Jeff Logan	7-9317
Marine Energy	Nic Lane	7-7905
Natural Gas	William Hederman	7-7738
Nuclear Energy and Loan Guarantees	Mark Holt	7-1704
Transmission and Electric Utilities	Amy Abel	7-7239

PUBLIC LAW 110–140—DEC. 19, 2007   121 STAT. 1783

"(B) PAYMENT.—Any company against which the Administrator assesses costs under this paragraph shall pay such costs.

"(2) DEPOSIT OF FUNDS.—Funds collected under this section shall be deposited in the account for salaries and expenses of the Administration.

"SEC. 394. MISCELLANEOUS.   15 USC 690m.

"To the extent such procedures are not inconsistent with the requirements of this part, the Administrator may take such action as set forth in sections 309, 311, 312, and 314 and an officer, director, employee, agent, or other participant in the management or conduct of the affairs of a Renewable Fuel Capital Investment company shall be subject to the requirements of such sections.

"SEC. 395. REMOVAL OR SUSPENSION OF DIRECTORS OR OFFICERS.   15 USC 690n.

"Using the procedures for removing or suspending a director or an officer of a licensee set forth in section 313 (to the extent such procedures are not inconsistent with the requirements of this part), the Administrator may remove or suspend any director or officer of any Renewable Fuel Capital Investment company.

"SEC. 396. REGULATIONS.   15 USC 690o.

"The Administrator may issue such regulations as the Administrator determines necessary to carry out the provisions of this part in accordance with its purposes.

"SEC. 397. AUTHORIZATIONS OF APPROPRIATIONS.   15 USC 690p.

"(a) IN GENERAL.—Subject to the availability of appropriations, the Administrator is authorized to make $15,000,000 in operational assistance grants under section 389 for each of fiscal years 2008 and 2009.

"(b) FUNDS COLLECTED FOR EXAMINATIONS.—Funds deposited under section 393(c)(2) are authorized to be appropriated only for the costs of examinations under section 393 and for the costs of other oversight activities with respect to the program established under this part.

"SEC. 398. TERMINATION.   15 USC 690q.

"The program under this part shall terminate at the end of the second full fiscal year after the date that the Administrator establishes the program under this part.".

SEC. 1208. STUDY AND REPORT.

The Administrator of the Small Business Administration shall conduct a study of the Renewable Fuel Capital Investment Program under part C of title III of the Small Business Investment Act of 1958, as added by this Act. Not later than 3 years after the date of enactment of this Act, the Administrator shall complete the study under this section and submit to Congress a report regarding the results of the study.

TITLE XIII—SMART GRID

SEC. 1301. STATEMENT OF POLICY ON MODERNIZATION OF ELEC-   15 USC 17381.
TRICITY GRID.

It is the policy of the United States to support the modernization of the Nation's electricity transmission and distribution system

121 STAT. 1784   PUBLIC LAW 110–140—DEC. 19, 2007

to maintain a reliable and secure electricity infrastructure that can meet future demand growth and to achieve each of the following, which together characterize a Smart Grid:

(1) Increased use of digital information and controls technology to improve reliability, security, and efficiency of the electric grid.

(2) Dynamic optimization of grid operations and resources, with full cyber-security.

(3) Deployment and integration of distributed resources and generation, including renewable resources.

(4) Development and incorporation of demand response, demand-side resources, and energy-efficiency resources.

(5) Deployment of "smart" technologies (real-time, automated, interactive technologies that optimize the physical operation of appliances and consumer devices) for metering, communications concerning grid operations and status, and distribution automation.

(6) Integration of "smart" appliances and consumer devices.

(7) Deployment and integration of advanced electricity storage and peak-shaving technologies, including plug-in electric and hybrid electric vehicles, and thermal-storage air conditioning.

(8) Provision to consumers of timely information and control options.

(9) Development of standards for communication and interoperability of appliances and equipment connected to the electric grid, including the infrastructure serving the grid.

(10) Identification and lowering of unreasonable or unnecessary barriers to adoption of smart grid technologies, practices, and services.

15 USC 17382.

SEC. 1302. SMART GRID SYSTEM REPORT.

The Secretary, acting through the Assistant Secretary of the Office of Electricity Delivery and Energy Reliability (referred to in this section as the "OEDER") and through the Smart Grid Task Force established in section 1303, shall, after consulting with any interested individual or entity as appropriate, no later than 1 year after enactment, and every 2 years thereafter, report to Congress concerning the status of smart grid deployments nationwide and any regulatory or government barriers to continued deployment. The report shall provide the current status and prospects of smart grid development, including information on technology penetration, communications network capabilities, costs, and obstacles. It may include recommendations for State and Federal policies or actions helpful to facilitate the transition to a smart grid. To the extent appropriate, it should take a regional perspective. In preparing this report, the Secretary shall solicit advice and contributions from the Smart Grid Advisory Committee created in section 1303; from other involved Federal agencies including but not limited to the Federal Energy Regulatory Commission ("Commission"), the National Institute of Standards and Technology ("Institute"), and the Department of Homeland Security; and from other stakeholder groups not already represented on the Smart Grid Advisory Committee.

15 USC 17383.

SEC. 1303. SMART GRID ADVISORY COMMITTEE AND SMART GRID TASK FORCE.

(a) SMART GRID ADVISORY COMMITTEE.—

PUBLIC LAW 110–140—DEC. 19, 2007   121 STAT. 1785

(1) ESTABLISHMENT.—The Secretary shall establish, within 90 days of enactment of this Part, a Smart Grid Advisory Committee (either as an independent entity or as a designated sub-part of a larger advisory committee on electricity matters). The Smart Grid Advisory Committee shall include eight or more members appointed by the Secretary who have sufficient experience and expertise to represent the full range of smart grid technologies and services, to represent both private and non-Federal public sector stakeholders. One member shall be appointed by the Secretary to Chair the Smart Grid Advisory Committee. Deadline.

(2) MISSION.—The mission of the Smart Grid Advisory Committee shall be to advise the Secretary, the Assistant Secretary, and other relevant Federal officials concerning the development of smart grid technologies, the progress of a national transition to the use of smart-grid technologies and services, the evolution of widely-accepted technical and practical standards and protocols to allow interoperability and inter-communication among smart-grid capable devices, and the optimum means of using Federal incentive authority to encourage such progress.

(3) APPLICABILITY OF FEDERAL ADVISORY COMMITTEE ACT.—The Federal Advisory Committee Act (5 U.S.C. App.) shall apply to the Smart Grid Advisory Committee.

(b) SMART GRID TASK FORCE.—

(1) ESTABLISHMENT.—The Assistant Secretary of the Office of Electricity Delivery and Energy Reliability shall establish, within 90 days of enactment of this Part, a Smart Grid Task Force composed of designated employees from the various divisions of that office who have responsibilities related to the transition to smart-grid technologies and practices. The Assistant Secretary or his designee shall be identified as the Director of the Smart Grid Task Force. The Chairman of the Federal Energy Regulatory Commission and the Director of the National Institute of Standards and Technology shall each designate at least one employee to participate on the Smart Grid Task Force. Other members may come from other agencies at the invitation of the Assistant Secretary or the nomination of the head of such other agency. The Smart Grid Task Force shall, without disrupting the work of the Divisions or Offices from which its members are drawn, provide an identifiable Federal entity to embody the Federal role in the national transition toward development and use of smart grid technologies. Deadline.

(2) MISSION.—The mission of the Smart Grid Task Force shall be to insure awareness, coordination and integration of the diverse activities of the Office and elsewhere in the Federal Government related to smart-grid technologies and practices, including but not limited to: smart grid research and development; development of widely accepted smart-grid standards and protocols; the relationship of smart-grid technologies and practices to electric utility regulation; the relationship of smart-grid technologies and practices to infrastructure development, system reliability and security; and the relationship of smart-grid technologies and practices to other facets of electricity supply, demand, transmission, distribution, and policy. The Smart Grid Task Force shall collaborate with the Smart Grid Advisory Committee and other Federal agencies and offices.

121 STAT. 1786   PUBLIC LAW 110–140—DEC. 19, 2007

The Smart Grid Task Force shall meet at the call of its Director as necessary to accomplish its mission.

(c) AUTHORIZATION.—There are authorized to be appropriated for the purposes of this section such sums as are necessary to the Secretary to support the operations of the Smart Grid Advisory Committee and Smart Grid Task Force for each of fiscal years 2008 through 2020.

42 USC 17384.   **SEC. 1304. SMART GRID TECHNOLOGY RESEARCH, DEVELOPMENT, AND DEMONSTRATION.**

(a) POWER GRID DIGITAL INFORMATION TECHNOLOGY.—The Secretary, in consultation with the Federal Energy Regulatory Commission and other appropriate agencies, electric utilities, the States, and other stakeholders, shall carry out a program—

(1) to develop advanced techniques for measuring peak load reductions and energy-efficiency savings from smart metering, demand response, distributed generation, and electricity storage systems;

(2) to investigate means for demand response, distributed generation, and storage to provide ancillary services;

(3) to conduct research to advance the use of wide-area measurement and control networks, including data mining, visualization, advanced computing, and secure and dependable communications in a highly-distributed environment;

(4) to test new reliability technologies, including those concerning communications network capabilities, in a grid control room environment against a representative set of local outage and wide area blackout scenarios;

(5) to identify communications network capacity needed to implement advanced technologies.

(6) to investigate the feasibility of a transition to time-of-use and real-time electricity pricing;

(7) to develop algorithms for use in electric transmission system software applications;

(8) to promote the use of underutilized electricity generation capacity in any substitution of electricity for liquid fuels in the transportation system of the United States; and

(9) in consultation with the Federal Energy Regulatory Commission, to propose interconnection protocols to enable electric utilities to access electricity stored in vehicles to help meet peak demand loads.

(b) SMART GRID REGIONAL DEMONSTRATION INITIATIVE.—

(1) IN GENERAL.—The Secretary shall establish a smart grid regional demonstration initiative (referred to in this subsection as the "Initiative") composed of demonstration projects specifically focused on advanced technologies for use in power grid sensing, communications, analysis, and power flow control. The Secretary shall seek to leverage existing smart grid deployments.

(2) GOALS.—The goals of the Initiative shall be—

(A) to demonstrate the potential benefits of concentrated investments in advanced grid technologies on a regional grid;

(B) to facilitate the commercial transition from the current power transmission and distribution system technologies to advanced technologies;

PUBLIC LAW 110–140—DEC. 19, 2007   121 STAT. 1787

(C) to facilitate the integration of advanced technologies in existing electric networks to improve system performance, power flow control, and reliability;

(D) to demonstrate protocols and standards that allow for the measurement and validation of the energy savings and fossil fuel emission reductions associated with the installation and use of energy efficiency and demand response technologies and practices; and

(E) to investigate differences in each region and regulatory environment regarding best practices in implementing smart grid technologies.

(3) DEMONSTRATION PROJECTS.—

(A) IN GENERAL.—In carrying out the initiative, the Secretary shall carry out smart grid demonstration projects in up to 5 electricity control areas, including rural areas and at least 1 area in which the majority of generation and transmission assets are controlled by a tax-exempt entity.

(B) COOPERATION.—A demonstration project under subparagraph (A) shall be carried out in cooperation with the electric utility that owns the grid facilities in the electricity control area in which the demonstration project is carried out.

(C) FEDERAL SHARE OF COST OF TECHNOLOGY INVESTMENTS.—The Secretary shall provide to an electric utility described in subparagraph (B) financial assistance for use in paying an amount equal to not more than 50 percent of the cost of qualifying advanced grid technology investments made by the electric utility to carry out a demonstration project.

(D) INELIGIBILITY FOR GRANTS.—No person or entity participating in any demonstration project conducted under this subsection shall be eligible for grants under section 1306 for otherwise qualifying investments made as part of that demonstration project.

(c) AUTHORIZATION OF APPROPRIATIONS.—There are authorized to be appropriated—

(1) to carry out subsection (a), such sums as are necessary for each of fiscal years 2008 through 2012; and

(2) to carry out subsection (b), $100,000,000 for each of fiscal years 2008 through 2012.

SEC. 1305. SMART GRID INTEROPERABILITY FRAMEWORK.   15 USC 17385.

(a) INTEROPERABILITY FRAMEWORK.—The Director of the National Institute of Standards and Technology shall have primary responsibility to coordinate the development of a framework that includes protocols and model standards for information management to achieve interoperability of smart grid devices and systems. Such protocols and standards shall further align policy, business, and technology approaches in a manner that would enable all electric resources, including demand-side resources, to contribute to an efficient, reliable electricity network. In developing such protocols and standards—

(1) the Director shall seek input and cooperation from the Commission, OEDER and its Smart Grid Task Force, the Smart Grid Advisory Committee, other relevant Federal and State agencies; and

121 STAT. 1788　　　PUBLIC LAW 110–140—DEC. 19, 2007

(2) the Director shall also solicit input and cooperation from private entities interested in such protocols and standards, including but not limited to the Gridwise Architecture Council, the International Electrical and Electronics Engineers, the National Electric Reliability Organization recognized by the Federal Energy Regulatory Commission, and National Electrical Manufacturer's Association.

(b) SCOPE OF FRAMEWORK.—The framework developed under subsection (a) shall be flexible, uniform and technology neutral, including but not limited to technologies for managing smart grid information, and designed—

(1) to accommodate traditional, centralized generation and transmission resources and consumer distributed resources, including distributed generation, renewable generation, energy storage, energy efficiency, and demand response and enabling devices and systems;

(2) to be flexible to incorporate—

(A) regional and organizational differences; and

(B) technological innovations;

(3) to consider the use of voluntary uniform standards for certain classes of mass-produced electric appliances and equipment for homes and businesses that enable customers, at their election and consistent with applicable State and Federal laws, and are manufactured with the ability to respond to electric grid emergencies and demand response signals by curtailing all, or a portion of, the electrical power consumed by the appliances or equipment in response to an emergency or demand response signal, including through—

(A) load reduction to reduce total electrical demand;

(B) adjustment of load to provide grid ancillary services; and

(C) in the event of a reliability crisis that threatens an outage, short-term load shedding to help preserve the stability of the grid; and

(4) such voluntary standards should incorporate appropriate manufacturer lead time.

(c) TIMING OF FRAMEWORK DEVELOPMENT.—The Institute shall begin work pursuant to this section within 60 days of enactment. The Institute shall provide and publish an initial report on progress toward recommended or consensus standards and protocols within 1 year after enactment, further reports at such times as developments warrant in the judgment of the Institute, and a final report when the Institute determines that the work is completed or that a Federal role is no longer necessary.

(d) STANDARDS FOR INTEROPERABILITY IN FEDERAL JURISDICTION.—At any time after the Institute's work has led to sufficient consensus in the Commission's judgment, the Commission shall institute a rulemaking proceeding to adopt such standards and protocols as may be necessary to insure smart-grid functionality and interoperability in interstate transmission of electric power, and regional and wholesale electricity markets.

(e) AUTHORIZATION.—There are authorized to be appropriated for the purposes of this section $5,000,000 to the Institute to support the activities required by this subsection for each of fiscal years 2008 through 2012.

PUBLIC LAW 110–140—DEC. 19, 2007   121 STAT. 1789

SEC. 1306. FEDERAL MATCHING FUND FOR SMART GRID INVESTMENT COSTS.   42 USC 17386.

(a) MATCHING FUND.—The Secretary shall establish a Smart Grid Investment Matching Grant Program to provide reimbursement of one-fifth (20 percent) of qualifying Smart Grid investments.

(b) QUALIFYING INVESTMENTS.—Qualifying Smart Grid investments may include any of the following made on or after the date of enactment of this Act:

(1) In the case of appliances covered for purposes of establishing energy conservation standards under part B of title III of the Energy Policy and Conservation Act of 1975 (42 U.S.C. 6291 et seq.), the documented expenditures incurred by a manufacturer of such appliances associated with purchasing or designing, creating the ability to manufacture, and manufacturing and installing for one calendar year, internal devices that allow the appliance to engage in Smart Grid functions.

(2) In the case of specialized electricity-using equipment, including motors and drivers, installed in industrial or commercial applications, the documented expenditures incurred by its owner or its manufacturer of installing devices or modifying that equipment to engage in Smart Grid functions.

(3) In the case of transmission and distribution equipment fitted with monitoring and communications devices to enable smart grid functions, the documented expenditures incurred by the electric utility to purchase and install such monitoring and communications devices.

(4) In the case of metering devices, sensors, control devices, and other devices integrated with and attached to an electric utility system or retail distributor or marketer of electricity that are capable of engaging in Smart Grid functions, the documented expenditures incurred by the electric utility, distributor, or marketer and its customers to purchase and install such devices.

(5) In the case of software that enables devices or computers to engage in Smart Grid functions, the documented purchase costs of the software.

(6) In the case of entities that operate or coordinate operations of regional electric grids, the documented expenditures for purchasing and installing such equipment that allows Smart Grid functions to operate and be combined or coordinated among multiple electric utilities and between that region and other regions.

(7) In the case of persons or entities other than electric utilities owning and operating a distributed electricity generator, the documented expenditures of enabling that generator to be monitored, controlled, or otherwise integrated into grid operations and electricity flows on the grid utilizing Smart Grid functions.

(8) In the case of electric or hybrid-electric vehicles, the documented expenses for devices that allow the vehicle to engage in Smart Grid functions (but not the costs of electricity storage for the vehicle).

(9) The documented expenditures related to purchasing and implementing Smart Grid functions in such other cases as the Secretary shall identify. In making such grants, the Secretary shall seek to reward innovation and early adaptation,

121 STAT. 1790   PUBLIC LAW 110–140—DEC. 19, 2007

even if success is not complete, rather than deployment of proven and commercially viable technologies.

(c) INVESTMENTS NOT INCLUDED.—Qualifying Smart Grid investments do not include any of the following:

(1) Investments or expenditures for Smart Grid technologies, devices, or equipment that are eligible for specific tax credits or deductions under the Internal Revenue Code, as amended.

(2) Expenditures for electricity generation, transmission, or distribution infrastructure or equipment not directly related to enabling Smart Grid functions.

(3) After the final date for State consideration of the Smart Grid Information Standard under section 1307 (paragraph (17) of section 111(d) of the Public Utility Regulatory Policies Act of 1978), an investment that is not in compliance with such standard.

(4) After the development and publication by the Institute of protocols and model standards for interoperability of smart grid devices and technologies, an investment that fails to incorporate any of such protocols or model standards.

(5) Expenditures for physical interconnection of generators or other devices to the grid except those that are directly related to enabling Smart Grid functions.

(6) Expenditures for ongoing salaries, benefits, or personnel costs not incurred in the initial installation, training, or start up of smart grid functions.

(7) Expenditures for travel, lodging, meals or other personal costs.

(8) Ongoing or routine operation, billing, customer relations, security, and maintenance expenditures.

(9) Such other expenditures that the Secretary determines not to be Qualifying Smart Grid Investments by reason of the lack of the ability to perform Smart Grid functions or lack of direct relationship to Smart Grid functions.

(d) SMART GRID FUNCTIONS.—The term "smart grid functions" means any of the following:

(1) The ability to develop, store, send and receive digital information concerning electricity use, costs, prices, time of use, nature of use, storage, or other information relevant to device, grid, or utility operations, to or from or by means of the electric utility system, through one or a combination of devices and technologies.

(2) The ability to develop, store, send and receive digital information concerning electricity use, costs, prices, time of use, nature of use, storage, or other information relevant to device, grid, or utility operations to or from a computer or other control device.

(3) The ability to measure or monitor electricity use as a function of time of day, power quality characteristics such as voltage level, current, cycles per second, or source or type of generation and to store, synthesize or report that information by digital means.

(4) The ability to sense and localize disruptions or changes in power flows on the grid and communicate such information instantaneously and automatically for purposes of enabling automatic protective responses to sustain reliability and security of grid operations.

(5) The ability to detect, prevent, communicate with regard to, respond to, or recover from system security threats, including cyber-security threats and terrorism, using digital information, media, and devices.

(6) The ability of any appliance or machine to respond to such signals, measurements, or communications automatically or in a manner programmed by its owner or operator without independent human intervention.

(7) The ability to use digital information to operate functionalities on the electric utility grid that were previously electro-mechanical or manual.

(8) The ability to use digital controls to manage and modify electricity demand, enable congestion management, assist in voltage control, provide operating reserves, and provide frequency regulation.

(9) Such other functions as the Secretary may identify as being necessary or useful to the operation of a Smart Grid.

(e) The Secretary shall—

    Procedures.
    Federal Register, publication.
    Deadline.

(1) establish and publish in the Federal Register, within 1 year after the enactment of this Act procedures by which applicants who have made qualifying Smart Grid investments can seek and obtain reimbursement of one-fifth of their documented expenditures;

(2) establish procedures to ensure that there is no duplication or multiple reimbursement for the same investment or costs, that the reimbursement goes to the party making the actual expenditures for Qualifying Smart Grid Investments, and that the grants made have significant effect in encouraging and facilitating the development of a smart grid;

    Records.

(3) maintain public records of reimbursements made, recipients, and qualifying Smart Grid investments which have received reimbursements;

(4) establish procedures to provide, in cases deemed by the Secretary to be warranted, advance payment of moneys up to the full amount of the projected eventual reimbursement, to creditworthy applicants whose ability to make Qualifying Smart Grid Investments may be hindered by lack of initial capital, in lieu of any later reimbursement for which that applicant qualifies, and subject to full return of the advance payment in the event that the Qualifying Smart Grid investment is not made; and

(5) have and exercise the discretion to deny grants for investments that do not qualify in the reasonable judgment of the Secretary.

(f) AUTHORIZATION OF APPROPRIATIONS.—There are authorized to be appropriated to the Secretary such sums as are necessary for the administration of this section and the grants to be made pursuant to this section for fiscal years 2008 through 2012.

SEC. 1307. STATE CONSIDERATION OF SMART GRID.

(a) Section 111(d) of the Public Utility Regulatory Policies Act of 1978 (16 U.S.C. 2621(d)) is amended by adding at the end the following:

"(16) CONSIDERATION OF SMART GRID INVESTMENTS.—

"(A) IN GENERAL.—Each State shall consider requiring that, prior to undertaking investments in nonadvanced grid technologies, an electric utility of the State demonstrate

121 STAT. 1792   PUBLIC LAW 110–140—DEC. 19, 2007

to the State that the electric utility considered an investment in a qualified smart grid system based on appropriate factors, including—

"(i) total costs;
"(ii) cost-effectiveness;
"(iii) improved reliability;
"(iv) security;
"(v) system performance; and
"(vi) societal benefit.

"(B) RATE RECOVERY.—Each State shall consider authorizing each electric utility of the State to recover from ratepayers any capital, operating expenditure, or other costs of the electric utility relating to the deployment of a qualified smart grid system, including a reasonable rate of return on the capital expenditures of the electric utility for the deployment of the qualified smart grid system.

"(C) OBSOLETE EQUIPMENT.—Each State shall consider authorizing any electric utility or other party of the State to deploy a qualified smart grid system to recover in a timely manner the remaining book-value costs of any equipment rendered obsolete by the deployment of the qualified smart grid system, based on the remaining depreciable life of the obsolete equipment.

"(17) SMART GRID INFORMATION.—

"(A) STANDARD.—All electricity purchasers shall be provided direct access, in written or electronic machine-readable form as appropriate, to information from their electricity provider as provided in subparagraph (B).

"(B) INFORMATION.—Information provided under this section, to the extent practicable, shall include:

"(i) PRICES.—Purchasers and other interested persons shall be provided with information on—

"(I) time-based electricity prices in the wholesale electricity market; and

"(II) time-based electricity retail prices or rates that are available to the purchasers.

"(ii) USAGE.—Purchasers shall be provided with the number of electricity units, expressed in kwh, purchased by them.

"(iii) INTERVALS AND PROJECTIONS.—Updates of information on prices and usage shall be offered on not less than a daily basis, shall include hourly price and use information, where available, and shall include a day-ahead projection of such price information to the extent available.

"(iv) SOURCES.—Purchasers and other interested persons shall be provided annually with written information on the sources of the power provided by the utility, to the extent it can be determined, by type of generation, including greenhouse gas emissions associated with each type of generation, for intervals during which such information is available on a cost-effective basis.

"(C) ACCESS.—Purchasers shall be able to access their own information at any time through the Internet and on other means of communication elected by that utility

PUBLIC LAW 110–140—DEC. 19, 2007   121 STAT. 1793

for Smart Grid applications. Other interested persons shall be able to access information not specific to any purchaser through the Internet. Information specific to any purchaser shall be provided solely to that purchaser.".

(b) COMPLIANCE.—

(1) TIME LIMITATIONS.—Section 112(b) of the Public Utility Regulatory Policies Act of 1978 (16 U.S.C. 2622(b)) is amended by adding the following at the end thereof:

"(6)(A) Not later than 1 year after the enactment of this paragraph, each State regulatory authority (with respect to each electric utility for which it has ratemaking authority) and each nonregulated utility shall commence the consideration referred to in section 111, or set a hearing date for consideration, with respect to the standards established by paragraphs (17) through (18) of section 111(d).

"(B) Not later than 2 years after the date of the enactment of this paragraph, each State regulatory authority (with respect to each electric utility for which it has ratemaking authority), and each nonregulated electric utility, shall complete the consideration, and shall make the determination, referred to in section 111 with respect to each standard established by paragraphs (17) through (18) of section 111(d).".

(2) FAILURE TO COMPLY.—Section 112(c) of the Public Utility Regulatory Policies Act of 1978 (16 U.S.C. 2622(c)) is amended by adding the following at the end:

"In the case of the standards established by paragraphs (16) through (19) of section 111(d), the reference contained in this subsection to the date of enactment of this Act shall be deemed to be a reference to the date of enactment of such paragraphs.".

(3) PRIOR STATE ACTIONS.—Section 112(d) of the Public Utility Regulatory Policies Act of 1978 (16 U.S.C. 2622(d)) is amended by inserting "and paragraphs (17) through (18)" before "of section 111(d)".

SEC. 1308. STUDY OF THE EFFECT OF PRIVATE WIRE LAWS ON THE DEVELOPMENT OF COMBINED HEAT AND POWER FACILITIES.

(a) STUDY.—

(1) IN GENERAL.—The Secretary, in consultation with the States and other appropriate entities, shall conduct a study of the laws (including regulations) affecting the siting of privately owned electric distribution wires on and across public rights-of-way.

(2) REQUIREMENTS.—The study under paragraph (1) shall include—

(A) an evaluation of—

(i) the purposes of the laws; and

(ii) the effect the laws have on the development of combined heat and power facilities;

(B) a determination of whether a change in the laws would have any operating, reliability, cost, or other impacts on electric utilities and the customers of the electric utilities; and

(C) an assessment of—

(i) whether privately owned electric distribution wires would result in duplicative facilities; and

121 STAT. 1794   PUBLIC LAW 110–140—DEC. 19, 2007

       (ii) whether duplicative facilities are necessary or desirable.

  (b) REPORT.—Not later than 1 year after the date of enactment of this Act, the Secretary shall submit to Congress a report that describes the results of the study conducted under subsection (a).

SEC. 1309. DOE STUDY OF SECURITY ATTRIBUTES OF SMART GRID SYSTEMS.

Deadline.
Reports.

  (a) DOE STUDY.—The Secretary shall, within 18 months after the date of enactment of this Act, submit a report to Congress that provides a quantitative assessment and determination of the existing and potential impacts of the deployment of Smart Grid systems on improving the security of the Nation's electricity infrastructure and operating capability. The report shall include but not be limited to specific recommendations on each of the following:

    (1) How smart grid systems can help in making the Nation's electricity system less vulnerable to disruptions due to intentional acts against the system.

    (2) How smart grid systems can help in restoring the integrity of the Nation's electricity system subsequent to disruptions.

    (3) How smart grid systems can facilitate nationwide, interoperable emergency communications and control of the Nation's electricity system during times of localized, regional, or nationwide emergency.

    (4) What risks must be taken into account that smart grid systems may, if not carefully created and managed, create vulnerability to security threats of any sort, and how such risks may be mitigated.

  (b) CONSULTATION.—The Secretary shall consult with other Federal agencies in the development of the report under this section, including but not limited to the Secretary of Homeland Security, the Federal Energy Regulatory Commission, and the Electric Reliability Organization certified by the Commission under section 215(c) of the Federal Power Act (16 U.S.C. 824o) as added by section 1211 of the Energy Policy Act of 2005 (Public Law 109–58; 119 Stat. 941).

Virginia Graeme Baker Pool and Spa Safety Act.
15 USC 8001 note.

TITLE XIV—POOL AND SPA SAFETY

SEC. 1401. SHORT TITLE.

  This title may be cited as the "Virginia Graeme Baker Pool and Spa Safety Act".

15 USC 8001.

SEC. 1402. FINDINGS.

  Congress finds the following:

    (1) Of injury-related deaths, drowning is the second leading cause of death in children aged 1 to 14 in the United States.

    (2) In 2004, 761 children aged 14 and under died as a result of unintentional drowning.

    (3) Adult supervision at all aquatic venues is a critical safety factor in preventing children from drowning.

    (4) Research studies show that the installation and proper use of barriers or fencing, as well as additional layers of protection, could substantially reduce the number of childhood residential swimming pool drownings and near drownings.

Congressional Research Service

Smart Grid Provisions in H.R. 6, 110th Congress

Amy Abel
Section Research Manager

February 13, 2008

Congressional Research Service

7-5700

www.crs.gov

RL34288

CRS Report for Congress

Prepared for Members and Committees of Congress

Summary

The term Smart Grid refers to a distribution system that allows for flow of information from a customer's meter in two directions: both inside the house to thermostats and appliances and other devices, and back to the utility. This could allow appliances to be turned off during periods of high electrical demand and cost, and give customers real-time information on constantly changing electric rates. Efforts are being made in both industry and government to modernize electric distribution to improve communications between utilities and the ultimate consumer. The goal is to use advanced, information-based technologies to increase power grid efficiency, reliability, and flexibility, and reduce the rate at which additional electric utility infrastructure needs to be built.

Both regulatory and technological barriers have limited the implementation of Smart Grid technology. At issue is whether a distinction for cost allocation purposes can be made between the impact of Smart Grid technology on the wholesale transmission system and its impact on the retail distribution system. Another issue limiting the deployment of this technology is the lack of consistent standards and protocols. There currently are no standards for these technologies. This limits the interoperability of Smart Grid technologies and limits future choices for companies that choose to install any particular type of technology.

H.R. 6, as signed by the President, contains provisions to encourage research, development, and deployment of Smart Grid technologies. Provisions include requiring the National Institute of Standards and Technology to be the lead agency to develop standards and protocols; creating a research, development, and demonstration program for Smart Grid technologies at the Department of Energy; and providing federal matching funds for portions of qualified Smart Grid investments.

Congressional Research Service

Smart Grid Provisions in H.R. 6, 110th Congress

Contents

Figures

Contacts

Introduction and Overview

The U.S. electric power system has historically operated at such a high level of reliability that any major outage, either caused by sabotage, weather, or operational errors, makes news headlines. As the August 14, 2003, Midwest and Northeast blackout demonstrated, a loss of electric power is very expensive and can entail considerable disruption to business, travel, government services, and daily life.

The electric utility industry operates as an integrated system of generation, transmission, and distribution facilities to deliver power to consumers. The electric power system in the United States consists of over 9,200 electric generating units with more than 950,000 megawatts of generating capacity connected to more than 300,000 miles of transmission lines; more than 210,000 miles of the transmission lines are rated at 230 kilovolts (kV) or higher (**Figure 1**).[1] In addition, approximately 150 control centers manage the flow of electricity through the system under normal operating conditions.

Figure 1. Electric Transmission Network

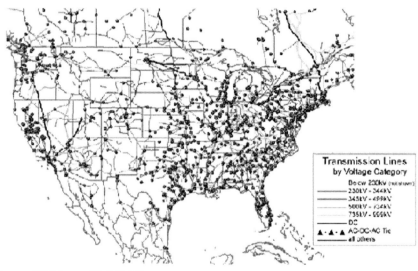

Source: GAO Report GAO 01 201

Most electricity in the United States is generated at power plants that use fossil fuels (oil, gas, coal), nuclear fission, or renewable energy (hydropower, geothermal, solar, wind, biomass). At the power plant, energy is converted into a set of three alternating electric currents, called three-phase power.[2] After power is generated, the first step in delivering electricity to the consumer is to transform the power from medium voltage (15-50 kilovolt (kV)) to high voltage (138-765 kV)

---

[1] North American Reliability Council. NERC 2007 Electricity Supply and Demand Database.

[2] The three currents are sinusoidal functions of time but with the same frequency (60 Hertz). In a three phase system, the phases are spaced equally, offset 120 degrees from each other. With three-phase power, one of the three phases is always nearing a peak.

alternating current (**Figure 2**).[3] This initial step-up of voltage occurs in a transformer located at transmission substations at the generating facilities. High voltages allow power to be moved long distances with the greatest efficiency, i.e. transmission line losses are minimized.[4] The three phases of power are carried over three wires that are connected to large transmission towers.[5] Close to the ultimate consumer, the power is stepped-down at another substation to lower voltages, typically less than 10 kV. At this point, the power is considered to have left the transmission system and entered the distribution system.

Figure 2. The Electric Power System

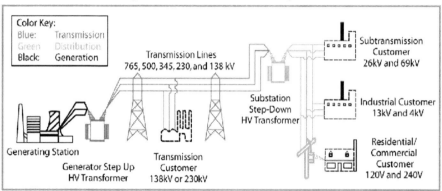

The transmission system continues to become more congested, and siting of transmission lines continues to be difficult. To try to maximize operation of existing infrastructure, efforts are being made in both industry and government to modernize electric distribution equipment to improve communications between utilities and the ultimate consumer. The goal is to use advanced, information-based technologies to increase power grid efficiency, reliability, and flexibility, and reduce the rate at which additional electric utility infrastructure needs to be built.

Some utilities have been using smart metering: meters that can be read remotely, primarily for billing purposes. However, these meters do not provide communication back to the utility with information on voltage, current levels, and specific usage. Similarly, these meters have very limited ability to allow the consumer the ability to either automatically or selectively change their usage patterns based on information provided by the utility.

The term Smart Grid refers to a distribution system that allows for flow of information from a customer's meter in two directions: both inside the house to thermostats and appliances and other devices, and back to the utility. It is expected that grid reliability will increase as additional information from the distribution system is available to utility operators. This will allow for better planning and operations during peak demand. For example, new technologies such as a Programmable Communicating Thermostat (PCT) could connect with a customer's meter

---

[3] 1kV=1,000 volts

[4] The loss of power on the transmission system is proportional to the square of the current (flow of electricity) while the current is inversely proportional to the voltage.

[5] Transmission towers also support a fourth wire running above the other three lines. This line is intended to attract lighting, so that the flow of electricity is not disturbed.

through a Home Area Network allowing the utility to change the settings on the thermostat based on load or other factors. PCTs are not commercially available, but are expected to be available within a year.[6] It is estimated that a 4% peak load reduction could be achieved using Smart Grid technologies.

Both regulatory and technological barriers have limited the implementation of Smart Grid technology. The Federal Energy Regulatory Commission (FERC) regulates the wholesale transmission system and the states regulate the distribution system. In general, the federal government has not interfered with state regulation of the electric distribution system. However, the Energy Policy Act of 2005 (EPACT05) required states to consider deploying smart meters for residential and small commercial customers.[7] At issue is whether a distinction for cost allocation purposes can be made between Smart Grid technologies' impact on the wholesale transmission system and retail distribution system. If FERC and the states cannot determine which costs should be considered transmission related (federally regulated) and which should be considered distribution related (state regulated) utilities may be reluctant to make large investments in Smart Grid technologies.

Another issue limiting the deployment of this technology is the lack of consistent standards and protocols. There currently are no standards for these technologies. Most systems are able to communicate only with technologies developed by the same manufacturer. This limits the interoperability of Smart Grid technologies and limits future choices for companies that choose to install any particular type of technology. The Department of Energy's (DOE's) Office of Electricity Delivery and Energy Reliability in partnership with industry is developing standards for advanced grid design and operations. In addition, DOE is funding research and development projects in this area.

Selected Utility Applications

Smart Grid technologies are currently being used by several utilities in small applications, mainly for testing purposes. However, the technologies within the customer's house or business cannot allow for dynamic control of thermostats, for instance, but rather use switches to either turn an appliance on or off depending on preset criteria. The following applications of Smart Grid technologies represent some of the largest installations.

Southern California Edison Company

The California Public Utility Commission as well as the California Energy Action Plan call for smart meters as part of the overall energy policy for California.[8] On July 31, 2007, Southern California Edison Company (SCE) filed an application with the Public Utility Commission of California for approval of advanced metering infrastructure (AMI) deployment activities and a cost recovery mechanism for the $1.7 billion in estimated costs.[9] Beginning in 2009, SCE

---

[6] Personal Communication. Tom Casey, CEO Current Technologies. August 2, 2007.

[7] P.L. 109-58, §1252.

[8] California Energy Commission. *Energy Action Plan II*, September 21, 2005. Available at http://www.energy.ca.gov/energy_action_plan/2005-09-21_EAP2_FINAL.PDF.

[9] Public Utilities Commission of the State of California. *Southern California Edison Company's (U 338-E)* (continued...)

proposes to install through its SmartConnect™ program advanced meters in all households and businesses under 200 kW throughout its service territory (approximately 5.3 million meters). It is expected that demand response at peak times could save SCE as much as 1,000 megawatts of capacity additions. Dynamic rates such as Time of Use and Critical Peak Pricing should provide incentives to customers to shift some of their electricity usage to off-peak hours. According to SCE's application before the California Public Utility Commission:

> Edison SmartConnect™ includes meter and indication functionality that (i) measures interval electricity usage and voltage; (ii) supports nonproprietary, open standard communication interfaces with technologies such as programmable communicating thermostats and device switches; (iii) improves reliability through remote outage detection at customer premises; (iv) improves service and reduces costs by remote service activation; (v) is capable of remote upgrades; (vi) is compatible with broadband over powerline used by third parties; (vii) supports contract gas and water meter reads; and (viii) incorporates industry-leading security capabilities.[10]

In its filing, SCE is requesting approval to recover the operation and maintenance and capital expenditures associated with deployment of Edison SmartConnect™.

SCE is planning to use three telecommunications elements in addition to a smart meter.[11] The telecommunications system will include a Home Area Network (HAN) that is a non-proprietary open standard two-way narrowband radio frequency mesh network interface from the meter to customer-owned smart appliances, displays, and thermostats. Second, there will be a Local Area Network (LAN) consisting of a proprietary two-way narrowband radio frequency network that will connect the meter to the electricity aggregator.[12] Finally, a Wide Area Network (WAN) will be installed using a non-proprietary open standard two-way broadband network that will be used to communicate between the aggregator and the utility back office systems.[13] The meter will integrate the LAN and HAN in order to provide electric usage measurements, service voltage measurements, and interval measurements for billing purposes. These meters will have net-metering capability to support measurement of solar and other distributed generation at the customer's location. In addition, the meters will have security that has sophisticated cryptographic capabilities.

For the consumer, benefits include load reduction and energy conservation, which could result in lower electric bills. Outage information will automatically be sent to the utilities so customers won't need to report these disturbances. SCE is expecting to achieve greater reliability over time as additional information from the system is available to manage operations. For the utility, manual meter reading will be eliminated as will field service to turn power on to new customers.

(...continued)

Application for Approval of Advanced Metering Infrastructure Deployment Activities and Cost Recovery Mechanism. Filed July 31, 2007.

[10] Ibid., p. 7.

[11] Email communication. Paul De Martini. Director Edison SmartConnect™. August 2, 2007.

[12] An electric aggregator purchases power at wholesale for resale to retail customers.

[13] The two-way broadband network could include cellular, WiMax, or broadband over powerline.

Pacific Northwest GridWise™ Demonstration

The Pacific Northwest National Laboratory (PNNL) is teaming with utilities in the states of Washington and Oregon to test new energy technologies designed to improve efficiency and reliability while at the same time increasing consumer choice and control.[14] The utilities involved in the demonstration projects include the Bonneville Power Administration, PacifiCorp, Portland General Electric, Mason County PUD #3, Clallam County PUD, and the City of Port Angeles, Washington. PNNL has received in-kind contributions from industrial collaborators, including Sears Kenmore dryers, and communications and market integration software from IBM.

Two demonstration projects involve 300 homes as well as some municipal and commercial customers. The first project on the Olympic Peninsula involves 200 homes that are receiving real-time price signals over the Internet and have demand-response thermostats and hot water heaters that can be programmed to respond automatically. The goal is to relieve congestion on the transmission and distribution grid during peak periods. These 200 homes will test a "home information gateway" that will allow smart appliances such as communicating thermostats, smart water heaters, and smart clothes dryers to respond to transmission congestion due to peak demand or when prices are high. In addition, consumers will be able to see the actual cost of producing and delivering electricity, and cash incentives will be used to motivate customers to reduce peak demand. Part of the demonstration will study how existing backup generators can be used to displace demand for electricity.

The second demonstration involves 50 homes on the Olympic Peninsula in Washington, 50 homes in Yakima, Washington, and 50 homes in Gresham, Oregon. Clothes dryers will be installed in 150 homes and water heaters will be installed in 50 homes to test the ability of PNNL-developed appliance controllers to detect fluctuations in frequency. Fluctuations in frequency can indicate that the grid is under stress, and the appliance controllers can quickly respond to that stress by reducing demand. The appliance controllers will automatically turn off some appliances for a few seconds or minutes, allowing grid operators to rebalance the system.

TXU Electric Delivery Company

In October 2006, TXU Electric Delivery entered into an agreement to purchase 400,000 advanced meters. TXU Electric Delivery plans to have 3 million automated meters installed primarily in the Dallas-Fort Worth area by 2011. As of December 31, 2006, TXU had installed 285,000 advanced meters, 10,000 of which had broadband over powerline (BPL) capabilities.[15] This system combines advanced meters manufactured by Landis+Gyr with BPL-enabled communications technology provided by CURRENT Technologies. TXU Electric Delivery in the near-term will primarily use the advanced meters for increased network reliability and power quality and to prevent, detect, and restore customer outages more effectively. It is expected that TXU electric delivery will eventually include time-of-use options and new billing methods to its consumers.

---

[14] http://gridwise.pnl.gov/

[15] TXU Electric Delivery Company Annual Report. Form 10-K filing to the Securities and Exchange Commission. March 7, 2007.

On May 10, 2007, the Public Utility Commission of Texas issued an order allowing for the cost recovery of advanced meters.[16]

Summary of H.R. 6 Smart Grid Provisions

H.R. 6, signed by the President, contains a provision on Smart Grid technologies to address some of the regulatory and technological barriers to widespread installation.[17] This section summarizes Title XIII.

Section 1301. Statement of Policy on Modernization of Electricity Grid

It is the policy of the United States to support the modernization of the electric transmission and distribution system to maintain reliability and infrastructure protection. The Smart Grid is defined to include: increasing the use of additional information controls to improve operation of the electric grid; optimizing grid operations and resources to reflect the changing dynamics of the physical infrastructure and economic markets, while ensuring cybersecurity; using and integrating distributed resources, including renewable resources; developing and integrating demand response, demand-side resources, and energy-efficiency resources; deploying smart technologies for metering, communications of grid operations and status, and distribution automation; integrating "smart" appliances and other consumer devices; deploying and integrating advanced electricity storage and peak-shaving technologies; transferring information to consumers in a timely manner to allow control decisions; developing standards for the communication and the interoperability of appliances and equipment connected to the electric grid; identifying and lowering of unreasonable or unnecessary barriers to adoption of smart grid technologies, practices, and services.

Section 1302. Smart Grid System Report

No later than one year after enactment, and every two years thereafter, the Secretary of Energy shall issue a report to Congress on the status of the deployment of smart grid technologies and any regulatory or government barriers to continued deployment.

Section 1303. Smart Grid Advisory Committee and Smart Grid Task Force

Within 90 days of enactment, the Secretary of Energy shall establish a Smart Grid Advisory Committee, whose mission is to advise the Secretary of Energy and other relevant federal officials on the development of smart grid technologies, the deployment of such technologies, and the development of widely-accepted technical and practical standards and protocols to allow

[16] Public Utility Commission of Texas. Project Number 31418. Rulemaking Related to Advanced Metering. May 10, 2007.

[17] P.L. 110-140, signed by President Bush on December 19, 2007.

interoperability and integration among Smart Grid capable devices, and the optimal means for using federal incentive authority to encourage such programs.

In addition, a Smart Grid Task Force shall be established within 90 days of enactment. This task force will be composed of employees of the Department of Energy, Federal Energy Regulatory Commission, and the National Institute of Standards and Technology. The mission of the Smart Grid Task Force is to ensure coordination and integration of activities among the federal agencies.

Section 1304. Smart Grid Technology Research, Development, and Demonstration

The Secretary of Energy, in consultation with appropriate agencies, electric utilities, the states, and other stakeholders, is directed to carry out a program, in part, to develop advanced measurement techniques to monitor peak load reductions and energy efficiency savings from smart metering, demand response, distributed generation, and electricity storage systems; to conduct research to advance the use of wide-area measurement and control networks; to test new reliability technologies; to investigate the feasibility of a transition to time-of-use and real-time electricity pricing; to promote the use of underutilized electricity generation capacity in any substitution of electricity for liquid fuels in the transportation system of the United States; and to propose interconnection protocols to enable electric utilities to access electricity stored in hybrid vehicles to help meet peak demand loads. The Secretary of Energy shall also establish a Smart Grid regional demonstration initiative focusing on projects using advanced technologies for use in power grid sensing, communications, analysis, and power flow control.

Section 1305. Smart Grid Interoperability Framework

The Director of the National Institute of Standards and Technology is primarily responsible for coordinating the development of a framework for protocols and model standards for information management to gain interoperability of smart grid devices and systems.

Section 1306. Federal Matching Funds for Smart Grid Investment Costs

The Secretary of Energy shall establish a program to reimburse 20% of qualifying Smart Grid investments.

Section 1307. State Consideration of Smart Grid

The Public Utility Regulatory Policies Act of 1978 (16 U.S.C. 2621 (d)) is amended to require each state to consider requiring electric utilities demonstrate that prior to investing in non-advanced grid technologies, Smart Grid technology is determined not to be appropriate. States must also consider regulatory standards that allow utilities to recover Smart Grid investments through rates.

Section 1308. Study of the Effect of Private Wire Laws on the Development of Combined Heat and Power Facilities

Within one year of enactment, the Secretary of Energy shall submit a report to Congress detailing a study of the laws and regulations affecting the siting of privately owned electric distribution wires on and across public rights-of-way. This study will assess whether privately owned electric distribution wires would result in duplicative facilities and whether duplicate facilities are necessary or desirable.

Section 1309. DOE Study of Security Attributes of Smart Grid Systems

Within 18 months of enactment, the Secretary of Energy shall report to Congress the results of a study which provides a quantitative assessment and determination of the existing and potential impacts of the deployment of Smart Grid systems on the security of the electricity infrastructure and its operating capability.

Author Contact Information

Amy Abel
Section Research Manager
aabel@crs.loc.gov, 7-7239

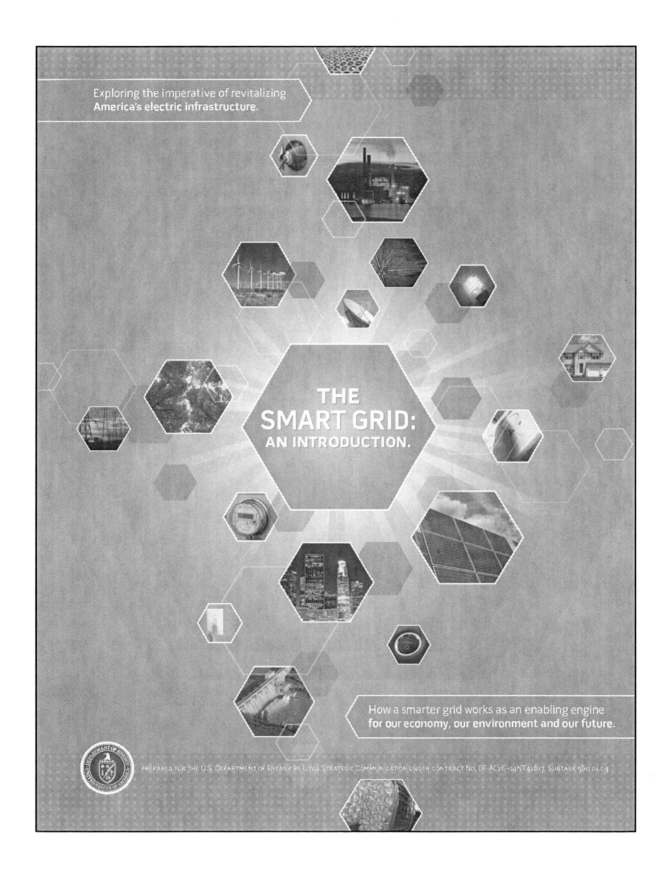

Exploring the imperative of revitalizing
America's electric infrastructure.

**THE
SMART GRID:
AN INTRODUCTION.**

How a smarter grid works as an enabling engine
for our economy, our environment and our future.

PREPARED FOR THE U.S. DEPARTMENT OF ENERGY BY LITOS STRATEGIC COMMUNICATION UNDER CONTRACT NO. DE-AC26-04NT41817, SUBTASK 560.01.04

PREFACE

IT IS A COLOSSAL TASK. BUT IT IS A TASK THAT MUST BE DONE.

The Department of Energy has been charged with orchestrating the wholesale modernization of our nation's electrical grid.

While it is running.

Full-tilt.

Heading this effort is the Office of Electricity Delivery and Energy Reliability. In concert with its cutting edge research and energy policy programs, the office's newly formed, multi-agency Smart Grid Task Force is responsible for coordinating standards development, guiding research and development projects, and reconciling the agendas of a wide range of stakeholders.

Equally critical to the success of this effort is the education of all interested members of the public as to the nature, challenges and opportunities surrounding the Smart Grid and its implementation.

It is to this mission that The Smart Grid: An Introduction is dedicated.

From the Department of Energy

The Smart Grid Introduction is intended primarily to acquaint non-technical yet interested readers about:

- the existence of, and benefits accruing from, a smarter electrical grid
- what the application of such intelligence means for our country
- how DOE is involved in helping to accelerate its implementation.

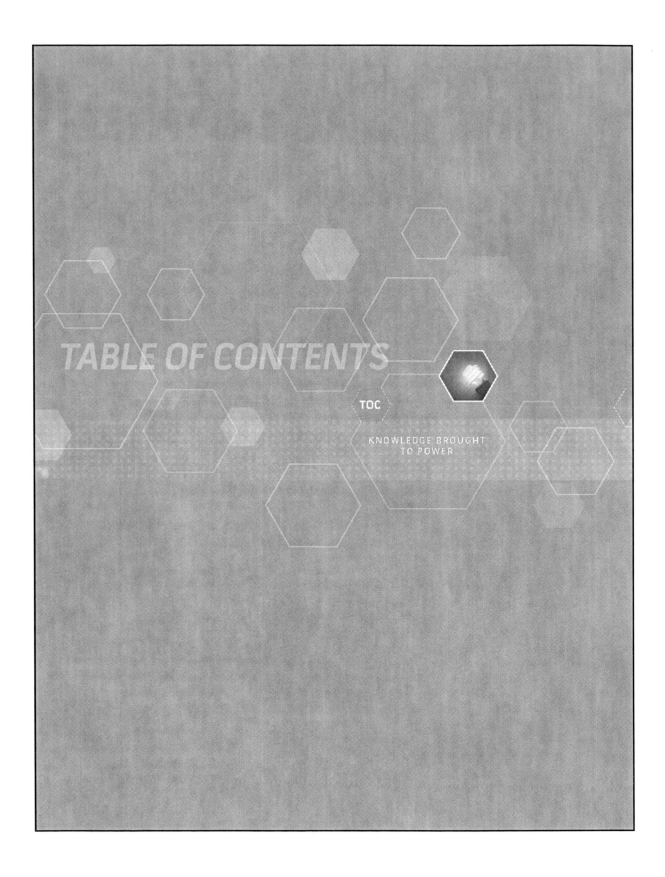

TABLE OF CONTENTS

TOC

KNOWLEDGE BROUGHT
TO POWER

1

123

SECTION ONE:
INTRODUCTION

WE DON'T HAVE MUCH TIME.

Our nation's electric power infrastructure that has served us so well for so long — also known as "the grid" — is rapidly running up against its limitations. Our lights may be on, but systemically, the risks associated with relying on an often overtaxed grid grow in size, scale and complexity every day. From national challenges like power system security to those global in nature such as climate change, our near-term agenda is formidable. Some might even say history-making.

Fortunately, we have a way forward.

There is growing agreement among federal and state policymakers, business leaders, and other key stakeholders, around the idea that a Smart Grid is not only needed but well within reach. Think of the Smart Grid as the internet brought to our electric system.

A tale of two timelines

There are in fact two grids to keep in mind as our future rapidly becomes the present.

The first — we'll call it "a smarter grid" — offers valuable technologies that can be deployed within the very near future or are already deployed today.

The second — the Smart Grid of our title — represents the longer-term promise of a grid remarkable in its intelligence and impressive in its scope, although it is universally considered to be a decade or more from realization. Yet given how a single "killer application" — e-mail — incited broad, deep and immediate acceptance of the internet, who is to say that a similar killer app in this space won't substantially accelerate that timetable?

In the short term, a smarter grid will function more efficiently, enabling it to deliver the level of service we've come to expect more affordably in an era of rising costs, while also offering considerable societal benefits — such as less impact on our environment.

Longer term, expect the Smart Grid to spur the kind of transformation that the internet has already brought to the way we live, work, play and learn.

124

*A smarter grid applies technologies, tools and techniques available now to bring knowledge to power —
knowledge capable of making the grid work far more efficiently...*

- Ensuring its reliability to degrees never before possible.
- Maintaining its affordability.
- Reinforcing our global competitiveness.
- Fully accommodating renewable and traditional energy sources.
- Potentially reducing our carbon footprint.
- Introducing advancements and efficiencies yet to be envisioned.

Transforming our nation's grid has been compared in significance with building the interstate highway system
or the development of the internet. These efforts, rightly regarded as revolutionary, were preceded by countless
evolutionary steps. Envisioned in the 1950s, the Eisenhower Highway System was not completed until the early
1980s. Similarly, the internet's lineage can be directly traced to the Advanced Research Projects Agency Network
(ARPANET) of the U.S. Department of Defense in the 60s and 70s, long before its appearance as a society-changing
technology in the 80s and 90s.

In much the same way, full implementation of the Smart Grid will evolve over time. However, countless positive
steps are being taken today, organizations energized and achievements realized toward reaching that goal. You
will learn about some of them here.

The purpose of this book is to give readers — in plain language — a fix on the current position of the Smart Grid and
its adoption. You will learn what the Smart Grid is — and what it is not. You will get a feel for the issues surrounding
it, the challenges ahead, the countless opportunities it presents and the benefits we all stand to gain.

Remember life before e-mail?
With every passing day, fewer and fewer people do.

With the appropriate application of ingenious ideas, advanced technology, entrepreneurial energy and political will,
there will also come a time when you won't remember life before the Smart Grid.

3

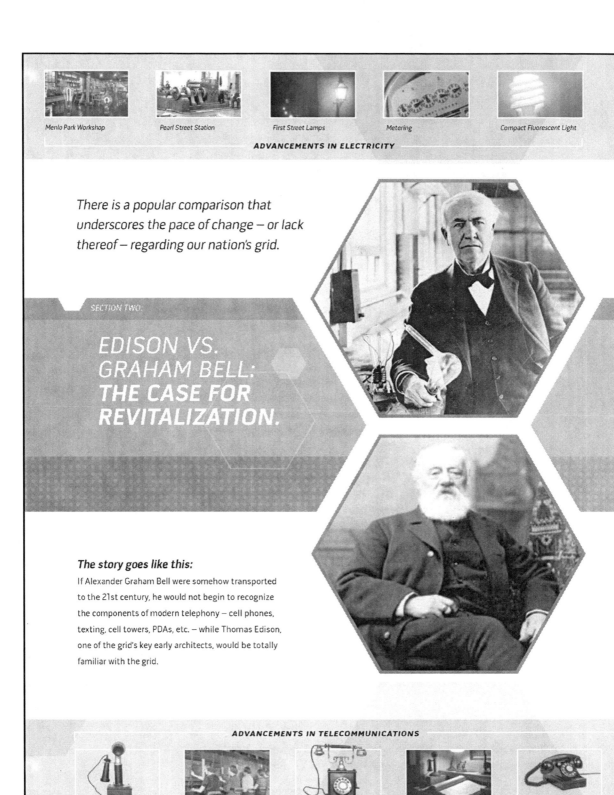

Menlo Park Workshop Pearl Street Station First Street Lamps Metering Compact Fluorescent Light

ADVANCEMENTS IN ELECTRICITY

There is a popular comparison that underscores the pace of change — or lack thereof — regarding our nation's grid.

SECTION TWO:

EDISON VS. GRAHAM BELL: THE CASE FOR REVITALIZATION.

The story goes like this:

If Alexander Graham Bell were somehow transported to the 21st century, he would not begin to recognize the components of modern telephony — cell phones, texting, cell towers, PDAs, etc. — while Thomas Edison, one of the grid's key early architects, would be totally familiar with the grid.

ADVANCEMENTS IN TELECOMMUNICATIONS

First Telephone Operator Switching Stations Rotary Dialing North American Numbering System Rotary Dial with Ringer and Handset

4

While this thought experiment speaks volumes about appearances, it is far from the whole story. Edison would be quite familiar with the grid's basic infrastructure and perhaps even an electromechanical connection or two, but he would be just as dazzled as Graham Bell with the technology behind the scenes.

Our century-old power grid is the largest interconnected machine on Earth, so massively complex and inextricably linked to human involvement and endeavor that it has alternately (and appropriately) been called an ecosystem. It consists of more than 9,200 electric generating units with more than 1,000,000 megawatts of generating capacity connected to more than 300,000 miles of transmission lines.

Given that the growth of the nation's global economic leadership over the past century has in many ways mirrored the trajectory of the grid's development, this choice is not surprising.

In many ways, the present grid works exceptionally well for what it was designed to do – for example, keeping costs down. Because electricity has to be used the moment it is generated, the grid represents the ultimate in just-in-time product delivery. Everything must work almost perfectly at all times – and does. Whenever an outage occurs in, say, Florida, there may well be repercussions up the Atlantic seaboard; however, due to the system's robustness and resultant reliability, very few outside the industry ever know about it.

> **POWER SYSTEM FACT**
>
> *Today's electricity system is 99.97 percent reliable, yet still allows for power outages and interruptions that cost Americans at least $150 billion each year — about $500 for every man, woman and child.*

In celebrating the beginning of the 21st century, the National Academy of Engineering set about identifying the single most important engineering achievement of the 20th century. The Academy compiled an estimable list of twenty accomplishments which have affected virtually everyone in the developed world. The internet took thirteenth place on this list, and "highways" eleventh. Sitting at the top of the list was electrification as made possible by the grid, "the most significant engineering achievement of the 20th Century."

Engineered and operated by dedicated professionals over decades, the grid remains our national engine. It continues to offer us among the highest levels of reliability in the world for electric power. Its importance to our economy, our national security, and to the lives of the hundreds of millions it serves cannot be overstated.

But we – all of us – have taken this marvelous machine for granted for far too long. As a result, our overburdened grid has begun to fail us more frequently and presents us with substantial risks.

Long Distance Calling

First Telecom Satellite

Touch-Tone Telephones

Cellular Communications

Phone Over the Internet

5

Since 1982, growth in peak demand for electricity – driven by population growth, bigger houses, bigger TVs, more air conditioners and more computers – has exceeded transmission growth by almost 25% every year. Yet spending on research and development – the first step toward innovation and renewal – is among the lowest of all industries.

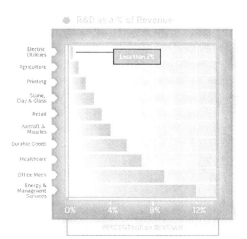

SECTION THREE:

THE GRID AS IT STANDS: *WHAT'S AT RISK?*

Even as demand has skyrocketed, there has been chronic underinvestment in getting energy where it needs to go through transmission and distribution, further limiting grid efficiency and reliability. While hundreds of thousands of high-voltage transmission lines course throughout the United States, only 668 additional miles of interstate transmission have been built since 2000. As a result, system constraints worsen at a time when outages and power quality issues are estimated to cost American business more than $100 billion on average each year.

In short, the grid is struggling to keep up.

● R&D as a % of Revenue

Electric Utilities	Less than 2%
Agriculture	
Printing	
Stone, Clay & Glass	
Retail	
Aircraft & Missiles	
Durable Goods	
Healthcare	
Office Mech	
Energy & Managment Services	

0%   4%   8%   12%

PERCENTAGE OF REVENUE

Based on 20TH century design requirements and having matured in an era when expanding the grid was the only option and visibility within the system was limited, the grid has historically had a single mission, i.e., keeping the lights on. As for other modern concerns...

Energy efficiency? A marginal consideration at best when energy was – as the saying went – "too cheap to meter."

Environmental impacts? Simply not a primary concern when the existing grid was designed.

Customer choice? What was that?

RELIABILITY: There have been five massive blackouts over the past 40 years, three of which have occurred in the past nine years. More blackouts and brownouts are occurring due to the slow response times of mechanical switches, a lack of automated analytics, and "poor visibility" – a "lack of situational awareness" on the part of grid operators. This issue of blackouts has far broader implications than simply waiting for the lights to come on. Imagine plant production stopped, perishable food spoiling, traffic lights dark, and credit card transactions rendered inoperable. Such are the effects of even a short regional blackout.

DID YOU KNOW

In many areas of the United States, the only way a utility knows there's an outage is when a customer calls to report it.

Today, the irony is profound: In a society where technology reigns supreme, America is relying on a centrally planned and controlled infrastructure created largely before the age of microprocessors that limits our flexibility and puts us at risk on several critical fronts:

EFFICIENCY: If the grid were just 5% more efficient, the energy savings would equate to permanently eliminating the fuel and greenhouse gas emissions from 53 million cars. Consider this, too: If every American household replaced just one incandescent bulb (Edison's pride and joy) with a compact fluorescent bulb, the country would conserve enough energy to light 3 million homes and save more than $600 million annually. Clearly, there are terrific opportunities for improvement.

POWER SYSTEM FACT

41% more outages affected 50,000 or more consumers in the second half of the 1990s than in the first half of the decade. The "average" outage affected 15 percent more consumers from 1996 to 2000 than from 1991 to 1995 (409,854 versus 355,204).

7

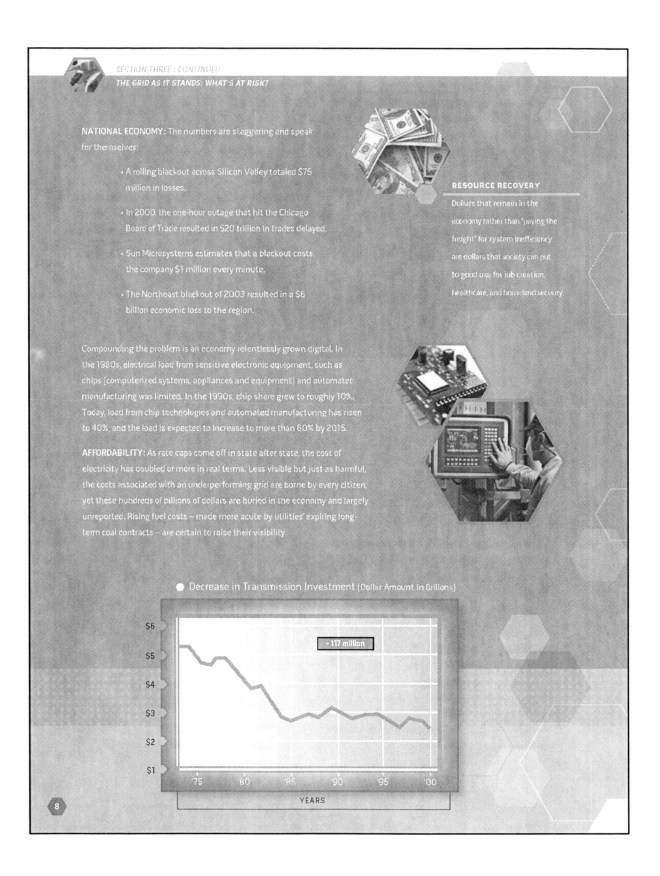

SECTION THREE : CONTINUED

THE GRID AS IT STANDS: WHAT'S AT RISK?

NATIONAL ECONOMY: The numbers are staggering and speak for themselves:

- A rolling blackout across Silicon Valley totaled $75 million in losses.

- In 2000, the one-hour outage that hit the Chicago Board of Trade resulted in $20 trillion in trades delayed.

- Sun Microsystems estimates that a blackout costs the company $1 million every minute.

- The Northeast blackout of 2003 resulted in a $6 billion economic loss to the region.

Compounding the problem is an economy relentlessly grown digital. In the 1980s, electrical load from sensitive electronic equipment, such as chips (computerized systems, appliances and equipment) and automated manufacturing was limited. In the 1990s, chip share grew to roughly 10%. Today, load from chip technologies and automated manufacturing has risen to 40%, and the load is expected to increase to more than 60% by 2015.

AFFORDABILITY: As rate caps come off in state after state, the cost of electricity has doubled or more in real terms. Less visible but just as harmful, the costs associated with an underperforming grid are borne by every citizen, yet these hundreds of billions of dollars are buried in the economy and largely unreported. Rising fuel costs — made more acute by utilities' expiring long-term coal contracts — are certain to raise their visibility.

RESOURCE RECOVERY

Dollars that remain in the economy rather than "paying the freight" for system inefficiency are dollars that society can put to good use for job creation, healthcare, and homeland security.

● Decrease in Transmission Investment (Dollar Amount In Billions)

- 117 million

$6
$5
$4
$3
$2
$1

'75   '80   '85   '90   '95   '00

YEARS

8

130

SECURITY: When the blackout of 2003 occurred — the largest in US history — those citizens not startled by being stuck in darkened, suffocating elevators turned their thoughts toward terrorism. And not without cause. The grid's centralized structure leaves us open to attack. In fact, the interdependencies of various grid components can bring about a domino effect — a cascading series of failures that could bring our nation's banking, communications, traffic, and security systems among others to a complete standstill.

ENVIRONMENT/CLIMATE CHANGE: From food safety to personal health, a compromised environment threatens us all. The United States accounts for only 4% of the world's population and produces 25% of its greenhouse gases. Half of our country's electricity is still produced by burning coal, a rich domestic resource but a major contributor to global warming. If we are to reduce our carbon footprint and stake a claim to global environmental leadership, clean, renewable sources of energy like solar, wind and geothermal must be integrated into the nation's grid. However, without appropriate enabling technologies linking them to the grid, their potential will not be fully realized.

GLOBAL COMPETITIVENESS: Germany is leading the world in the development and implementation of photo-voltaic solar power. Japan has similarly moved to the forefront of distribution automation through its use of advanced battery-storage technology. The European Union has an even more aggressive "Smart Grids" agenda, a major component of which has buildings functioning as power plants. Generally, however, these countries don't have a "legacy system" on the order of the grid to consider or grapple with.

How will a smarter grid address these risks and others? Read on.

U.S. Share of World Population Compared to Its Production of Greenhouse Gases

4%

The U.S. accounts for 4% of the world's population while contributing 25% of its greenhouse gases.

25%

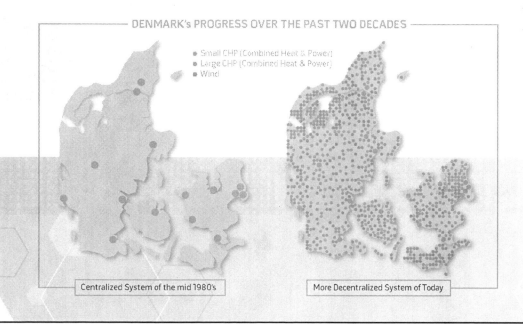

DENMARK's PROGRESS OVER THE PAST TWO DECADES

- Small CHP (Combined Heat & Power)
- Large CHP (Combined Heat & Power)
- Wind

Centralized System of the mid 1980's

More Decentralized System of Today

9

Prepare for an electric system that is cleaner and more efficient, reliable, resilient and responsive — a smarter grid.

THE SMART GRID: WHAT IT IS. WHAT IT ISN'T.

PART 1: **WHAT IT IS.**

The electric industry is poised to make the transformation from a centralized, producer-controlled network to one that is less centralized and more consumer-interactive. The move to a smarter grid promises to change the industry's entire business model and its relationship with all stakeholders, involving and affecting utilities, regulators, energy service providers, technology and automation vendors and all consumers of electric power.

A smarter grid makes this transformation possible by bringing the philosophies, concepts and technologies that enabled the internet to the utility and the electric grid. More importantly, it enables the industry's best ideas for grid modernization to achieve their full potential.

Concepts in action.
It may surprise you to know that many of these ideas are already in operation. Yet it is only when they are empowered by means of the two-way digital communication and plug-and-play capabilities that exemplify a smarter grid that genuine breakthroughs begin to multiply.

Because this interaction occurs largely "in the background," with minimal human intervention, there's a dramatic savings on energy that would otherwise be consumed.

This type of program has been tried in the past, but without Smart Grid tools such as enabling technologies, interoperability based on standards, and low-cost communication and electronics, it possessed none of the potential that it does today.

Visualization technology. Consider grid visualization and the tools associated with it. Already used for real-time load monitoring and load-growth planning at the utility level, such tools generally lack the ability to integrate information from a variety of sources or display different views to different users. The result: Limited

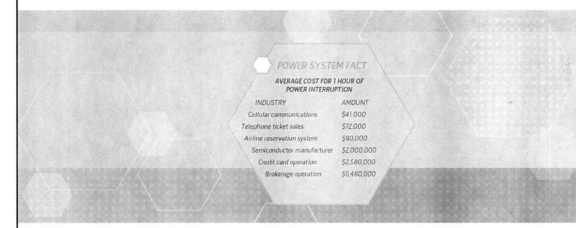

POWER SYSTEM FACT

AVERAGE COST FOR 1 HOUR OF
POWER INTERRUPTION

INDUSTRY	AMOUNT
Cellular communications	$41,000
Telephone ticket sales	$72,000
Airline reservation system	$90,000
Semiconductor manufacturer	$2,000,000
Credit card operation	$2,580,000
Brokerage operation	$6,480,000

Advanced Metering Infrastructure (AMI) is an approach to integrating consumers based upon the development of open standards. It provides consumers with the ability to use electricity more efficiently and provides utilities with the ability to detect problems on their systems and operate them more efficiently.

AMI enables consumer-friendly efficiency concepts like "Prices to Devices" to work like this: Assuming that energy is priced on what it costs in near real-time — a Smart Grid imperative — price signals are relayed to "smart" home controllers or end-consumer devices like thermostats, washer/dryers and refrigerators — the home's major energy-users. The devices, in turn, process the information based on consumers' learned wishes and power accordingly. The house or office responds to the occupants, rather than vice-versa.

situational awareness. This condition will grow even more acute as customer-focused efficiency and demand-response programs increase, requiring significantly more data as well as the ability to understand and act on that data.

Next-generation visualization is on its way. Of particular note is VERDE, a project in development for DOE at the Oak Ridge National Laboratory. VERDE (Visualizing Energy Resources Dynamically on Earth) will provide wide-area grid awareness, integrating real-time sensor data, weather information and grid modeling with geographical information. Potentially, it will be able to explore the state of the grid at the national level and switch within seconds to explore specific details at the street level. It will provide rapid information about blackouts and power quality as well as insights into system operation for utilities. With a platform built on Google Earth, it can also take advantage of content generated by Google Earth's user community.

11

133

Just who's running the grid?

Formed at the recommendation of the Federal Energy Regulatory Commission (FERC), an Independent System Operator (ISO) or Regional Transmission Organization (RTO) is a profit-neutral organization in charge of reconciling supply and demand as it coordinates, controls and monitors the operation of the power system. The ISO's control area can encompass one state or several.

The role of these organizations is significant in making the Smart Grid real. ISOs and RTOs will use the smart distribution system as another resource for managing a secure and most economic transmission system. "Lessons learned" from their experiences in building processes and technologies, etc., will be directly applicable to efforts in grid transformation, both short-term and long-term.

Phasor Measurement Units.

Popularly referred to as the power system's "health meter," Phasor Measurement Units (PMU) sample voltage and current many times a second at a given location, providing an "MRI" of the power system compared to the "X-Ray" quality available from earlier Supervisory Control and Data Acquisition (SCADA) technology. Offering wide-area situational awareness, phasors work to ease congestion and bottlenecks and mitigate – or even prevent – blackouts.

Typically, measurements are taken once every 2 or 4 seconds offering a steady-state view into the power system behavior. Equipped with Smart Grid communications technologies, measurements taken are precisely time-synchronized and taken many times a second (i.e., 30 samples/second) offering dynamic visibility into the power system.

Adoption of the Smart Grid will enhance every facet of the electric delivery system, including generation, transmission, distribution and consumption. It will energize those utility initiatives that encourage consumers to modify patterns of electricity usage, including the timing and level of electricity demand. It will increase the possibilities of distributed generation, bringing generation closer to those it serves (think: solar panels on your roof rather than some distant power station). The shorter the distance from generation to consumption, the more efficient, economical and "green" it may be. It will empower consumers to become active participants in their energy choices to a degree never before possible. And it will offer a two-way visibility and control of energy usage.

SMART DEFINITION: DISTRIBUTED GENERATION

Distributed generation is the use of small-scale power generation technologies located close to the load being served, capable of lowering costs, improving reliability, reducing emissions and expanding energy options.

12

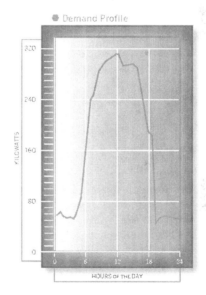

An automated, widely distributed energy delivery network, the Smart Grid will be characterized by a two-way flow of electricity and information and will be capable of monitoring everything from power plants to customer preferences to individual appliances. It incorporates into the grid the benefits of distributed computing and communications to deliver real-time information and enable the near-instantaneous balance of supply and demand at the device level.

The problem with peak.

While supply and demand is a bedrock concept in virtually all other industries, it is one with which the current grid struggles mightily because, as noted, electricity must be consumed the moment it's generated.

Without being able to ascertain demand precisely, at a given time, having the 'right' supply available to deal with every contingency is problematic at best. This is particularly true during episodes of peak demand, those times of greatest need for electricity during a particular period.

● Demand Profile

HOURS OF THE DAY

13

Imagine that it is a blisteringly hot summer afternoon. With countless commercial and residential air conditioners cycling up to maximum, demand for electricity is being driven substantially higher, to its "peak." Without a greater ability to anticipate, without knowing *precisely* when demand will peak or how high it will go, grid operators and utilities must bring generation assets called peaker plants online to ensure reliability and meet peak demand. Sometimes older and always difficult to site, peakers are expensive to operate — requiring fuel bought on the more volatile "spot" market. But old or not, additional peakers generate additional greenhouse gases, degrading the region's air quality. Compounding the inefficiency of this scenario is the fact that peaker plants are generation assets that typically sit idle for most of the year without generating revenue but must be paid for nevertheless.

In making real-time grid response a reality, a smarter grid makes it possible to reduce the high cost of meeting peak demand. It gives grid operators far greater visibility into the system at a finer "granularity," enabling them to control loads in a way that minimizes the need for traditional peak capacity. In addition to driving down costs, it may even eliminate the need to use existing peaker plants or build new ones — to save everyone money and give our planet a breather.

PART 2: **WHAT THE SMART GRID ISN'T.**

People are often confused by the terms Smart Grid and smart meters. Are they not the same thing? Not exactly. Metering is just one of hundreds of possible applications that constitute the Smart Grid; a smart meter is a good example of an enabling technology that makes it possible to extract value from two-way communication in support of distributed technologies and consumer participation.

As much as "smart technologies" can enhance this familiar device, it's not the same thing as the Smart Grid.

14

136

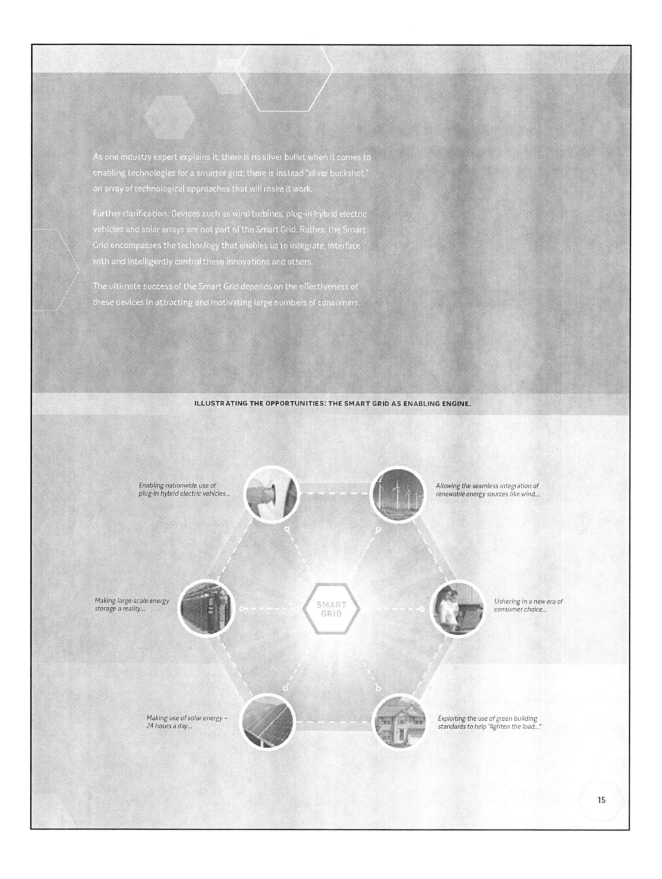

As one industry expert explains it, there is no silver bullet when it comes to enabling technologies for a smarter grid; there is instead "silver buckshot," an array of technological approaches that will make it work.

Further clarification: Devices such as wind turbines, plug-in hybrid electric vehicles and solar arrays are not part of the Smart Grid. Rather, the Smart Grid encompasses the technology that enables us to integrate, interface with and intelligently control these innovations and others.

The ultimate success of the Smart Grid depends on the effectiveness of these devices in attracting and motivating large numbers of consumers.

ILLUSTRATING THE OPPORTUNITIES: THE SMART GRID AS ENABLING ENGINE.

Enabling nationwide use of plug-in hybrid electric vehicles...

Allowing the seamless integration of renewable energy sources like wind...

Making large-scale energy storage a reality...

SMART GRID

Ushering in a new era of consumer choice...

Making use of solar energy – 24 hours a day...

Exploiting the use of green building standards to help "lighten the load..."

15

The Smart Grid transforms the current grid to one that functions more cooperatively, responsively and organically.

SECTION FIVE:

COMPARE AND CONTRAST: A GRID WHERE EVERYTHING IS POSSIBLE.

SMART GRID FACT

Made possible by a smarter grid, DOE's Solar Energy Grid Integration Systems (SEGIS) is a suite of tools, techniques and technologies designed to achieve a high penetration of photovoltaic (PV) systems into homes and businesses.

16

IN TERMS OF OVERALL VISION, THE SMART GRID IS:

Intelligent – capable of sensing system overloads and rerouting power to prevent or minimize a potential outage; of working autonomously when conditions require resolution faster than humans can respond...and cooperatively in aligning the goals of utilities, consumers and regulators

Efficient – capable of meeting increased consumer demand without adding infrastructure

Accommodating – accepting energy from virtually any fuel source including solar and wind as easily and transparently as coal and natural gas; capable of integrating any and all better ideas and technologies – energy storage technologies, for example – as they are market-proven and ready to come online

Motivating – enabling real-time communication between the consumer and utility so consumers can tailor their energy consumption based on individual preferences, like price and/or environmental concerns

Opportunistic – creating new opportunities and markets by means of its ability to capitalize on plug-and-play innovation wherever and whenever appropriate

Quality-focused – capable of delivering the power quality necessary – free of sags, spikes, disturbances and interruptions – to power our increasingly digital economy and the data centers, computers and electronics necessary to make it run

Resilient – increasingly resistant to attack and natural disasters as it becomes more decentralized and reinforced with Smart Grid security protocols

"Green" – slowing the advance of global climate change and offering a genuine path toward significant environmental improvement

---

Applied across various key constituencies, the benefits of creating a smarter grid are drawn in even sharper relief.

The Smart Grid as it applies to utilities.

Whether they're investor-owned, cooperatively owned or public, utilities are dedicated to providing for the public good – i.e., taking care of society's electricity needs – by operating, maintaining and building additional electric infrastructure. The costs associated with such tasks can run to billions of dollars annually and the challenges associated with them are enormous.

For a smarter grid to benefit society, it must reduce utilities' capital and/or operating expenses today – or reduce costs in the future. It is estimated that Smart Grid enhancements will ease congestion and increase utilization (of full capacity), sending 50% to 300% more electricity through existing energy corridors.

The more efficient their systems, the less utilities need to spend.

Given our nation's population growth and the exponential increase in the number of power-hungry digital components in our digital economy, additional infrastructure must be built – Smart or not. According to The Brattle Group, investment totaling approximately $1.5 trillion will be required between 2010 and 2030 to pay for this infrastructure. The Smart Grid holds the potential to be the most affordable alternative to "building out" by building less, and saving more energy. It will clearly require investments that are not typical for utilities. But the overall benefits of such efforts will outweigh the costs, as some utilities are already discovering.

17

SECTION FIVE CONTINUED
COMPARE AND CONTRAST: A GRID WHERE EVERYTHING IS POSSIBLE

One afternoon in early 2008, the wind stopped blowing in Texas.
A leader in this renewable energy, the state experienced a sudden, unanticipated and dramatic drop in wind power — 1300 Mw in just three hours. An emergency demand response program was initiated in which large industrial and commercial users restored most of the lost generation within ten minutes, acting as a buffer for fluctuations in this intermittent resource. Smart Grid principles in action.

The Smart Grid as it applies to consumers.
For most consumers, energy has long been considered a passive purchase. After all, what choice have they been given? The typical electric bill is largely unintelligible to consumers and delivered days after the consumption actually occurs — giving consumers no visibility into decisions they could be making regarding their energy consumption.

However, it pays to look at electric bills closely if for no other reason than this: they also typically include a hefty "mortgage payment" to pay for the infrastructure needed to generate and deliver power to consumers.

A surprisingly substantial portion of your electric bill — between 33% – 50% — is currently assigned to funding our "infrastructure mortgage," our current electric infrastructure. This item is non-negotiable because that infrastructure — power plants, transmission lines, and everything else that connects them — must be maintained to keep the grid running as reliably as it does. In fact, the transmission and distribution charge on the electric bill is specifically for infrastructure.

With demand estimated to double by 2050 — and more power plants, transmission lines, transformers and substations to be built — the costs of this "big iron" will also show up on your bill in one way or another. (The only difference this time is that global demand for the iron, steel, and concrete required to build this infrastructure will make these commodities far more costly; in fact, the cost of many raw materials and grid components has more than tripled since 2006.)

> POWER SYSTEM FACT
>
> In the United States, the average generating station was built in the 1960s using even older technology. Today, the average age of a substation transformer is 42, two years more than their expected life span.

SMART DEFINITION: REAL-TIME PRICING — These are energy prices that are set for a specific time period on an advance or forward basis and which may change according to price changes in the market. Prices paid for energy consumed during these periods are typically established and known to consumers a day ahead ("day-ahead pricing") or an hour ahead ("hour-ahead pricing") in advance of such consumption, allowing them to vary their demand and usage in response to such prices and manage their energy costs by shifting usage to a lower cost period, or reducing consumption overall.

18

Now for the good news. The Smart Grid connects consumers to the grid in a way that is beneficial to both, because it turns out there's a lot that average consumers can do to help the grid.

Simply by connecting to consumers – by means of the right price signals and smart appliances, for example – a smarter grid can reduce the need for some of that infrastructure while keeping electricity reliable and affordable. As noted, during episodes of peak demand, stress on the grid threatens its reliability and raises the probability of widespread blackouts.

By enabling consumers to automatically reduce demand for brief periods through new technologies and motivating mechanisms like real-time pricing, the grid remains reliable – and consumers are compensated for their help.

Enabling consumer participation also provides tangible results for utilities which are experiencing difficulty in siting new transmission lines and power plants. Ultimately, tapping the collaborative power of millions of consumers to shed load will put significant brakes on the need for new infrastructure at any cost. Instead, utilities will have time to build more cost-efficiencies into their siting and building plans.

Consumers are more willing to be engaged.

Consumers are advocating for choice in market after market, from telecom to entertainment. Already comfortable with the concept of time-differentiated service thanks to time-dependent cell phone rates and airline fares, it follows that they just might want insight and visibility into the energy choices they are making, too. Enabled by Smart Grid technology and dynamic pricing, consumers will have the opportunity to see what price they are paying for energy before they buy – a powerful motivator toward managing their energy costs by reducing electric use during peak periods.

Currently, recognition of the time-dependent cost of energy varies by region. In areas where costs are low and specialized rates to this point non-existent, there is little interest or economic incentive on the part of the consumer to modify usage or even think about energy having an hourly cost. In California, on a hot afternoon, consumers are well aware of the possibility of a blackout driven by peak demand and familiar with adjusting their energy usage accordingly.

Efficiency is the way.

10% of all generation assets and 25% of distribution infrastructure are required less than 400 hours per year, roughly 5% of the time. While Smart Grid approaches can't completely displace the need to build new infrastructure, they will enable new, more persistent forms of demand response that will succeed in deferring or avoiding some of it.

The rewards of getting involved.

Smart Grid consumer mantra: Ask not what the grid can do for you. Ask what you can do for the grid – and prepare to get paid for it.

19

Given new awareness, understanding, tools and education made possible by a smarter grid, all consumers will be able to make choices that save money, enhance personal convenience, improve the environment – or all three.

The message from consumers about the Smart Grid: Keep It Simple.

Research indicates that consumers are ready to engage with the Smart Grid as long as their interface with the Smart Grid is simple, accessible and in no way interferes with how they live their lives. Consumers are not interested in sitting around for an hour a day to change how their house uses energy; what they will do is spend two hours per year to set their comfort, price and environmental preferences – enabling collaboration with the grid to occur automatically on their behalf and saving money each time.

At the residential level, Smart Grid must be simple, "set-it-and-forget-it" technology, enabling consumers to easily adjust their own energy use. Equipped with rich, useful information, consumers can help manage load on-peak to save money and energy for themselves and, ultimately, all of us.

The Smart Grid as it applies to our environment.

While the nation's transportation sector emits 20% of all the carbon dioxide we produce, the generation of electricity emits 40% – clearly presenting an enormous challenge for the electric power industry in terms of global climate change. Smart Grid deployment is a key tool in addressing the challenges of climate change, ultimately and significantly reducing greenhouse gases and criteria pollutants such as NOx, SOx and particulates.

For the growing number of environmentally-aware consumers, a smarter grid finally provides a "window" for them to assess and react to their personal environmental impacts. Already, some utilities are informing consumers about their carbon footprint alongside their energy costs. In time, the Smart Grid will enable consumers to react in near real-time to lessen their impacts.

POWER SYSTEM FACT

From 1988-98, U.S. electricity demand rose by nearly 30 percent, while the transmission network's capacity grew by only 15%. Summer peak demand is expected to increase by almost 20% during the next 10 years.

SMART DEFINITION: CRITERIA POLLUTANTS - Criteria pollutants are six common air pollutants that the scientific community has established as being harmful to our health and welfare when present at specified levels. They include nitrogen dioxide (NOx), carbon monoxide, ozone, lead, sulfur dioxide (SOx) and particulate matter, which includes dirt, soot, car and truck exhaust, cigarette smoke, spray paint droplets, and toxic chemical compounds.

20

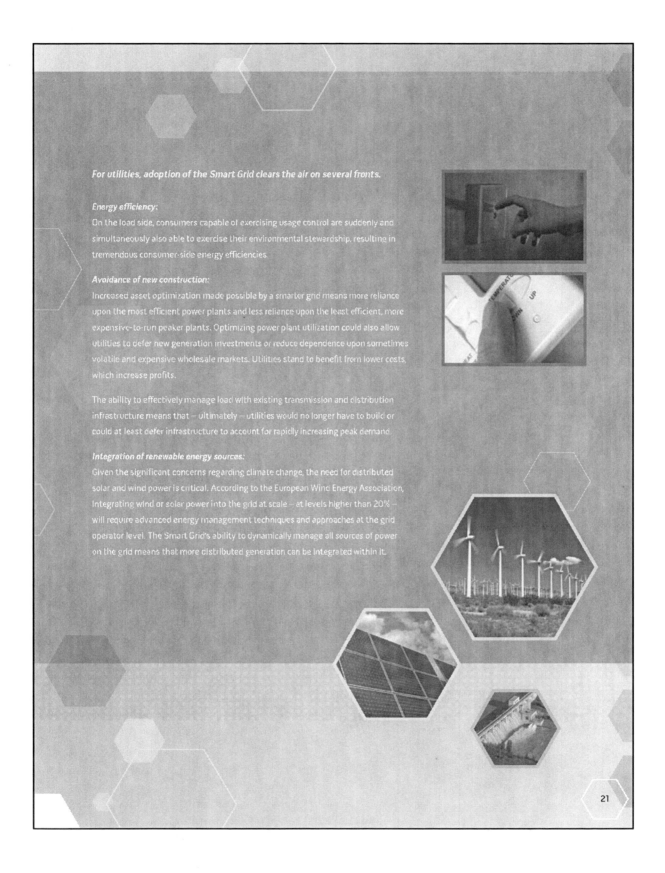

For utilities, adoption of the Smart Grid clears the air on several fronts.

Energy efficiency:
On the load side, consumers capable of exercising usage control are suddenly and simultaneously also able to exercise their environmental stewardship, resulting in tremendous consumer-side energy efficiencies.

Avoidance of new construction:
Increased asset optimization made possible by a smarter grid means more reliance upon the most efficient power plants and less reliance upon the least efficient, more expensive-to-run peaker plants. Optimizing power plant utilization could also allow utilities to defer new generation investments or reduce dependence upon sometimes volatile and expensive wholesale markets. Utilities stand to benefit from lower costs, which increase profits.

The ability to effectively manage load with existing transmission and distribution infrastructure means that – ultimately – utilities would no longer have to build or could at least defer infrastructure to account for rapidly increasing peak demand.

Integration of renewable energy sources:
Given the significant concerns regarding climate change, the need for distributed solar and wind power is critical. According to the European Wind Energy Association, integrating wind or solar power into the grid at scale – at levels higher than 20% – will require advanced energy management techniques and approaches at the grid operator level. The Smart Grid's ability to dynamically manage all sources of power on the grid means that more distributed generation can be integrated within it.

21

SECTION FIVE : CONTINUED

COMPARE AND CONTRAST: *A GRID WHERE EVERYTHING IS POSSIBLE*

Preparation for the future:

A smarter grid is also a necessity for plugging in the next generation of automotive vehicles – including plug-in hybrid electric vehicles (PHEVs) – to provide services supporting grid operation. Such ancillary services hold the potential for storing power and selling it back to the grid when the grid requires it.

Enabled by new technologies, plug-in hybrid vehicles – currently scheduled for showroom floors by 2010 – may dramatically reduce our nation's foreign oil bill. According to the Pacific Northwest National Laboratory, existing U.S. power plants could meet the electricity needs of 73% of the nation's light vehicles (i.e., cars and small trucks) if the vehicles were replaced by plug-ins that recharged at night. Such a shift would reduce oil consumption by 6.2 million barrels per day, eliminating 52% of current imports.

However, there is a lot more to realizing this potential than simply plugging in.

Without an integrated communications infrastructure and corresponding price signals, handling the increased load of plug-in hybrids and electric vehicles would be exceedingly difficult and inefficient. Smart Chargers, however – enabled by the Smart Grid – will help manage this new energy device on already constrained grids and avoid any unintended consequences on the infrastructure.

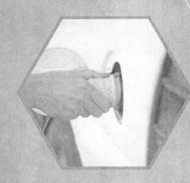

To get a greener grid, you need a Smart Grid.

Solar and wind power are necessary and desirable components of a cleaner energy future. To make the grid run cleaner, it will take a grid capable of dealing with the variable nature of these renewable resources.

SMART DEFINITION : OFF PEAK

A period of relatively low system demand, often occurring in daily, weekly, and seasonal patterns. Off-peak periods differ for each individual electric utility.

22

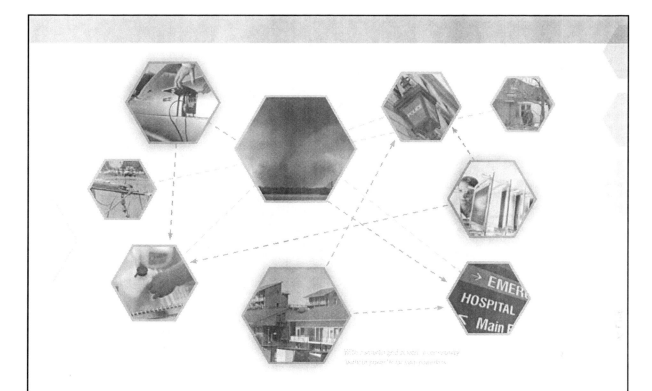

With a smart grid power is no longer dependent on its sources.

What might the longer-term future look like?

It is a decade from now.

An unusually destructive storm has isolated a community or region. Ten years ago, the wait for the appearance of a utility's "trouble trucks" would begin. The citizens would remain literally in the dark, their food spoiling, their security compromised and their families at risk.

Instead, with full Smart Grid deployment, this future community is not waiting. Instead, it's able immediately to take advantage of distributed resources and standards that support a Smart Grid concept known as "islanding." Islanding is the ability of distributed generation to continue to generate power even when

power from a utility is absent. Combining distributed resources of every description — rooftop PV (solar), fuel cells, electric vehicles — the community can generate sufficient electricity to keep the grocery store, the police department, traffic lights, the phone system and the community health center up and running.

While it may take a week to restore the lines, the generation potential resident in the community means that citizens still have sufficient power to meet their essential needs.

THIS IS POWER FROM THE PEOPLE.

And it is coming.

23

Getting from Point A to Point B – from our present grid to the Smart Grid – requires a brief examination of the history and culture of the industry's primary custodians; namely, utilities and regulators.

SECTION SIX:

FIRST THINGS FIRST: TEASING OUT THE COMPLEXITIES.

When electricity's regulatory compact was first struck in the 1930s, a nation with little appetite for monopolies recognized the provision of electricity as a "natural monopoly" service, one best accomplished by a single entity, whether it was investor-owned, a municipal utility or a co-op.

Under the terms of the compact, in exchange for providing electric service to all consumers within the utility's service territory, utilities were provided a return on their investments plus a return on those investments commensurate with risks they take in ensuring service and reliability. State regulatory commissions were charged with determining whether the investments made were prudent and what a reasonable return on those investments should be.

SMART GRID FACT

The American Public Power Association (APPA) has launched a task force to develop a framework for deploying Smart Grid technologies in a public-power environment.

Over the ensuing decades, much hard work was done on both sides of the compact as much of the grid as we know it was built.

Within utilities, efforts toward this objective were typically segmented or "siloed." This division of labor worked well for utilities, providing efficiencies within the organization for quick execution and maintenance of system reliability.

Meanwhile, regulators focus on their respective states as a matter of law, an understandable circumstance given that each state must answer first and foremost to its citizens and their unique set of needs, resources and agendas.

Until relatively recently, this statutory arrangement has resulted in little regulatory action among the states and little reason to engage in collective action on a national basis, although they work at common purposes through regional associations.

Similarly, regulated utilities have traditionally been reactive, with no need or incentive to be proactive on a national level. Well aligned for utility operations, they are not necessarily well positioned for integrated strategic initiatives like the Smart Grid although they have collectively and forcefully advocated in the past on issues such as security and climate change.

STATES TAKING ACTION:

30 states have developed and adopted renewable portfolio standards, which require a pre-determined amount of a state's energy portfolio (up to 20%) to come exclusively from renewable sources by as early as 2010.

STATE	AMOUNT	YEAR	RPS ADMINISTRATOR
Arizona	15%	2025	Arizona Corporation Commission
California	20%	2010	California Energy Commission
Colorado	20%	2020	Colorado Public Utilities Commission
Connecticut	23%	2020	Department of Public Utility Control
District of Columbia	11%	2022	DC Public Service Commission
Delaware	20%	2019	Delaware Energy Office
Hawaii	20%	2020	Hawaii Strategic Industries Division
Iowa	105 MW		Iowa Utilities Board
Illinois	25%	2025	Illinois Department of Commerce
Massachusetts	4%	2009	Massachusetts Division of Energy Resources
Maryland	9.5%	2022	Maryland Public Service Commission
Maine	10%	2017	Maine Public Utilities Commission
Minnesota	25%	2025	Minnesota Department of Commerce
Missouri*	11%	2020	Missouri Public Service Commission
Montana	15%	2015	Montana Public Service Commission
New Hampshire	16%	2025	New Hampshire Office of Energy and Planning
New Jersey	22.5%	2021	New Jersey Board of Public Utilities
New Mexico	20%	2020	New Mexico Public Regulation Commission
Nevada	20%	2015	Public Utilities Commission of Nevada
New York	24%	2013	New York Public Service Commission
North Carolina	12.5%	2021	North Carolina Utilities Commission
Oregon	25%	2025	Oregon Energy Office
Pennsylvania	18%	2020	Pennsylvania Public Utility Commission
Rhode Island	15%	2020	Rhode Island Public Utilities Commission
Texas	5,880 MW	2015	Public Utility Commission of Texas
Utah*	20%	2025	Utah Department of Environmental Quality
Vermont*	10%	2013	Vermont Department of Public Service
Virginia*	12%	2022	Virginia Department of Mines, Minerals, and Energy
Washington	15%	2020	Washington Secretary of State
Wisconsin	10%	2015	Public Service Commission of Wisconsin

*Four states, Missouri, Utah, Vermont, & Virginia, have set voluntary goals for adopting renewable energy instead of portfolio standards with binding targets.

25

SECTION SIX : CONTINUED

FIRST THINGS FIRST. *TEASING OUT THE COMPLEXITIES*

With growing consensus around the crucial need for Smart Grid deployment, the cultures of these entities are now changing dramatically.

For their part, regulators are actively sharing ideas and information with other states. Acting with an eye toward national agreement, twenty-nine states have also developed and adopted renewable portfolio standards, which require a pre-determined amount of a state's energy portfolio (up to 20%) to come exclusively from renewable sources by as early as 2010.

Regulators on both the state and federal level are stepping up their dialog. State regulators represented by the National Association of Regulatory Utility Commissions (NARUC) are exploring options for expediting Smart Grid implementation with their federal counterpart, the Federal Energy Regulatory Commission (FERC). Meanwhile, DOE is providing leadership with the passing into law of the Energy Independence and Security Act of 2007 (EISA), which codifies a research, development and demonstration program for Smart Grid technologies.

Thanks to these and other efforts, many regulators are moving toward new regulations designed to incentivize utility investment in the Smart Grid. Among these are dynamic pricing, selling energy back to the grid, and policies that guarantee utilities cost recovery and/or favorable depreciation on new Smart Grid investments and legacy systems made obsolete by the switch to "smart meters" and other Smart Grid investments.

26

148

As for utilities, an increasing number of them are taking a more integrated view of a smarter grid, particularly when there are areas of overlap that can be leveraged for cost reduction or benefit increase. There are regulatory implications here as well; if utilities are to argue for cost recovery project by project rather than by single integrated plan, some beneficial aspects of deployment of a smarter grid could be lost.

Integrated plans are being proposed and considered. In California, smart meters only became economic when the commission considered non-utility benefits — benefits to consumers from lower bills.

To an industry historically regulated for prior investment, the transformation to regulation for value delivery promises to stimulate substantial progress and alignment around the Smart Grid vision and implementation. Keep in mind, though, that regulators will continue to require a showing that the value of the investments to consumers — whatever they may be — ultimately exceeds the costs.

SMART GRID FACT

To advance the modernization of our nation's electric grid, DOE has entered into public/private partnerships with leading champions of the Smart Grid which include the GridWise Alliance, EPRI/Intelligrid, and the Galvin Electricity Initiative.

27

Open architecture. Internet protocol. Plug and play. Common technology standards. Non-proprietary. Interoperability.

Fine concepts all, yet one of the reasons the electric industry has been slow to take advantage of common technology standards – which would speed Smart Grid adoption – is a lack of agreement on what those standards should be and who should issue them.

SECTION SEVEN:

HOW THINGS WORK: CREATING THE PLATFORM FOR THE SMART GRID.

The industry is not without its role models in this regard.

Consider the ATM. It is available virtually anywhere. Every unit features a similar user interface, understandable whether or not you know the local language. Users don't give it a second thought. It simply works. Yet the fact that the ATM exists at all was made possible only by industry-wide agreement on a multitude of common standards, from communication to security to business rules.

Fortunately, the agendas of utilities, regulators and automation vendors are rapidly aligning and movement toward identifying and adopting Smart Grid standards is gaining velocity.

28

150

DOE lists five fundamental technologies that will drive the Smart Grid:

- Integrated communications, connecting components to open architecture for real-time information and control, allowing every part of the grid to both 'talk' and 'listen'

- Sensing and measurement technologies, to support faster and more accurate response such as remote monitoring, time-of-use pricing and demand-side management

- Advanced components, to apply the latest research in superconductivity, storage, power electronics and diagnostics

- Advanced control methods, to monitor essential components, enabling rapid diagnosis and precise solutions appropriate to any event

- Improved interfaces and decision support, to amplify human decision-making, transforming grid operators and managers quite literally into visionaries when it come to seeing into their systems

KILLER APP

Will the PHEV be the Smart Grid's "killer app," the outward expression of the Smart Grid that consumers adopt en masse as they did e-mail? There are plenty of experts who think so.

The National Institute of Standards and Technology (NIST), an agency of the U.S. Department of Commerce, has been charged under EISA (Energy Independence and Security Act) with identifying and evaluating existing standards, measurement methods, technologies, and other support in service to Smart Grid adoption. Additionally, they will be preparing a report to Congress recommending areas where standards need to be developed.

The GridWise Architecture Council is an important resource for NIST. The Council, representing a wide array of utility and technology stakeholders and underwritten by DOE, has been working closely with NIST to develop common principles and an interoperability framework spanning the entire electricity delivery chain. Already, the work of the GridWise Architecture Council and other organizations such as ANSI (American National Standards Institute), IEEE (Institute of Electrical and Electronics Engineers) and the ZigBee Alliance have enabled a smarter grid to readily accept innovation across a wide spectrum of applications.

29

Integration in practice.

On Washington's Olympic Peninsula, a DOE demonstration project set in motion a sophisticated system that responded to simple instructions set in place by a consumer in his or her preference profile. Meanwhile, in the background, energy was managed on the consumer's behalf to save money and reduce the impact on the grid.

Consumers saved approximately 10% on their bills. More significantly, peak load was reduced by 15%, bringing the constrained regional grid another 3-5 years of peak load growth and enabling the installation of cleaner, more efficient technologies for supply.

Across the nation, companies are developing new Smart Grid technologies for utility-scale deployments that are progressively raising the bar on what is possible and practical.

Steps toward a common "language."

The independent, non-profit Electric Power Research Institute (EPRI) is also conducting research on key issues facing the electric power industry and working towards the development of open standards for the Smart Grid. The International ElectroTechnical Commission (IEC) recently published EPRI's IntelliGrid Methodology for Developing Requirements for Energy Systems as a publicly available specification.

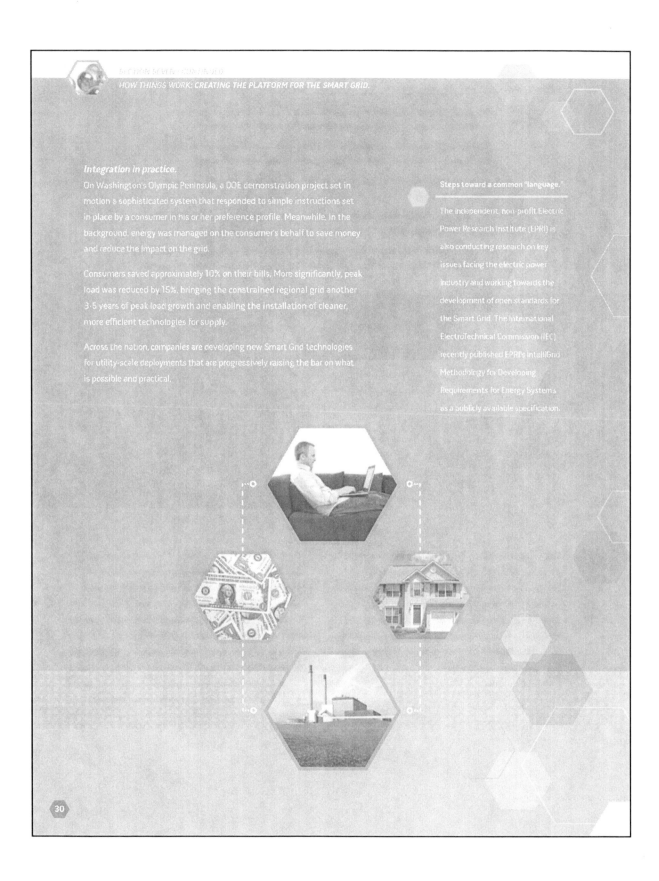

Another look at the future: PHEV (Plug-in Hybrid Electric Vehicles)
Assuming customer acceptance regarding price, performance and longevity, these
vehicles offer consumers the opportunity to shift use of oil and gasoline to electricity
— and to power a car for the equivalent of $.90 per gallon. (As inefficient as the grid
is today, it is cleaner on balance than oil and gasoline.) Consumers get far more
affordable transportation. Relying more on electricity for transportation and less
on fossil fuels increases our energy independence as well as our environmental
prospects. PHEVs take advantage of lower cost and off-peak capacity and can provide
grid support during the peak periods.

ADVANCEMENTS ALSO IN DEVELOPMENT...

Zero-net energy commercial buildings:
Whether measured by cost, energy, or carbon emissions, structures equipped with Smart
Grid technologies capable of balancing energy generation and energy conservation.

Superconducting power cables:
Capable of reducing line losses and carrying 3-5 times more power in a smaller
right of way than traditional copper-based cable.

Energy storage:
While electricity cannot be economically stored, energy can be — with the application
of Smart Grid technologies. Thermal storage, sometimes called hybrid air conditioning,
holds promising potential for positively affecting peak load today. Also of note is the
near-term potential of lithium-ion batteries for PHEV applications.

Advanced sensors:
Monitoring and reporting line conditions in real time, advanced sensors enable more
power to flow over existing lines.

31

153

The Department of Energy is actively engaged in supporting a wide variety of Smart Grid projects. The role of DOE is to act as an objective facilitator, allowing the best ideas to prove themselves. Smart Grid efforts are well underway on several key fronts, from forward-thinking utilities to the 50th state.

SECTION EIGHT:

PROGRESS NOW!: A LOOK AT CURRENT SMART GRID EFFORTS AND HOW THEY'RE SUCCEEDING.

SMART GRID FACT

States such as Texas, California, Ohio, New Jersey, Illinois, New York and others are already actively exploring ways to increase the use of tools and technologies toward the realization of a smarter grid.

Distribution Management System (DMS) Platform by the University of Hawaii

The integrated energy management platform will be developed, featuring advanced functions for home energy management by consumers and for improved distribution system operations by utilities. This platform will integrate AMI as a home portal for demand response; home automation for energy conservation; optimal dispatch of distributed generation, storage, and loads in the distribution system, and controls to make the distribution system a dispatchable entity to collaborate with other entities in the bulk grid.

Home energy management of this type will enable consumers to take control, automating energy conservation and demand response practices based on their personal preferences.

The home automation will be based on the SmartMeter and ecoDashboard products from General Electric. The SmartMeter with a ZigBee network will communicate with household appliances, and the dashboard will automate controls of their operations. In addition, this platform will provide ancillary services to the local utility such as spinning reserve, load-following regulation, and intermittency management for wind and solar energy. This platform will be deployed at the Maui Lani Substation in Maui, Hawaii.

Perfect Power by Illinois Institute of Technology (IIT)

A "Perfect Power" system is defined as: An electric system that cannot fail to meet the electric needs of the individual end-user. A Perfect Power system has the flexibility to supply the power required by various types of end-users and their needs without fail. The functionalities of such a system will be enabled by the Smart Grid.

This project will design a Perfect Power prototype that leverages advanced technology to create microgrids responding to grid conditions and providing increased reliability and demand reduction. This prototype model will be demonstrated at the IIT campus to showcase its operations to the industry. The model is designed to be replicable in any municipality-sized system where customers can participate in electric market opportunities.

33

West Virginia Super Circuit by Allegheny Energy

The super circuit project is designed to demonstrate an advanced distribution circuit with improved reliability and security through integration of distributed resources and advanced monitoring, control, and protection technologies. This circuit will integrate biodiesel generation and energy storage with the AMI and a mesh-based Wi-Fi communications network for rapid fault anticipation and location and rapid fault restoration with minimized impact to customers.

Currently during a circuit fault, all customers on this circuit are being affected with power loss or with power quality issues. The super circuit will demonstrate an ability to dynamically reconfigure the circuit to allow isolation of the faulted segment, transfer uninterrupted services to "unfaulted" segments, and tap surplus capacity from adjacent feeders to optimize consumer service.

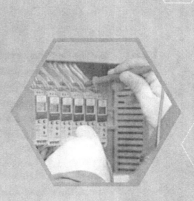

Beach Cities MicroGrid by San Diego Gas & Electric

As its name implies, a microgrid resembles our current grid although on a much smaller scale. It is unique in its ability – during a major grid disturbance – to isolate from the utility seamlessly with little or no disruption to the loads within it and seamlessly reconnect later.

The Beach Cities Microgrid Project will be demonstrated at an existing substation identified as "Beach City Substation." It is intended to offer a blueprint to all distribution utilities – proving the effectiveness of integrating multiple distributed energy resources with advanced controls and communications. It seeks to improve reliability and reduce peak loads on grid components such as distribution feeders and substations.

Both utility-owned and customer-owned generation, i.e., photovoltaic (PV) systems and biodiesel-fueled generators, and energy storage will be integrated along with advanced metering infrastructure (AMI) into the real-world substation operations with a peak load of approximately 50 MW.

Beach Cities will serve as a guide for improved asset use as well as for operating the entire distribution network in the future. Successfully "building" such capabilities will enable customer participation in reliability- and price-driven load management practices, both of which are key to the realization of a smarter grid.

High Penetration of Clean Energy Technologies by The City of Fort Collins

The city and its city-owned Fort Collins Utility support a wide variety of clean energy initiatives, including the establishment of a Zero Energy District within the city (known as FortZED).

One such initiative seeks to modernize and transform the electrical distribution system by developing and demonstrating an integrated system of mixed distributed resources to increase the penetration of renewables – such as solar and wind – while delivering improved efficiency and reliability.

These and other distributed resources will be fully integrated into the electrical distribution system to support achievement of a Zero Energy District. In fact, this DOE-supported project involves the integration of a mix of nearly 30 distributed generation, renewable energy, and demand response resources across 5 customer locations for an aggregated capacity of more than 3.5 Megawatts.

The resources being integrated include:

- *photovoltaic (PV)*

- *microturbines (small combustion turbines that produce between 25 kW and 500 kW of power)*

- *dual-fuel combined power and heat (CHP) systems (utilizing the by-product methane generated from a water treatment plant operation)*

- *reciprocating (or internal combustion) engines*

- *backup generators*

- *wind*

- *plug-in hybrid electric vehicles (PHEV) in an ancillary-services role*

- *fuel cells*

This project will help determine the maximum degree of penetration of distributed resources based on system performance and economics.

35

When Smart Grid implementation becomes reality, everyone wins — and what were once our risks become our strengths.

SECTION NINE:

EDISON UNBOUND: WHAT'S YOUR STAKE IN ALL THIS?

POWER

36

LET'S REVISIT THAT LIST:

EFFICIENCY: It is estimated that tens of billions of dollars will be saved thanks to demand-response programs that provide measurable, persistent savings and require no human intervention or behavior change. The dramatically reduced need to build more power plants and transmission lines will help, too.

RELIABILITY: A Smart Grid that anticipates, detects and responds to problems rapidly reduces wide-area blackouts to

"We are being presented with unprecedented opportunity and challenge across our industry. By coming together around a shared vision of a smarter grid, we have an equally unprecedented opportunity and challenge for shaping our industry's and our nation's future."

STEVEN G HAUSER PRESIDENT, THE GRIDWISE ALLIANCE

near zero (and will have a similarly diminishing effect on the lost productivity).

AFFORDABILITY: Energy prices will rise; however, the trajectory of future cost increases will be far more gradual post-Smart Grid. Smart Grid technologies, tools, and techniques will also provide customers with new options for managing their own electricity consumption and controlling their own utility bills.

SECURITY: The Smart Grid will be more resistant to attack and natural disasters. So fortified, it will also move us

toward energy independence from foreign energy sources, which themselves may be targets for attack, outside of our protection and control.

ENVIRONMENT/CLIMATE CHANGE: Clean, renewable sources of energy like solar, wind, and geothermal can easily be integrated into the nation's grid. We reduce our carbon footprint and stake a claim to global environmental leadership.

NATIONAL ECONOMY: Opening the grid to innovation will enable markets to grow unfettered and innovation to flourish. For comparison's sake, consider the market-making effect of the opening of the telephone industry in the 1980s. With revenues of $33 billion at the time, the ensuing proliferation of consumer-centric products and services transformed it into a $117 billion market as of 2006.

GLOBAL COMPETITIVENESS: Regaining our early lead in solar and wind will create an enduring green-collar economy.

37

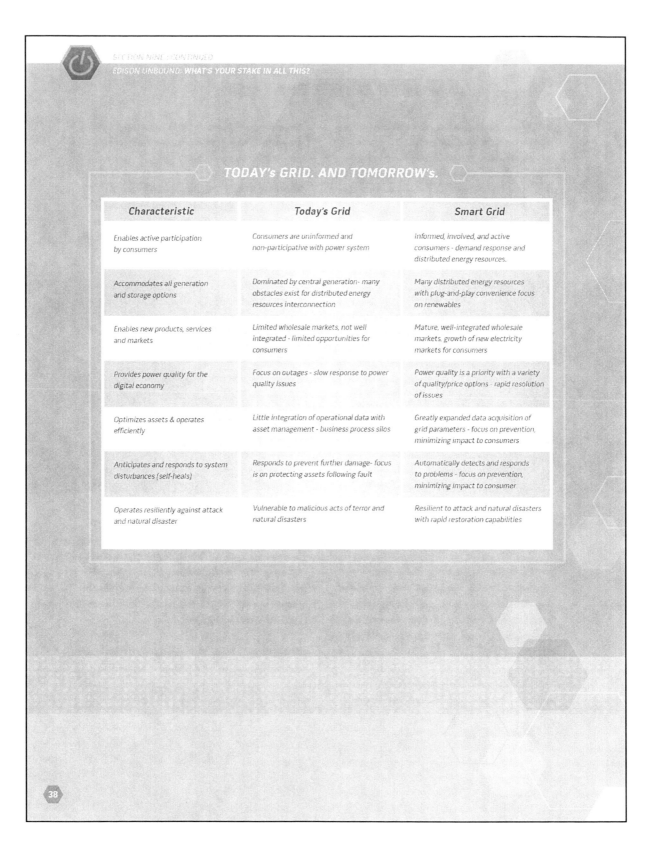

TODAY's GRID. AND TOMORROW's.

Characteristic	Today's Grid	Smart Grid
Enables active participation by consumers	Consumers are uninformed and non-participative with power system	Informed, involved, and active consumers - demand response and distributed energy resources.
Accommodates all generation and storage options	Dominated by central generation- many obstacles exist for distributed energy resources interconnection	Many distributed energy resources with plug-and-play convenience focus on renewables
Enables new products, services and markets	Limited wholesale markets, not well integrated - limited opportunities for consumers	Mature, well-integrated wholesale markets, growth of new electricity markets for consumers
Provides power quality for the digital economy	Focus on outages - slow response to power quality issues	Power quality is a priority with a variety of quality/price options - rapid resolution of issues
Optimizes assets & operates efficiently	Little integration of operational data with asset management - business process silos	Greatly expanded data acquisition of grid parameters - focus on prevention, minimizing impact to consumers
Anticipates and responds to system disturbances (self-heals)	Responds to prevent further damage- focus is on protecting assets following fault	Automatically detects and responds to problems - focus on prevention, minimizing impact to consumer
Operates resiliently against attack and natural disaster	Vulnerable to malicious acts of terror and natural disasters	Resilient to attack and natural disasters with rapid restoration capabilities

38

The Smart Grid creates value up and down the value chain, much like the internet has. As we've experienced with the internet, affordable, rapid and universal communication can enable sophisticated transactions, create entirely new business models and sweep across society with surprising speed.

Consider for a moment your iPod, YouTube, internet banking...

Prior to the internet's adoption, markets didn't have the ability to operate as cost-effectively and productively as they do today. Few predicted that people would engage as seriously with the internet as they have. And no one could have predicted the revolutionary advancements it has fostered.

Similarly, we had no idea that the internet would revolutionize so many aspects of our lives.

The Smart Grid represents the relatively simple extension of this movement to power consumption.

Thomas Edison, The Wizard of Menlo Park, would approve of the enterprise and innovation that drive the Smart Grid. He might even ask what took us so long. New technologies and public policies, economic incentives and regulations are aligning to bring the Smart Grid to full implementation. Its success is imperative to the economic growth and vitality of America far into the future.

We hope that **The Smart Grid: An Introduction** has given you a clearer understanding of the need for immediate and concerted action in the transformation of our nation's electrical grid. To learn more, please visit the websites listed on the following page.

"If we all did the things we are capable of doing, we would literally astound ourselves."

THOMAS A. EDISON (1847-1931)

39

RESOURCES

2007 INTERGRAPH ROCKET CITY GEOSPATIAL CONFERENCE:
Presentation: http://www.directionsmag.com/images/RCG/Main/Damon%20Dougherty.ppt#257,3,What%20is%20Smart%20Grid

COLUMBIA UNIVERSITY: http://www.ldeo.columbia.edu/res/pi/4d4/testbeds/Smart-Grid-White-Paper.pdf

EEI :
http://www.eei.org/industry_issues/electricity_policy/advanced_metering_infrastructure.htm
http://www.eei.org/magazine/editorial_content/nonav_stories/2007-09-01-Smart.pdf
http://www.eei.org/magazine/editorial_content/nonav_stories/2005-01-01-Smarter.pdf

ENERGY FUTURE COALITION :
Smart Grid Working Group: http://www.energyfuturecoalition.org/preview.cfm?catID=13
Reports: http://www.energyfuturecoalition.org/pubs/app_smart_grid.pdf
http://energyfuturecoalition.org/pubs/PJMsmartgrid.pdf

ELSTER:
http://www.eei.org/meetings/nonav_2008-02-06-ja/Gray_Elster.pdf
http://www.eei.org/industry_issues/electricity_policy/federal_legislation/deciding_on_smart_meters.pdf

GRIDPOINT: http://www.electricitydeliveryforum.org/pdfs/Gridpoint_SmartGrid.pdf

NATIONAL ELECTRICAL MANUFACTURERS ASSOCIATION:
Presentation: http://www.nema.org/gov/energy/smartgrid/upload/Presentation-Smart-Grid.pdf

NATIONAL ENERGY TECHNOLOGIES LABORATORY:
Presentation: http://www.energetics.com/supercon07/pdfs/NETL_Synergies_of_the_SmartGrid_and_Superconducitivity_Pullins.pdf

THE PEW CENTER ON GLOBAL CLIMATE CHANGE:
Workshop Proceedings: http://www.pewclimate.org/docUploads/10-50_Anderson_120604_120713.pdf

SAN DIEGO SMART GRID STUDY: http://www.sandiego.edu/epic/publications/documents/061017_SDSmartGridStudyFINAL.pdf

SMART GRID NEWS: http://www.smartgridnews.com/artman/publish/index.html

XCEL ENERGY:
Smart Grid City Web site: http://www.xcelenergy.com/XLWEB/CDA/0,3080,1-1-1_15531_43141_46932-39884-0_0_0-0,00.html

EPRI INTELLIGRID: http://intelligrid.epri.com/

PNNL GRIDWISE: http://gridwise.pnl.gov/

SMART GRID TASK FORCE: http://www.oe.energy.gov/smartgrid_taskforce.htm

SMART GRID: http://www.oe.energy.gov/smartgrid.htm

GRIDWISE ALLIANCE: www.gridwise.org

GRID WEEK: www.gridweek.com

SOURCES

Sources for this book include the Department of Energy, the GridWise Alliance, the Galvin Electricity Initiative and EPRI/Intelligrid.

GLOSSARY: COMING TO TERMS WITH THE SMART GRID

AMI: Advanced Metering Infrastructure is a term denoting electricity meters that measure and record usage data at a minimum, in hourly intervals, and provide usage data to both consumers and energy companies at least once daily.

AMR: Automated Meter Reading is a term denoting electricity meters that collect data for billing purposes only and transmit this data one way, usually from the customer to the distribution utility.

ANCILLARY SERVICES: Services that ensure reliability and support the transmission of electricity from generation sites to customer loads. Such services may include: load regulation, spinning reserve, non-spinning reserve, replacement reserve, and voltage support.

APPLIANCE: A piece of equipment, commonly powered by electricity, used to perform a particular energy-driven function. Examples of common appliances are refrigerators, clothes washers and dishwashers, conventional ranges/ovens and microwave ovens, humidifiers and dehumidifiers, toasters, radios, and televisions. Note: Appliances are ordinarily self-contained with respect to their function. Thus, equipment such as central heating and air conditioning systems and water heaters, which are connected to distribution systems inherent to their purposes, are not considered appliances.

CAPITAL COST: The cost of field development and plant construction and the equipment required for industry operations.

CARBON DIOXIDE (CO₂): A colorless, odorless, non-poisonous gas that is a normal part of Earth's atmosphere. Carbon dioxide is a product of fossil-fuel combustion as well as other processes. It is considered a greenhouse gas as it traps heat (infrared energy) radiated by the Earth into the atmosphere and thereby contributes to the potential for global warming. The global warming potential (GWP) of other greenhouse gases is measured in relation to that of carbon dioxide, which by international scientific convention is assigned a value of one (1).

CLIMATE CHANGE: A term used to refer to all forms of climatic inconsistency, but especially to significant change from one prevailing climatic condition to another. In some cases, "climate change" has been used synonymously with the term "global warming"; scientists, however, tend to use the term in a wider sense inclusive of natural changes in climate, including climatic cooling.

CONGESTION: A condition that occurs when insufficient transfer capacity is available to implement all of the preferred schedules for electricity transmission simultaneously.

DSM: This Demand-Side Management category represents the amount of consumer load reduction at the time of system peak due to utility programs that reduce consumer load during many hours of the year. Examples include utility rebate and shared savings activities for the installation of energy efficient appliances, lighting and electrical machinery, and weatherization materials. In addition, this category includes all other Demand-Side Management activities, such as thermal storage, time-of-use rates, fuel substitution, measurement and evaluation, and any other utility-administered Demand-Side Management activity designed to reduce demand and/or electricity use.

DISTRIBUTED GENERATOR: A generator that is located close to the particular load that it is intended to serve. General, but non-exclusive, characteristics of these generators include: an operating strategy that supports the served load; and interconnection to a distribution or sub-transmission system.

DISTRIBUTION: The delivery of energy to retail customers.

DISTRIBUTION SYSTEM: The portion of the transmission and facilities of an electric system that is dedicated to delivering electric energy to an end-user.

ELECTRIC GENERATION INDUSTRY: Stationary and mobile generating units that are connected to the electric power grid and can generate electricity. The electric generation industry includes the "electric power sector" (utility generators and independent power producers) and industrial and commercial power generators, including combined-heat-and-power producers, but excludes units at single-family dwellings.

ELECTRIC GENERATOR: A facility that produces only electricity, commonly expressed in kilowatthours (kWh) or megawatthours (MWh). Electric generators include electric utilities and independent power producers.

ELECTRIC POWER: The rate at which electric energy is transferred. Electric power is measured by capacity and is commonly expressed in megawatts (MW).

ELECTRIC POWER GRID: A system of synchronized power providers and consumers connected by transmission and distribution lines and operated by one or more control centers. In the continental United States, the electric power grid consists of three systems: the Eastern Interconnect, the Western Interconnect, and the Texas Interconnect. In Alaska and Hawaii, several systems encompass areas smaller than the State (e.g., the interconnect serving Anchorage, Fairbanks, and the Kenai Peninsula; individual islands).

ELECTRIC SYSTEM RELIABILITY: The degree to which the performance of the elements of the electrical system results in power being delivered to consumers within accepted standards and in the amount desired. Reliability encompasses two concepts, adequacy and security. Adequacy implies that there are sufficient generation and transmission resources installed and available to meet projected electrical demand plus reserves for contingencies. Security implies that the system will remain intact operationally (i.e., will have sufficient available operating capacity) even after outages or other equipment failure. The degree of reliability may be measured by the frequency, duration, and magnitude of adverse effects on consumer service.

41

GLOSSARY (CONT'D)

ELECTRIC UTILITY: Any entity that generates, transmits, or distributes electricity and recovers the cost of its generation, transmission or distribution assets and operations, either directly or indirectly, through cost-based rates set by a separate regulatory authority (e.g., State Public Service Commission), or is owned by a governmental unit or the consumers that the entity serves. Examples of these entities include: investor-owned entities, public power districts, public utility districts, municipalities, rural electric cooperatives, and State and Federal agencies.

ELECTRICITY CONGESTION: A condition that occurs when insufficient transmission capacity is available to implement all of the desired transactions simultaneously.

ELECTRICITY DEMAND: The rate at which energy is delivered to loads and scheduling points by generation, transmission, and distribution facilities.

ENERGY EFFICIENCY, ELECTRICITY: Refers to programs that are aimed at reducing the energy used by specific end-use devices and systems, typically without affecting the services provided. These programs reduce overall electricity consumption (reported in megawatthours), often without explicit consideration for the timing of program-induced savings. Such savings are generally achieved by substituting technologically more advanced equipment to produce the same level of end-use services (e.g. lighting, heating, motor drive) with less electricity. Examples include high-efficiency appliances, efficient lighting programs, high-efficiency heating, ventilating and air conditioning (HVAC) systems or control modifications, efficient building design, advanced electric motor drives, and heat recovery systems.

ENERGY SAVINGS: A reduction in the amount of electricity used by end users as a result of participation in energy efficiency programs and load management programs.

ENERGY SERVICE PROVIDER: An energy entity that provides service to a retail or end-use customer.

FEDERAL ENERGY REGULATORY COMMISSION (FERC): The Federal agency with jurisdiction over interstate electricity sales, wholesale electric rates, hydroelectric licensing, natural gas pricing, oil pipeline rates, and gas pipeline certification. FERC is an independent regulatory agency within the Department of Energy and is the successor to the Federal Power Commission.

FUEL CELL: A device capable of generating an electrical current by converting the chemical energy of a fuel (e.g., hydrogen) directly into electrical energy. Fuel cells differ from conventional electrical cells in that the active materials such as fuel and oxygen are not contained within the cell but are supplied from outside. It does not contain an intermediate heat cycle, as do most other electrical generation techniques.

GENERATION: The process of producing electric energy by transforming other forms of energy; also, the amount of electric energy produced, expressed in kilowatthours.

GLOBAL WARMING: An increase in the near surface temperature of the Earth. Global warming has occurred in the distant past as the result of natural influences, but the term is today most often used to refer to the warming some scientists predict will occur as a result of increased anthropogenic emissions of greenhouse gases.

GREENHOUSE GASES: Those gases, such as water vapor, carbon dioxide, nitrous oxide, methane, hydrofluorocarbons (HFCs), perfluorocarbons (PFCs) and sulfur hexafluoride, that are transparent to solar (short-wave) radiation but opaque to long-wave (infrared) radiation, thus preventing long-wave radiant energy from leaving Earth's atmosphere. The net effect is a trapping of absorbed radiation and a tendency to warm the planet's surface.

INTERMITTENT ELECTRIC GENERATOR OR INTERMITTENT RESOURCE: An electric generating plant with output controlled by the natural variability of the energy resource rather than dispatched based on system requirements. Intermittent output usually results from the direct, non-stored conversion of naturally occurring energy fluxes such as solar energy, wind energy, or the energy of free-flowing rivers (that is, run-of-river hydroelectricity).

INTERRUPTIBLE LOAD: This Demand-Side Management category represents the consumer load that, in accordance with contractual arrangements, can be interrupted at the time of annual peak load by the action of the consumer at the direct request of the system operator. This type of control usually involves large-volume commercial and industrial consumers. Interruptible Load does not include Direct Load Control.

LINE LOSS: Electric energy lost because of the transmission of electricity. Much of the loss is thermal in nature.

LOAD (ELECTRIC): The amount of electric power delivered or required at any specific point or points on a system. The requirement originates at the energy-consuming equipment of the consumers.

LOAD CONTROL PROGRAM: A program in which the utility company offers a lower rate in return for having permission to turn off the air conditioner or water heater for short periods of time by remote control. This control allows the utility to reduce peak demand.

OFF PEAK: Period of relatively low system demand. These periods often occur in daily, weekly, and seasonal patterns; these off-peak periods differ for each individual electric utility.

ON PEAK: Periods of relatively high system demand. These periods often occur in daily, weekly, and seasonal patterns; these on-peak periods differ for each individual electric utility.

OUTAGE: The period during which a generating unit, transmission line, or other facility is out of service.

PEAK DEMAND OR PEAK LOAD: The maximum load during a specified period of time.

PEAKER PLANT OR PEAK LOAD PLANT: A plant usually housing old, low-efficiency steam units, gas turbines, diesels, or pumped-storage hydroelectric equipment normally used during the peak-load periods.

PEAKING CAPACITY: Capacity of generating equipment normally reserved for operation during the hours of highest daily, weekly, or seasonal loads. Some generating equipment may be operated at certain times as peaking capacity and at other times to serve loads on an around-the-clock basis.

RATE BASE: The value of property upon which a utility is permitted to earn a specified rate of return as established by a regulatory authority. The rate base generally represents the value of property used by the utility in providing service and may be calculated by any one or a combination of the following accounting methods: fair value, prudent investment, reproduction cost, or original cost. Depending on which method is used, the rate base includes cash, working capital, materials and supplies, deductions for accumulated provisions for depreciation, contributions in aid of construction, customer advances for construction, accumulated deferred income taxes, and accumulated deferred investment tax credits.

RATE CASE: A proceeding, usually before a regulatory commission, involving the rates to be charged for a public utility service.

RATE FEATURES: Special rate schedules or tariffs offered to customers by electric and/or natural gas utilities.

RATE OF RETURN: The ratio of net operating income earned by a utility is calculated as a percentage of its rate base.

RATE OF RETURN ON RATE BASE: The ratio of net operating income earned by a utility, calculated as a percentage of its rate base.

RATE SCHEDULE (ELECTRIC): A statement of the financial terms and conditions governing a class or classes of utility services provided to a customer. Approval of the schedule is given by the appropriate rate-making authority.

RATEMAKING AUTHORITY: A utility commission's legal authority to fix, modify, approve, or disapprove rates as determined by the powers given the commission by a State or Federal legislature.

RATES: The authorized charges per unit or level of consumption for a specified time period for any of the classes of utility services provided to a customer.

RELIABILITY (ELECTRIC SYSTEM): A measure of the ability of the system to continue operation while some lines or generators are out of service. Reliability deals with the performance of the system under stress.

RENEWABLE ENERGY RESOURCES: Energy resources that are naturally replenishing but flow-limited. They are virtually inexhaustible in duration but limited in the amount of energy that is available per unit of time. Renewable energy resources include: biomass, hydro, geothermal, solar, wind, ocean thermal, wave action, and tidal action.

SOLAR ENERGY: The radiant energy of the sun, which can be converted into other forms of energy, such as heat or electricity.

TARIFF: A published volume of rate schedules and general terms and conditions under which a product or service will be supplied.

THERMAL ENERGY STORAGE: The storage of heat energy during utility off-peak times at night, for use during the next day without incurring daytime peak electric rates.

THERMAL LIMIT: The maximum amount of power a transmission line can carry without suffering heat-related deterioration of line equipment, particularly conductors.

TIME-OF-DAY PRICING: A special electric rate feature under which the price per kilowatthour depends on the time of day.

TIME-OF-DAY RATE: The rate charged by an electric utility for service to various classes of customers. The rate reflects the different costs of providing the service at different times of the day.

TRANSMISSION AND DISTRIBUTION LOSS: Electric energy lost due to the transmission and distribution of electricity. Much of the loss is thermal in nature.

TRANSMISSION (ELECTRIC) (VERB): The movement or transfer of electric energy over an interconnected group of lines and associated equipment between points of supply and points at which it is transformed for delivery to consumers or is delivered to other electric systems. Transmission is considered to end when the energy is transformed for distribution to the consumer.

UTILITY GENERATION: Generation by electric systems engaged in selling electric energy to the public.

UTILITY-SPONSORED CONSERVATION PROGRAM: Any program sponsored by an electric and/or natural gas utility to review equipment and construction features in buildings and advise on ways to increase the energy efficiency of buildings. Also included are utility-sponsored programs to encourage the use of more energy-efficient equipment. Included are programs to improve the energy efficiency in the lighting system or building equipment or the thermal efficiency of the building shell.

WIND ENERGY: Kinetic energy present in wind motion that can be converted to mechanical energy for driving pumps, mills, and electric power generators.

43

www.energy.gov

Printed on
recycled paper

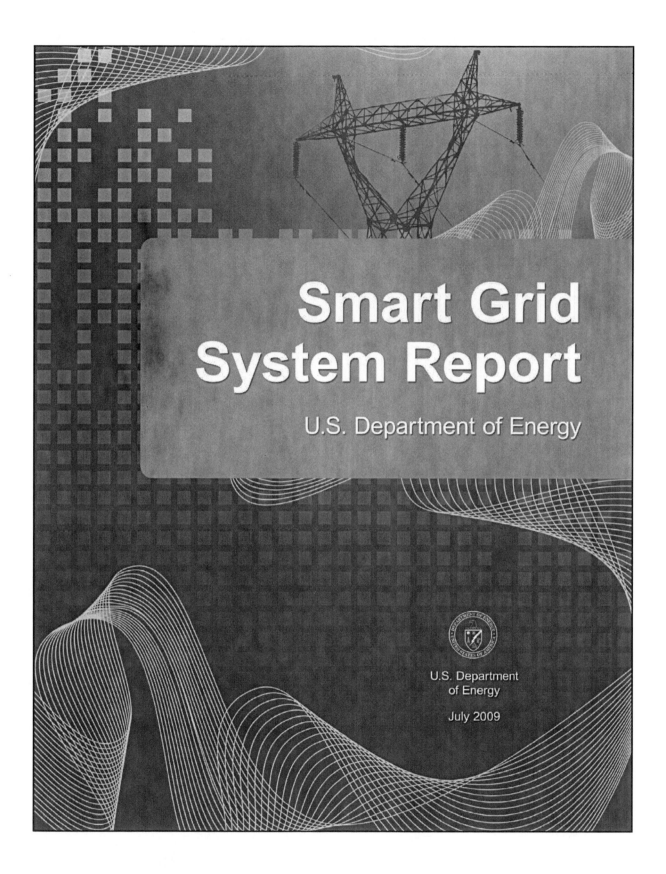

Smart Grid System Report

U.S. Department of Energy

U.S. Department
of Energy

July 2009

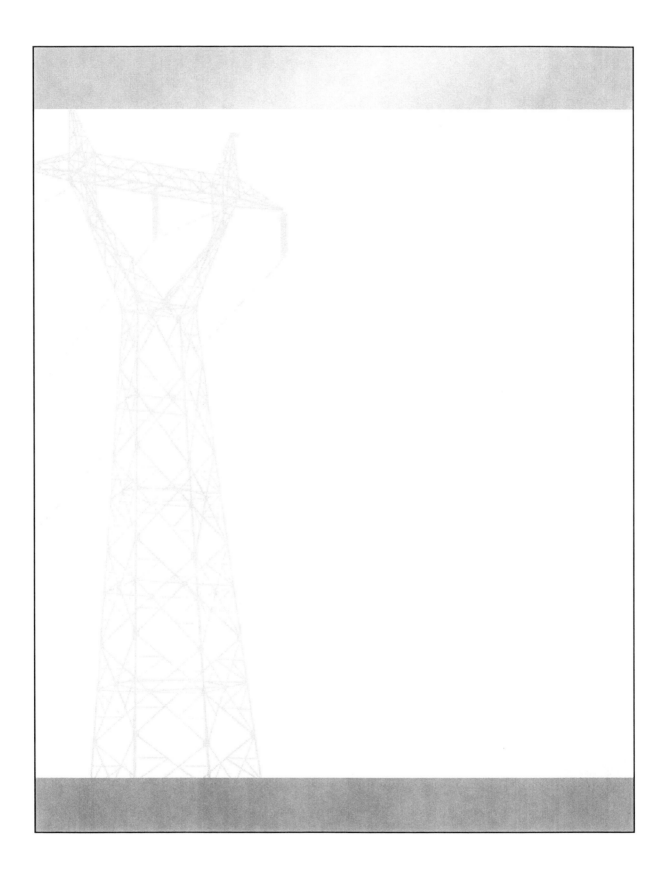

Smart Grid
System Report

U.S. Department of Energy

July 2009

169

SEC. 1302.
SMART GRID SYSTEM REPORT

The Secretary, acting through the Assistant Secretary of the Office of Electricity Delivery and Energy Reliability (referred to in this section as the "OEDER") and through the Smart Grid Task Force established in section 1303, shall, after consulting with any interested individual or entity as appropriate, no later than one year after enactment and every two years thereafter, report to Congress concerning the status of smart grid deployments nationwide and any regulatory or government barriers to continued deployment. The report shall provide the current status and prospects of smart grid development, including information on technology penetration, communications network capabilities, costs, and obstacles. It may include recommendations for State and Federal policies or actions helpful to facilitate the transition to a smart grid. To the extent appropriate, it should take a regional perspective. In preparing this report, the Secretary shall solicit advice and contributions from the Smart Grid Advisory Committee created in section 1303; from other involved Federal agencies including but not limited to the Federal Energy Regulatory Commission ("Commission",) the National Institute of Standards and Technology ("Institute"), and the Department of Homeland Security; and from other stakeholder groups not already represented on the Smart Grid Advisory Committee.

—Energy Independence and Security Act of 2007, December 19, 2007

Executive Summary

Section 1302 of Title XIII of the Energy Independence and Security Act of 2007 directs the Secretary of Energy to "…report to Congress concerning the status of smart grid deployments nationwide and any regulatory or government barriers to continued deployment." This document satisfies this directive and represents the first installment of this report to Congress, which is to be updated biennially.

The state of smart grid deployment covers a broad array of electric system capabilities and services enabled through pervasive communications and information technology, with the objective to improve reliability, operating efficiency, resiliency to threats, and our impact to the environment. By collecting information from a workshop, interviews, and research of existing smart grid literature and studies, this report attempts to present a balanced view of progress toward a smart grid across many fronts. The Department of Energy sponsored a workshop, "Implementing the Smart Grid," that engaged stakeholders from utilities, reliability coordinators, electricity market operators, end users, suppliers, trade organizations, and state and federal regulators, as well as the National Institute of Standards and Technology and the Federal Energy Regulatory Commission. The workshop's outcomes provide a foundation for the metrics identified in this report. In addition, the Department's Energy Advisory Committee and their Smart Grid Subcommittee were consulted along with the inter-agency Smart Grid Task Force that includes representatives from NIST, FERC, the Department of Homeland Security, and the Environmental Protection Agency among others. While future reports will improve the measurement and perspective of this progress, the investigation done for this first report reveals the following key findings.

Key Findings

- Distributed energy resources: The ability to connect distributed generation, storage, and renewable resources is becoming more standardized and cost effective. While the penetration level remains low, the area is experiencing high growth. Several other concepts associated with a smart grid are in a nascent phase of deployment these include the integration of microgrids, electric vehicles, and demand response initiatives, including grid-sensitive appliances.

- Electricity infrastructure: Those smart grid areas that fit within the traditional electricity utility business and policy model have a history of automation and advanced communication deployment to build upon. Advanced metering infrastructure is taking automated meter reading approaches to a new level, and is seen as a necessary step to enabling dynamic pricing and consumer participation mechanisms. Though penetration of these systems is still low, the growth and attention by businesses and policymakers is strong. Transmission substation automation remains strong with greater levels of information exchanged with control centers. Cost/benefit thresholds are now encouraging greater levels of automation at the distribution substation level. While reliability indices show some slight degradation, generation and electricity transport efficiencies are improving.

- Business and policy: The business cases, financial resources, paths to deployment, and models for enabling governmental policy are only now emerging with experimentation. This is true of the regulated and non-regulated aspects of the electric system. Understanding and articulating the environmental and consumer perspectives also

iii

Smart Grid System Report — July 2009

remains in its infancy, though recent reports and deliberations indicate that significant attention is beginning to be given to these issues.

- **High-tech culture change**: A smart grid is socially transformational. As with the Internet or cell phone communications, our experience with electricity will change dramatically. To successfully integrate high levels of automation requires cultural change. The integration of automation systems within and between the electricity delivery infrastructure, distributed resources, and end-use systems needs to evolve from specialized interfaces to embrace solutions that recognize well-accepted principles, methodology, and tools that are commonly recognized by communications, information technology, and related disciplines that enable interactions within all economic sectors and individual businesses. The solutions to improving physical and cyber security, information privacy, and interoperability (conveniently connect and work within a collaborative system) require disciplines and best practices that are subscribed to by all stakeholders. A cross-disciplinary change that instills greater interaction among all the stakeholders is a necessary characteristic as we advance toward a smart grid. Progress in areas such as cyber security and interoperability is immature and difficult to measure, though improved approaches for future measurements are proposed.

The Scope of a Smart Grid

A smart grid uses digital technology to improve reliability, security, and efficiency (both economic and energy) of the electric system from large generation, through the delivery systems to electricity consumers and a growing number of distributed-generation and storage resources (DOE/OEDER 2008a[1]). The information networks that are transforming our economy in other areas are also being applied to applications for dynamic optimization of electric system operations, maintenance, and planning. Resources and services that were separately managed are now being integrated and rebundled as we address traditional problems in new ways, adapt the system to tackle new challenges, and discover new benefits that have transformational potential.

Areas of the electric system that cover the scope of a smart grid include the following:

- the delivery infrastructure (e.g., transmission and distribution lines, transformers, switches),

- the end-use systems and related distributed-energy resources (e.g., building and factory loads, distributed generation, storage, electric vehicles),

- management of the generation and delivery infrastructure at the various levels of system coordination (e.g., transmission and distribution control centers, regional reliability coordination centers, national emergency response centers),

- the information networks themselves (e.g., remote measurement and control communications networks, inter- and intra-enterprise communications, public Internet), and

- the financial and regulatory environment that fuels investment and motivates decision makers to procure, implement, and maintain all aspects of the system (e.g., stock and bond markets, government incentives, regulated or non-regulated rate-of-return on investment).

Some aspect of the electricity system touches every person in the Nation.

(1) Items in parentheses such as this indicate source material listed in Section 6.0 References.

iv

Some aspect of the electricity system touches every person in the Nation. The stakeholder landscape for a smart-grid is complex. The lines of distinction are not always crisp, as corporations and other organizations can take on the characteristics and responsibilities of multiple functions. Stakeholders include the following: end users (consumers), electric-service retailers, distribution and transmission service providers, balancing authorities, wholesale-electricity traders/brokers/markets, reliability coordinators, product and service suppliers, energy policymakers, regulators, and advocates, standards organizations, and the financial community.

The State of Smart Grid Deployments

The report looks across a spectrum of smart-grid concerns and identifies 20 metrics for measuring the status of smart-grid deployment and impacts. Across the vast scope of smart-grid deployment, many things can be measured. The approach is to identify key indicators that can provide a sense of smart-grid progress while balancing detail and complexity. The metrics are derived from a Department of Energy-sponsored workshop on "Implementing the Smart Grid" (DOE/OEDER 2008b). At this workshop a cross-section of the stakeholder representation worked to identify over 50 areas for measurement. This list was distilled to remove areas of overlap and metrics that, while appropriate, were deemed too hard to measure. The result is the list shown in Table ES.1, *Summary of Smart Grid Metrics and Status*. The list covers the various areas of concern in a smart grid. Finding accurate measurements is difficult in a few cases, but recommendations for improving the metric or measurement capture are included in the report.

> *The approach is to identify key indicators that can provide a sense of smart-grid progress while balancing detail and complexity.*

Besides describing the 20 metrics, the table summarizes a broad indication of the penetration level (if a build metric) or maturity level (if a value metric), as well as the trending direction. Build metrics describe attributes that are built in support of a smart grid, while value metrics describe the value that may derive from achieving a smart grid. While build metrics tend to be quantifiable, value metrics can be influenced by many developments and therefore generally require more qualifying discussion. The indication levels used in the table for build metrics refer to level of penetration: nascent (very low and just emerging), low, or moderate. Because smart grid activity is relatively new, there are no high penetration levels to report on these metrics. The value metrics indicate whether the present state is nascent or mature. The trend (recent past and near-term projection) is indicated for either type of metric as declining, flat, or growing at nascent, low, moderate, or high, levels.

As smart-grid concerns are future-looking, several of the metrics represent areas where deployment activities are just being explored. Finding baseline status information for these areas is difficult. This is the case with several metrics in the Distributed Resource Technology area and Information Networks. Microgrids, electric or hybrid vehicles, and grid-responsive, non-generating demand-side equipment fall into this category. In the Delivery Infrastructure area, dynamic line limit technology deployment is also emerging; though the concept and pilots have been around for several years, the value proposition keeps it at the nascent level. Cyber security and open architecture/standards metrics are also listed as nascent. Even though attention is being paid to these areas, the application of a disciplined approach to cyber security and interoperability issues is new. A development and operational culture that addresses these concerns needs to mature and better methods are needed to measure progress.

Other smart-grid metrics are in areas that have been receiving attention for several years, and while the technology deployment may be low or moderate, implementation paths have had

Smart Grid System Report — July 2009

time to mature. The area of Delivery (T&D) Infrastructure is a good example. Substation automation has a long history of progress in the transmission area and is now beginning to penetrate the distribution system. Advanced metering infrastructure has a low level of penetration today, but the attention given to this area by utilities and regulators has resulted in significant investments with large near-term deployment growth projections. Though not as dramatic, advanced measurement systems, such as synchro-phasor technology deployment, are also growing.

In the Area, Regional, and National Coordination area, the metrics indicate a moderate level of activity, though the penetration level is low and regional progress is diverse. Policy makers are naturally cautious in making dynamic-pricing and rate-recovery decisions as value propositions remain immature, untested, and therefore, risky. Distributed-resource interconnection rules and procedures vary significantly across different service regions. Distributed energy resources, including renewable and non-renewable generation, storage, and grid-responsive load are being integrated, but at low penetration level.

The report monitors the following value metrics that pertain to the Delivery Infrastructure area: reliability, capacity, operating efficiency, and power quality. While operating efficiency has seen some improvement, the trends for the other indices have been flat or deteriorating over recent history. Though these values are difficult to cleanly associate with steps toward a smart grid, we expect improvement of these trends.

Venture capital funding of startups grew at an average annual rate of 27%

To convey the present situation of smart-grid deployment, the report uses a set of six characteristics derived from the seven characteristics of the National Energy Technology Laboratory (NETL) Modern Grid Strategy project and documented in "Characteristics of the Modern Grid" (NETL 2008).[1] The metrics listed in Table ES.1 provide insights into progress toward these characteristics. Nearly all of the metrics contribute information to understanding multiple characteristics. The main findings are summarized below:

- **Enables Informed Participation by Customers:** Supporting the bi-directional flow of information and energy is a foundation for enabling participation by consumer resources. Advanced metering infrastructure (AMI) is receiving the most attention in terms of planning and investment. Currently AMI comprises about 4.7% of all electric meters being used for demand response. Approximately 52 million more meters are projected to be installed by 2012. A large number of the meters installed are not being used for demand response activities [Metric 12]. Pricing signals can provide valuable information for consumers (and the automation systems that reflect their preferences) to decide on how to react to grid conditions. A FERC study found that in 2008 slightly over 1% of all customers received a dynamic pricing tariff [Metric 1], with nearly the entire amount represented by time-of-use tariffs. Lastly, the amount of load participating based on grid conditions is beginning to show a shift from traditional interruptible demand at industrial plants toward demand-response programs that either allow an energy-service provider to perform direct load control or provide financial incentives for customer-responsive demand at homes and businesses [Metric 5].

- **Accommodates All Generation and Storage Options:** Distributed energy resources and interconnection standards to accommodate generation capacity appear to be moving in positive directions. Accommodating a large number of disparate generation and storage resources requires anticipation of intermittency and unavailability, while balancing costs, reliability, and environmental emissions. Distributed generation (carbon-based and renewable) and storage, although a small fraction (1.6%) of total

---

(1) The sixth characteristic is a merger of the Modern Grid Initiative's characteristics a) Self-Heals and b) Resists Attack. The same metrics substantially contribute to both of these concerns.

vi

summer peak, appear to be increasing rapidly [Metric 7]. In addition, 31 states have interconnection standards in place, with 11 states progressing toward a standard, one state with some elements in place, and only 8 states with none. Unfortunately, only 15 states had interconnection standards that were deemed to be favorable to the integration of these resources [Metric 3].

- **Enables New Products, Services, and Markets:** Companies with new smart-grid concepts are receiving a significant injection of money. Venture-capital funding of startups grew from $58.4 million in 2002 to $194.1 million in 2007, yielding an average annual rate increase of 27% [Metric 20]. Electric utilities are finding some incentives from regulatory rulings that allow them rate recovery for smart-grid investments [Metric 4]. Some of these rulings have allowed AMI deployments to move forward and more information is being obtained to characterize the consumer benefits from the emerging new products and services. Great interest and investment in electric vehicles, including plug-in hybrids, is changing the future complexion of transportation and represents a significant demand for new products and services, including bi-directional information flow as being supported in AMI systems and smart charging systems. Today only 0.02% of light-duty vehicles are grid-connected, but most forecasts estimate ultimate penetration of this market at 8-16%, with some aggressive estimates at 37%, by 2020 [Metric 8]. A smart grid will also include consumer-oriented "smart" equipment, such as thermostats, space heaters, clothes dryers, and water heaters that communicate to enable demand participation. This smart equipment and related demand participation program offerings are just emerging, primarily in pilot programs [Metric 9].

- **Provides the Power Quality for the Range of Needs:** Not all customers have the same power-quality requirements, though traditionally these requirements and the costs to provide them have been shared. While the state of power quality has been difficult to quantify, the number of customer complaints has been rising slightly [Metric 17]. Smart grid solutions range from local control of your power needs in a microgrid [Metric 6] and supporting distributed generation [Metric 7], to more intelligent operation of the delivery system through technology such as is used in substation automation [Metric 11] (see next bullet). As mentioned earlier, distributed energy resource deployment is trending upward, while microgrid parks are just emerging and are mostly represented in pilot programs.

- **Optimizes Asset Utilization and Operating Efficiently:** Gross annual measures of operating efficiency have been improving slightly as energy lost in generation dropped 0.6 % to 67.7% in 2007 and transmission and distribution losses also improved slightly [Metric 15]. The summer peak capacity factor declined slightly to 80.8% while overall annual average capacity factor is projected to increase slightly to 46.5% [Metric 14]. Contributions to these measures include substation automation deployments. While transmission substations have considerable instrumentation and coordination, the value proposition for distribution-substation automation is now receiving more attention. Presently about 31% of substations have some form of automation, with the number expected to rise to 40% by 2010 [Metric 11]. The deployment of dynamic line rating technology is also expected to increase asset utilization and operating efficiency; however, implementations thus far have had very limited penetration levels [Metric 16].

- **Operates Resiliently to Disturbances, Attacks, and Natural Disasters:** The national averages for reliability indices (outage duration and frequency measures SAIDI, SAIFI, and MAIFI) appear to be trending upward [Metric 10]. Smart-grid directions, such as demand-side resource and distributed-generation participation in system operations

Smart Grid System Report — July 2009

discussed in the first two bullets are expected to more elegantly respond to disturbances and emergencies. Within the delivery-system field operations, substation automation (discussed in the previous bullet) is showing progress [Metric 11]. At the regional system operations level, advanced measurement equipment is being deployed within the delivery infrastructure to support situational awareness and enhance reliability coordination. Deployment numbers for one technology, synchro-phasor measurements, have increased from 100 units in 2006 to roughly 150 in 2008. Lastly, cyber-security challenges are beginning to be addressed with a more disciplined approach. NERC Critical Infrastructure Protection security assessments are more common with about 95% of companies interviewed for this report indicating that they have conducted at least one security assessment of their operations [Metric 18]. This characteristic is a merger of the Modern Grid Initiative's characteristics a) Self-Heals and b) Resists Attack. The same metrics substantially contribute to both of these concerns.

Different areas of the country have distinctions with regard to their generation resources, their business economy, climate, topography, environmental concerns, and public policy. These distinctions influence the picture for smart-grid deployment in each region, provide different incentives, and pose different obstacles for development. Where appropriate, the report discusses progress and issues associated with the state of smart-grid deployment measures on a regional basis.

Challenges to Smart Grid Deployments

A smart grid challenge is the uncertainty of the path that its development will take.

Among the significant challenges facing development of a smart grid are the cost of implementing a smart grid, with estimates for just the electric utility advanced metering capability ranging up to $27 billion (Kuhn 2008), and the regulations that allow recovery of such investments. For perspective, the Brattle Group estimates that it may take as much as $1.5 trillion to update the grid by 2030 (Chupka et al. 2008). Ensuring interoperability of smart-grid standards is another hurdle state and federal regulators will need to leap. Major technical barriers include developing economical storage systems; these storage systems can help solve other technical challenges, such as integrating distributed renewable-energy sources with the grid, addressing power-quality problems that would otherwise exacerbate the situation, and enhancing asset utilization. Without a smart grid, high penetrations of variable renewable resources may become more difficult and expensive to manage due to the greater need to coordinate these resources with dispatchable generation and demand.

Another challenge facing a smart grid is the uncertainty of the path that its development will take over time with changing technology, changing energy mixes, changing energy policy, and developing climate change policy. Trying to legislate or regulate the development of a smart grid or its related technologies can severely diminish the benefits of the virtual, flexible, and transparent energy market it strives to provide. Conversely, with the entire nation's energy grid potentially at risk, some may see the introduction of a smart grid in the United States as too important to allow laissez-faire evolution. Thus, the challenge of development becomes an issue of providing flexible regulation that leverages desired and developing technology through goal-directed and business-case-supported policy that promotes a positive economic outcome.

Policy Questions for Future Reports

Many policy questions continue to be raised surrounding smart-grid systems. Listed below are policy questions to consider related to reporting on smart grid deployments.

- As the first in a series of biennial smart-grid status reports, consideration should be given to the information gathered for this report as a framework and measurement baseline for future reports. The metrics identified are indicators of smart grid deployment progress that facilitate discussion regarding the main characteristics of a smart grid, but they are not comprehensive measures of all smart grid concerns. Because of this, should they be reviewed for continued relevance and appropriate emphasis of major smart grid attributes?

- Should the status of smart grid deployment project as balanced a view as possible across the diverse stakeholder perspectives related to the electric system? Should workshops, interviews, and research into smart grid related literature reflect a complete cross-section of the stakeholders? Should future reports review the stakeholder landscape to ensure coverage of these perspectives?

- Given the time period for developing the report, investigation was restricted to existing literature research and interviews with 21 electricity-service providers, representing a cross section of organizations by type, size, and location. Is further research needed to better gage the metrics and gain insights into deployment directions, as well as engage the other stakeholder groups? Will a more extensive interview process facilitate gathering this information?

- Should coordination with other smart grid information collection activities be supported? Should future reports require the development of assessment models that support those metrics that are difficult to measure, particularly regarding progress on cyber-security and automation system interoperability related to open architecture and standards?

- How comprehensive should a review related to smart grid deployment be? Should the number of metrics proliferate beyond the current number? In deciding if a new metric is merited, should consideration be given to how it fits with the other metrics, if a previous metric can be retired, and the strength of a metric's contribution to explaining the smart grid progress regarding the identified characteristics?

Report Content

The Smart-Grid System Report is organized into a main body and two supporting annexes. The main body discusses the metrics chosen to provide insight into the progress of smart-grid deployment nationally. The measurements resulting from research into the metrics are used to convey the state of smart-grid progress according to six characteristics derived from the NETL Modern Grid Strategy's work in this area. The main body of the report concludes with a summary of the challenges to smart-grid deployment including technical, business and financial challenges.

The first of two annexes presents a discussion of each of the metrics chosen to help measure the progress of smart-grid deployment. The second summarizes the results of interviews with 21 electricity-service providers chosen to represent a cross-section of the nation in terms of size, location, and type of organization (e.g., public or private company, rural electric cooperative, etc.). The interview questions were designed to support many of the identified metrics and the results are incorporated into the metric write-ups to support measurement estimates.

ix

Table ES.1. Summary of Smart Grid Metrics and Status

#	Metric Title	Type	Penetration/ Maturity	Trend
	Area, Regional, and National Coordination Regime			
1	Dynamic Pricing: fraction of customers and total load served by RTP, CPP, and TOU tariffs	build	low	moderate
2	Real-time System Operations Data Sharing: Total SCADA points shared and fraction of phasor measurement points shared.	build	moderate	moderate
3	Distributed-Resource Interconnection Policy: percentage of utilities with standard distributed-resource interconnection policies and commonality of such policies across utilities.	build	moderate	moderate
4	Policy/Regulatory Progress: weighted-average percentage of smart grid investment recovered through rates (respondents' input weighted based on total customer share).	build	low	moderate
	Distributed-Energy-Resource Technology			
5	Load Participation Based on Grid Conditions: fraction of load served by interruptible tariffs, direct load control, and consumer load control with incentives.	build	low	low
6	Load Served by Microgrids: the percentage total grid summer capacity.	build	nascent	low
7	Grid-Connected Distributed Generation (renewable and non-renewable) and Storage: percentage of distributed generation and storage.	build	low	high
8	EVs and PHEVs: percentage shares of on-road, light-duty vehicles comprising of EVs and PHEVs.	build	nascent	low
9	Grid-Responsive Non-Generating Demand-Side Equipment: total load served by smart, grid-responsive equipment.	build	nascent	low
	Delivery (T&D) Infrastructure			
10	T&D System Reliability: SAIDI, SAIFI, MAIFI.	value	mature	declining
11	T&D Automation: percentage of substations using automation.	build	moderate	high
12	Advanced Meters: percentage of total demand served by advanced metered (AMI) customers	build	low	high
13	Advanced System Measurement: percentage of substations possessing advanced measurement technology.	build	low	moderate
14	Capacity Factors: yearly average and peak-generation capacity factor	value	mature	flat
15	Generation and T&D Efficiencies: percentage of energy consumed to generate electricity that is not lost.	value	mature	improving
16	Dynamic Line Ratings: percentage miles of transmission circuits being operated under dynamic line ratings.	build	nascent	low
17	Power Quality: percentage of customer complaints related to power quality issues, excluding outages.	value	mature	declining
	Information Networks and Finance			
18	Cyber Security: percent of total generation capacity under companies in compliance with the NERC Critical Infrastructure Protection standards.	build	nascent	nascent
19	Open Architecture/Standards: Interoperability Maturity Level – the weighted average maturity level of interoperability realized among electricity system stakeholders	build	nascent	nascent
20	Venture Capital: total annual venture-capital funding of smart-grid startups located in the U.S.	value	nascent	high

x

Acronyms and Abbreviations

AMI	advanced metering infrastructure
AMR	automated meter reading
AMS	advanced metering system
CA	Control Area
CAIDI	Customer Average Interruption Duration Index
CBL	customer baseline load
CHP	combined heat and power
CIP	critical infrastructure protection
CPP	critical peak pricing
CPUC	California Public Utilities Commission
DC	District of Columbia
DER	distributed energy resources
DG	distributed generation
DLR	dynamic line ratings
DMS	distribution management systems
DOE	Department of Energy
eGRID	Emissions and Generation Resource Integrated Database
EIA	Energy Information Administration
EIOC	Electricity Infrastructure Operations Center
EMS	energy management systems
EPA	Environmental Protection Agency
EPAct	Energy Policy Act
EPRI	Electric Power Research Institute
ERCOT	Electric Reliability Council of Texas
EV	electric vehicle
FERC	Federal Energy Regulatory Commission
GW	gigawatt, billion watts
GWAC	GridWise* Architecture Council
HAN	home area network
IED	intelligent electronic device
IEEE	Institute of Electrical and Electronic Engineers
IOU	investor-owned utilities
ISO	independent system operator
IT	information technology
KVAR	KilovoltAmpere Reactive
kWh	kilowatt-hours
LBNL	Lawrence Berkeley National Laboratory

xi

MAIFI	Momentary Average Interruption Frequency Index
MDMS	meter data management system
MW	megawatts
MWh	megawatt-hours
NASPI	North American Synchro-Phasor Initiative
NERC	North American Electric Reliability Corporation
NETL	National Energy Technology Laboratory
NIST	National Institute of Standards and Technology
NRDC	National Resources Defense Council
OEM	original equipment manufacturer
OWL	Web Ontology Language
PG&E	Pacific Gas and Electric Company
PHEV	plug-in hybrid electric vehicles
PMU	phasor measurement units
PQ	power quality
PSC	public service commission
PUC	public utility commission
RCC	Reliability Coordination Center
RFC	Reliability First Corporation
RFP	request for proposal
RTO	regional transmission operator
RTP	real-time pricing
SAIDI	system average interruption duration index
SAIFI	system average interruption frequency index
SCADA	supervisory control and data acquisition
SDG&E	San Diego Gas & Electric Company
SERC	Southeastern Electric Reliability Company
SPP	Southwest Power Pool
T&D	transmission and distribution
TCP	Transmission Control Protocol
TOU	time-of-use pricing
TW	terawatt, trillion watts
TWh	terawatt-hours
UML	Unified Modeling Language⁺
VAR	volt-amps reactive
WAMS	Wide Area Measurement System
WECC	Western Electricity Coordination Council
XML	Extensible Markup Language

Contents

xiii

Figures

Tables

1.0 Introduction

Section 1302 of Title XIII of the Energy Independence and Security Act of 2007 directs the Secretary of Energy to, "…report to Congress concerning the status of smart grid deployments nationwide and any regulatory or government barriers to continued deployment." The first report is to occur no later than one year after enactment. This is the first installment of this report to Congress, which is to be updated biennially. Please note that this report does not include impacts related to the American Recovery and Reinvestment Act of 2009.

1.1 Objectives

The objective of Title XIII is to support the modernization of the Nation's electricity system to maintain a reliable and secure infrastructure that can meet future load growth and achieve the characteristics of a smart grid. The Smart Grid System Report is to provide the current status of smart-grid development, the prospects for its future, and the obstacles to progress. In addition to providing the state of smart-grid deployments, the legislation includes the following requirements:

1. report the prospects of smart-grid development including costs and obstacles;

2. identify regulatory or government barriers;

3. may provide recommendations for state and federal policies or actions; and

4. take a regional perspective.

In the process of developing this report, the advice of the Electricity Advisory Committee and its Subcommittee on Smart Grid has been solicited, in addition to the advice from the Federal Smart Grid Task Force, an inter-agency group that includes representation from U.S. Department of Energy (DOE), Federal Energy Regulatory Commission (FERC), National Institute for Standards and Technology (NIST), U.S. Department of Homeland Security (DHS), U.S. Environmental Protection Agency (EPA), and other involved federal agencies.

As the first in a series of biennial smart-grid status reports, aspects of this report are expected to form the framework for future reports. However, such future reports will be able to go into greater assessment detail using this framework and measure smart-grid-related progress based on the baseline established in this report.

1.2 Scope of a Smart Grid

A smart grid uses digital technology to improve reliability, security, and efficiency of the electric system. Due to the vast number of stakeholders and their various perspectives, there has been debate on a definition of a smart grid that addresses the special emphasis desired by each participant. To define the scope of a smart grid for this report, we reviewed application areas throughout the electric system related to dynamic optimization of system operations, maintenance, and planning. Figure 1.1 provides a pictorial view of the many aspects of the electric system touched by smart grid concerns.

A smart grid uses digital technology to improve reliability, security, and efficiency of the electric system.

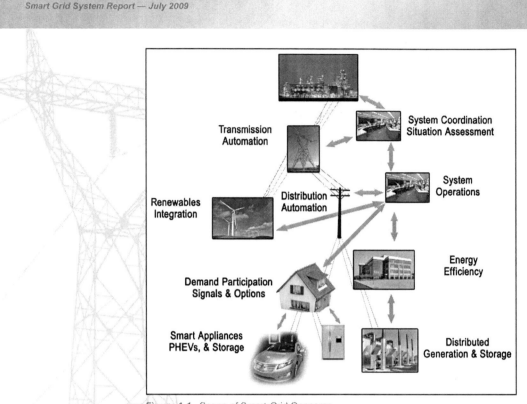

Figure 1.1. Scope of Smart-Grid Concerns

The following areas arguably represent a reasonable partitioning of the electric system that covers the scope of smart grid concerns. To describe the progress being made in moving toward a smart grid, one must also consider the interfaces between elements within each area and the systemic issues that transcend areas. The areas of the electric system that cover the scope of a smart grid include the following:

- Area, regional and national coordination regimes: A series of interrelated, hierarchical coordination functions exists for the economic and reliable operation of the electric system. These include balancing areas, independent system operators (ISOs), regional transmission operators (RTOs), electricity market operations, and government emergency-operation centers. Smart-grid elements in this area include collecting measurements from across the system to determine system state and health, and coordinating actions to enhance economic efficiency, reliability, environmental compliance, or response to disturbances.

- Distributed-energy resource technology: Arguably, the largest "new frontier" for smart grid advancements, this area includes the integration of distributed-generation, storage, and demand-side resources for participation in electric-system operation. Consumer products such as smart appliances and electric vehicles are expected to become important components of this area as are renewable-generation components such as those derived from solar and wind sources. Aggregation mechanisms of distributed-energy resources are also considered.

- **Delivery (transmission and distribution [T&D]) infrastructure:** T&D represents the delivery part of the electric system. Smart-grid items at the transmission level include substation automation, dynamic limits, relay coordination, and the associated sensing, communication, and coordinated action. Distribution-level items include distribution automation (such as feeder-load balancing, capacitor switching, and restoration) and advanced metering (such as meter reading, remote-service enabling and disabling, and demand-response gateways).

- **Central generation:** Generation plants already contain sophisticated plant automation systems because the production-cost benefits provide clear signals for investment. While technological progress is related to the smart grid, change is expected to be incremental rather than transformational, and therefore, this area in not emphasized as part of this report.

- **Information networks and finance:** Information technology and pervasive communications are cornerstones of a smart grid. Though the information networks requirements (capabilities and performance) will be different in different areas, their attributes tend to transcend application areas. Examples include interoperability and the ease of integration of automation components as well as cyber-security concerns. Information technology related standards, methodologies, and tools also fall into this area. In addition, the economic and investment environment for procuring smart-grid-related technology is an important part of the discussion concerning implementation progress.

Some aspect of the electricity system touches every person in the Nation.

Section 1301 of the legislation identifies characteristics of a smart grid. The NETL Modern Grid Initiative provides a list of smart-grid attributes in "Characteristics of the Modern Grid" (NETL 2008). These characteristics were used to help organize a Department of Energy-sponsored workshop on "Implementing the Smart Grid" (DOE/OEDER 2008b). The results of that workshop are used to organize the reporting of smart grid progress around six characteristics. The sixth characteristic is a merger of the Modern Grid Initiative's characteristics a) Self- Heals and b) Resists Attack. The same metrics substantially contribute to both of these concerns.

- Enabling Informed Participation by Customers
- Accommodating All Generation and Storage Options
- Enabling New Products, Services, and Markets
- Providing the Power Quality for the Range of Needs
- Optimizing Asset Utilization and Operating Efficiently
- Operating Resiliently: Disturbances, Attacks, and Natural Disasters

1.3 Stakeholder Landscape

Some aspect of the electricity system touches every person in the Nation. The stakeholder landscape for a smart-grid is complex (see Figure 1.2). The lines of distinction are not always

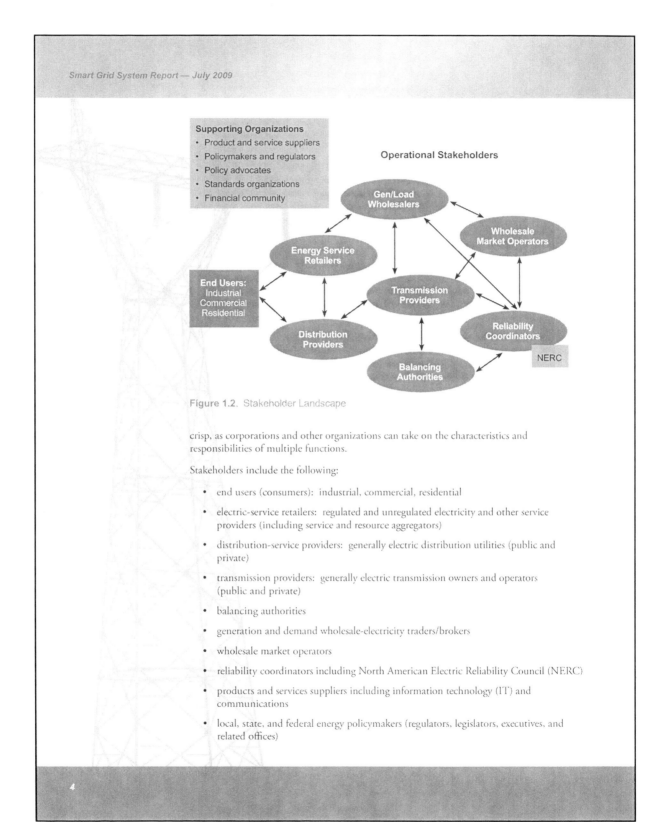

Smart Grid System Report — July 2009

Supporting Organizations
• Product and service suppliers
• Policymakers and regulators
• Policy advocates
• Standards organizations
• Financial community

Operational Stakeholders

Gen/Load Wholesalers

Wholesale Market Operators

Energy Service Retailers

End Users: Industrial Commercial Residential

Transmission Providers

Reliability Coordinators

NERC

Distribution Providers

Balancing Authorities

Figure 1.2. Stakeholder Landscape

crisp, as corporations and other organizations can take on the characteristics and responsibilities of multiple functions.

Stakeholders include the following:

• end users (consumers): industrial, commercial, residential

• electric-service retailers: regulated and unregulated electricity and other service providers (including service and resource aggregators)

• distribution-service providers: generally electric distribution utilities (public and private)

• transmission providers: generally electric transmission owners and operators (public and private)

• balancing authorities

• generation and demand wholesale-electricity traders/brokers

• wholesale market operators

• reliability coordinators including North American Electric Reliability Council (NERC)

• products and services suppliers including information technology (IT) and communications

• local, state, and federal energy policymakers (regulators, legislators, executives, and related offices)

- policy advocates (consumer groups, trade organizations, environmental advocates)
- standards organizations
- financial community

The major stakeholder groups are referenced throughout the report as appropriate to the topic in question.

1.4 Regional Influences

Different areas of the country have distinctions with regard to their fuel and generation resources, their business economy, climate, topography, environmental concerns, demography (e.g., rural versus urban), consumer values, and public policy. These distinctions influence the picture for smart-grid deployment in each region and service territory. They provide different incentives and pose different obstacles for development. The result is a transformation toward a smart grid that will vary across the nation. The major regions of the country can be divided into the 10 NERC reliability regions (see Figure 1.3) (EPA 2008a). The EPA further subdivides these into 26 subregions (see EPA map, Figure 1.4), and each of these regions has their distinctive state and local governments. Regional factors are woven into various aspects of the report including the smart-grid deployment metrics, deployment attributes, trends, and obstacles. The primary regional influences focus on the states and major NERC reliability regions; however, other regional aspects are presented as appropriate.

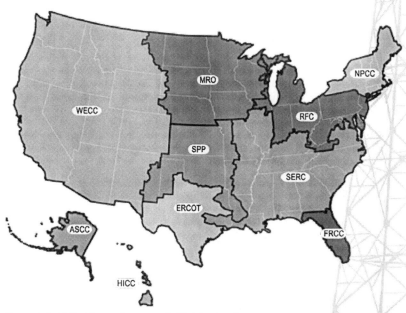

Figure 1.3. United States Portions of NERC Region Representation Map

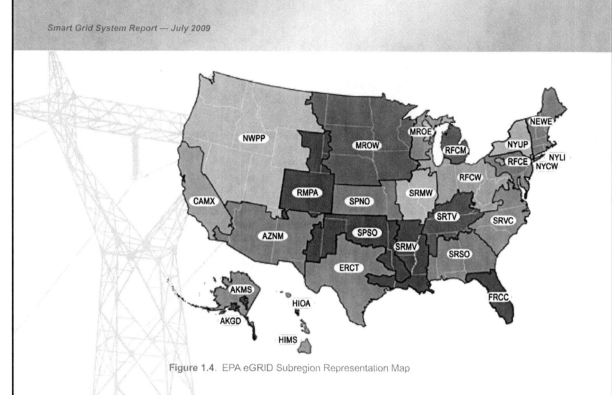

Figure 1.4. EPA eGRID Subregion Representation Map

1.5 About This Document

The Smart-Grid System Report is organized into a main body and two supporting annexes. The main body discusses the metrics chosen to provide insight into the progress of smart-grid deployment nationally. The measurements resulting from research into the metrics are used to convey the state of smart-grid progress according to six characteristics derived from the NETL Modern Grid Strategy's work in this area and discussions at the DOE Smart Grid Implementation Workshop. The main body of the report concludes with a summary of the challenges to smart-grid deployment including technical, business and financial challenges, and implications for state and federal policy.

The main body of the report is supported by two annexes. The first presents a discussion of each of the metrics chosen to help measure the progress of smart-grid deployment. The second summarizes the results of interviews with 21 electricity-service providers chosen to represent a cross section of the nation in terms of size, location, and type of organization (e.g., public or private company, rural electric cooperative, etc.). The interview questions were designed to support many of the identified metrics and the results are incorporated into the metric write-ups to support measurement estimates.

Different regions will deploy smart grid capabilities to varying degrees.

As mentioned earlier, throughout this discussion, special concerns of stakeholders and regional influences are described. As this is the first edition of this biennial report to Congress, recommendations for further investigation and improvements for future reports are also provided.

6

190

2.0 Deployment Metrics and Measurements

The scope of a smart grid extends across the electricity system and its supply chain. To measure the status of smart-grid deployments, a set of metrics has been chosen as indicators for discussing smart-grid progress. Though these metrics do not comprehensively cover all aspects of a smart grid, they were chosen to address a balance of coverage in significant functional areas and to support the communication of its status through a set of smart-grid characteristics that have been formed through workshop engagements with industry.

2.1 Smart-Grid Metrics

On June 19-20, 2008, the U.S. Department of Energy brought together 140 experts, representing the various stakeholder groups associated with a smart grid, at a workshop in Washington, D.C. The objective of the workshop was to identify a set of metrics for measuring progress toward implementation of smart-grid technologies, practices, and services. Breakout sessions for the workshop were organized around seven major smart-grid characteristics as developed through another set of industry workshops sponsored by the NETL Modern Grid Strategy (NETL 2008). The results of the workshop document submissions of over 50 metrics for measuring smart-grid progress (DOE/OEDER 2008b). Having balanced participation across the diverse electric system stakeholders is important to deriving appropriate metrics and was an important objective for selecting individuals to invite to the workshop. While most aspects of the stakeholder landscape described in Section 1.3 were well represented, the following groups arguably deserve to have greater representation in the future: end users (industrial, commercial, and residential) and their consumer advocates, environmental groups, as well as the financial community, including venture capitalists.

The workshop described two types of metrics: build metrics that describe attributes that are built in support of a smart grid, and value metrics that describe the value that may derive from achieving a smart grid. While build metrics tend to be quantifiable, value metrics can be influenced by many developments and therefore generally require more qualifying discussion. Build metrics generally lead the value that is eventually provided, while value metrics generally lag in reflecting the contributions that accrue from implementations. A metric's type is specifically identified in the discussion below. Both build and value metrics are important to describe the status of smart grid implementation.

In reviewing the workshop results, one finds several similar metrics identified by different breakout groups; the overlap arises because a metric may be an indicator of progress in more than one characteristic of a smart grid. The list of metrics in Table 2.1 results from a distillation of the recorded ideas into a relatively small number of metrics and augmented to provide a reasonable prospect of measurement or assessment. These 20 metrics are used in this report to describe the state of smart grid deployment. A detailed investigation of the measurements for each metric can be found in Annex A of this report.

7

191

Table 2.1. Summary of Smart Grid Metrics and Status

#	Metric Title	Type	Penetration/ Maturity	Trend
Area, Regional, and National Coordination Regime				
1	Dynamic Pricing: fraction of customers and total load served by RTP, CPP, and TOU tariffs	build	low	moderate
2	Real-time System Operations Data Sharing: Total SCADA points shared and fraction of phasor measurement points shared.	build	moderate	moderate
3	Distributed-Resource Interconnection Policy: percentage of utilities with standard distributed-resource interconnection policies and commonality of such policies across utilities.	build	moderate	moderate
4	Policy/Regulatory Progress: weighted-average percentage of smart grid investment recovered through rates (respondents' input weighted based on total customer share).	build	low	moderate
Distributed-Energy-Resource Technology				
5	Load Participation Based on Grid Conditions: fraction of load served by interruptible tariffs, direct load control, and consumer load control with incentives.	build	low	low
6	Load Served by Microgrids: the percentage total grid summer capacity.	build	nascent	low
7	Grid-Connected Distributed Generation (renewable and non-renewable) and Storage: percentage of distributed generation and storage.	build	low	high
8	EVs and PHEVs: percentage shares of on-road, light-duty vehicles comprising of EVs and PHEVs.	build	nascent	low
9	Grid-Responsive Non-Generating Demand-Side Equipment: total load served by smart, grid-responsive equipment.	build	nascent	low
Delivery (T&D) Infrastructure				
10	T&D System Reliability: SAIDI, SAIFI, MAIFI.	value	mature	declining
11	T&D Automation: percentage of substations using automation.	build	moderate	high
12	Advanced Meters: percentage of total demand served by advanced metered (AMI) customers	build	low	high
13	Advanced System Measurement: percentage of substations possessing advanced measurement technology.	build	low	moderate
14	Capacity Factors: yearly average and peak-generation capacity factor	value	mature	flat
15	Generation and T&D Efficiencies: percentage of energy consumed to generate electricity that is not lost.	value	mature	improving
16	Dynamic Line Ratings: percentage miles of transmission circuits being operated under dynamic line ratings.	build	nascent	low
17	Power Quality: percentage of customer complaints related to power quality issues, excluding outages.	value	mature	declining
Information Networks and Finance				
18	Cyber Security: percent of total generation capacity under companies in compliance with the NERC Critical Infrastructure Protection standards.	build	nascent	nascent
19	Open Architecture/Standards: Interoperability Maturity Level – the weighted average maturity level of interoperability realized among electricity system stakeholders	build	nascent	nascent
20	Venture Capital: total annual venture-capital funding of smart-grid startups located in the U.S.	value	nascent	high

The table includes two columns to indicate the metric's state (penetration level/maturity) and trend. The intent is to provide a high level, simplified perspective to a complicated picture. If it is a build metric, the penetration level is indicated as nascent (very low and just emerging), low, or moderate. Because smart grid activity is relatively new, there are no high penetration levels to report on these metrics. If it is a value metric, the maturity of the system with respect to this metric is indicated as either nascent or mature. The trend (recent past and near-term projection) is indicated for either type of metric as declining, flat, or growing at nascent, low, moderate, or high, levels.

Other observations about selecting metrics follow:

- Metrics can be combined in various ways to provide potentially interesting insights into smart-grid progress. The same metric is used multiple times in this report to explain progress with respect to smart grid characteristics.

- The selection process strove to identify fundamental metrics that can support more complex combinations.

- Though the list of metrics is flat, headings are used to group metrics into logically-related areas that support balanced coverage of smart-grid concerns.

- For each metric, serious consideration is required regarding how measurements can be obtained. In some situations (particularly with value metrics), qualifying statements tend to dominate how a smart grid may influence the measurement.

- Wherever appropriate, metrics should be expressed on a proportional basis.

2.2 Smart-Grid Characteristics

The metrics identified above are used in Section 3 to describe deployment status as organized around six major characteristics of a smart grid, as described in Table 2.2. These characteristics are derived from the seven characteristics in the Modern Grid Strategy work described earlier and augmented slightly in the organization of the metrics workshop. The sixth characteristic in the table is a merger of the workshop characteristics a) Addresses and Responds to System Disturbances in a Self-Healing Manner and b) Operates Resiliently Against Physical and Cyber Attacks and Natural Disasters. The same metrics substantially contribute to both of these concerns.

Metrics can be combined in various ways to provide potentially interesting insights into smart-grid progress.

9

Table 2.2   Smart-Grid Characteristics

Characteristic	Description
1. Enables Informed Participation by Customers	Consumers become an integral part of the electric power system. They help balance supply and demand and ensure reliability by modifying the way they use and purchase electricity. These modifications come as a result of consumers having choices that motivate different purchasing patterns and behavior. These choices involve new technologies, new information about their electricity use, and new forms of electricity pricing and incentives.
2. Accommodates All Generation and Storage Options	A smart grid accommodates not only large, centralized power plants, but also the growing array of distributed energy resources (DER). DER integration will increase rapidly all along the value chain, from suppliers to marketers to customers. Those distributed resources will be diverse and widespread, including renewables, distributed generation and energy storage.
3. Enables New Products, Services, and Markets	Correctly-designed and -operated markets efficiently reveal cost-benefit tradeoffs to consumers by creating an opportunity for competing services to bid. A smart grid accounts for all of the fundamental dynamics of the value/cost relationship. Some of the independent grid variables that must be explicitly managed are energy, capacity, location, time, rate of change, and quality. Markets can play a major role in the management of these variables. Regulators, owners/operators, and consumers need the flexibility to modify the rules of business to suit operating and market conditions.
4. Provides the Power Quality for the Range of Needs	Not all commercial enterprises, and certainly not all residential customers, need the same quality of power. A smart grid supplies varying grades of power and supports variable pricing accordingly. The cost of premium power-quality (PQ) features can be included in the electrical service contract. Advanced control methods monitor essential components, enabling rapid diagnosis and precise solutions to PQ events, such as arise from lightning, switching surges, line faults and harmonic sources. A smart grid also helps buffer the electrical system from irregularities caused by consumer electronic loads.
5. Optimizes Asset Utilization & Operating Efficiency	A smart grid applies the latest technologies to optimize the use of its assets. For example, optimized capacity can be attainable with dynamic ratings, which allow assets to be used at greater loads by continuously sensing and rating their capacities. Maintenance efficiency involves attaining a reliable state of equipment or "optimized condition." This state is attainable with condition-based maintenance, which signals the need for equipment maintenance at precisely the right time. System-control devices can be adjusted to reduce losses and eliminate congestion. Operating efficiency increases when selecting the least-cost energy-delivery system available through these adjustments of system-control devices.
6. Operates Resiliently to Disturbances, Attacks, & Natural Disasters	Resiliency refers to the ability of a system to react to events such that problematic elements are isolated while the rest of the system is restored to normal operation. These self-healing actions result in reduced interruption of service to consumers and help service providers better manage the delivery infrastructure. A smart grid responds resiliently to attacks, whether organized by others or the result of natural disasters. These threats include physical attacks and cyber attacks. A smart grid addresses security from the outset, as a requirement for all the elements, and ensures an integrated and balanced approach across the system.

2.3 Mapping Metrics to Characteristics

Section 3 of the report describes the status of smart-grid deployment using the six characteristics discussed above. A map of how the 20 metrics support the 6 characteristics is shown in Table 2.3. Notice that nearly every metric contributes to multiple characteristics. To reduce the repetition of statements about the metrics, each metric was assigned a primary

2.0 Deployment Metrics and Measurements

characteristic for emphasis. The table indicates the characteristic where a metric is emphasized as "emphasis." The other characteristic cells where a metric plays an important role are indicated by "mention." This should not be interpreted to be of secondary importance, only that a metric finding is mentioned under that characteristic in order to reduce redundancy of material in explaining the status of smart grid deployment.

The interviews with 21 electric-service providers also provide insight into a measure of the metrics and how they relate to the smart-grid characteristics. The interview questions were designed to gather information related to the metrics of interest. The interview results are presented in Annex B and the information gained from these interviews is woven into the metric write-ups in Annex A as well as the smart-grid status descriptions presented for each characteristic in the next section.

Table 2.3. Map of Metrics to Smart-Grid Characteristics

Metric No.	Metric Name	Enables Informed Participation by Customers	Accom-modates All Generation & Storage Options	Enables New Products, Services, & Markets	Provides Power Quality for the Range of Needs	Optimizes Asset Utilization & Efficient Operation	Operates Resiliently to Disturbances, Attacks, & Natural Disasters
1	Dynamic Pricing	Emphasis	Mention	Mention			Mention
2	Real-Time Data Sharing					Mention	Emphasis
3	DER Interconnection	Mention	Emphasis	Mention		Mention	
4	Regulatory Policy			Emphasis			
5	Load Participation	Emphasis			Mention	Mention	Mention
6	Microgrids		Mention	Mention	Emphasis		Mention
7	DG & Storage	Mention	Emphasis	Mention	Mention	Mention	Mention
8	Electric Vehicles	Mention	Mention	Emphasis			Mention
9	Grid-responsive Load	Mention	Mention	Mention	Mention		Emphasis
10	T&D Reliability						Emphasis
11	T&D Automation				Mention	Emphasis	Mention
12	Advanced Meters	Emphasis	Mention	Mention			Mention
13	Advanced Sensors					Mention	Emphasis
14	Capacity Factors					Emphasis	
15	Generation, T&D Efficiency					Emphasis	
16	Dynamic Line Rating					Emphasis	Mention
17	Power Quality			Mention	Emphasis		
18	Cyber Security						Emphasis
19	Open Architecture/Stds			Emphasis			
20	Venture Capital			Emphasis			

11

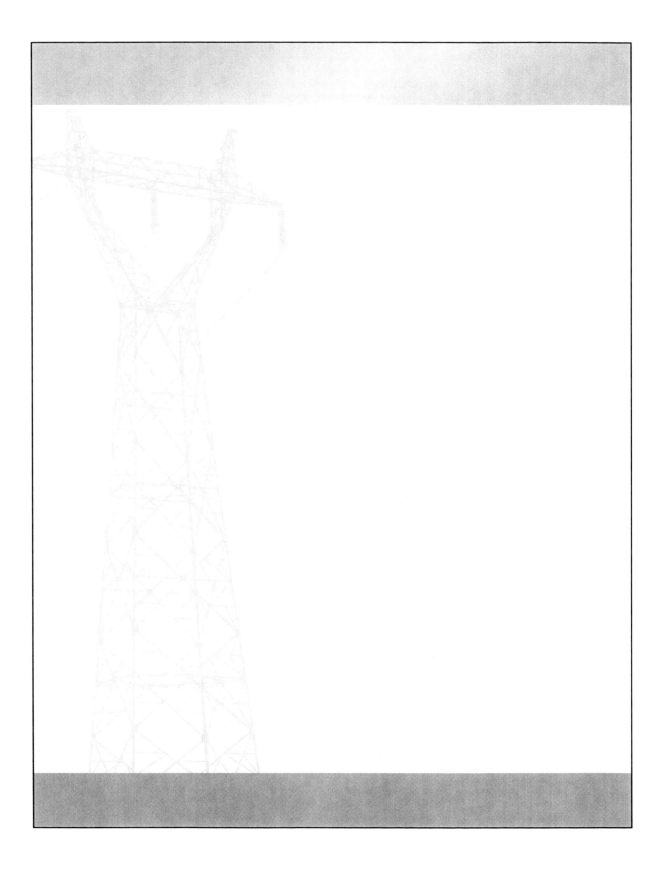

3.0 Deployment Trends and Projections

Deploying a smart grid is a journey that has been underway for some time, but will accelerate because of the Energy Independence and Security Act of 2007, and the recognition of characteristics and benefits collected and emphasized under the term "smart grid." Though there has been much debate over the exact definition, a smart grid comprises a broad range of technology solutions that optimize the energy value chain. Depending on where and how specific participants operate within that chain, they can benefit from deploying certain parts of a smart grid solution set (EAC 2008). Based on the identification of deployment metrics, this section of the report presents recent deployment trends. In addition, it reviews plans of the stakeholders relevant to smart-grid deployment to provide insight about near-term and future directions.

The status of smart-grid deployment expressed in this section is supported by an investigation of 20 metrics obtained through available research, such as advanced metering and T&D substation-automation assessment reports, penetration rates for energy resources, and capability enabled by a smart grid. An important contribution to that investigation is information collected through interviews of a diverse set of 21 electric service providers. In each subsection that follows, the metrics contributing to explaining the state of the smart grid characteristic are called out so the reader may review more detailed information in Annex A. The metrics emphasized to explain the status of a characteristic are so highlighted with a asterik (*).

3.1 Enables Informed Participation by Customers

A part of the vision of a smart grid is its ability to enable informed participation by customers, making them an integral part of the electric power system. With bi-directional flows of energy and coordination through communication mechanisms, a smart grid should help balance supply and demand and enhance reliability by modifying the manner in which customers use and purchase electricity. These modifications can be the result of consumer choices that motivate shifting patterns of behavior and consumption. These choices involve new technologies, new information regarding electricity use, and new pricing and incentive programs.

A smart grid adds consumer demand as another manageable resource, joining power generation, grid capacity, and energy storage. From the standpoint of the consumer, energy management in a smart-grid environment involves making economic choices based on the variable cost of electricity, the ability to shift load, and the ability to store or sell energy.

Consumers who are presented with a variety of options when it comes to energy purchases and consumption are enabled to:

- respond to price signals and other economic incentives to make better-informed decisions regarding when to purchase electricity, when to generate energy using distributed generation, and whether to store and re-use it later with distributed storage.

- make informed investment decisions regarding more efficient and smarter appliances, equipment, and control systems.

Related Metrics
1*, 3, 5*, 7, 8, 9, 12*

13

3.1.1 Grid-Enabled Bi-Directional Communication and Energy Flows

A smart grid system relies on the accurate, up to date, and predictable delivery of data between the customer and utility company. A conduit through which this information may be exchanged is advanced metering infrastructure (AMI). AMI, unlike conventional metering systems, relies on fixed, digital network technologies. At the most basic level, AMI serves as a middleman between a consumer's energy consumption and the utility that provides electricity, by reading household energy consumption at some predetermined requested interval (e.g., hourly) and then storing and transmitting the data via a wired or wireless network to the service provider. This basic level supports automated meter reading (AMR). At higher levels, AMI technology can incorporate bi-directional communication, including transmitting real-time price and consumption data between the household and utility, and coordinating with a Home Area Network. Figure 3.1 presents an overview of an AMI interface enabling the bi-directional flow of information [Metric 12].[1] Currently, AMI composes about 4.7%, or 6.7 million, of total U.S. electric meters (FERC 2007). The number of installed meters has been projected to grow by another 52 million by 2012.

When customers are motivated by economic incentives through dynamic pricing structures or other programs, their investments in "smart" devices could facilitate reductions or shifts in energy consumption. "Smart" devices (e.g., communicating thermostats, clothes washers and dryers, microwaves, hot water heaters, refrigerators) use signaling software or firmware to communicate with the grid [Metric 9]. For example, a "smart" water heater could be equipped with a device that coordinates with a facility's energy-management system to adjust temperature controls, within specified limits, based on energy prices.

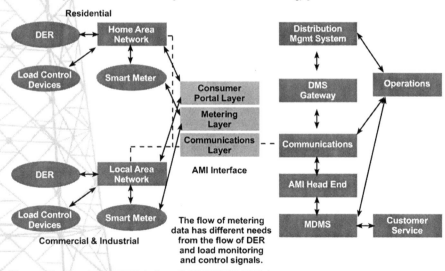

Figure 3.1. Overview of AMI Interface (DOE/OEDER 2008c)

(1) DMS: Distribution management system. MDMS: Meter data management systems.

One recent experiment that focused on the "smart" device concept in combination with smart-grid technologies and dynamic pricing was conducted in the Olympic Peninsula of Washington State by the U.S. Department of Energy (Hammerstrom 2007). In the experiment, thermostats, washers, dryers, and water heaters were fitted with smart grid-responsive equipment and were programmed to respond to peak loads. The results of the experiment were promising; the smart devices reduced load fluctuations, decreased peak loads, and significantly reduced energy costs.

The technology exists today to implement grid-responsive equipment. Industrial plants and modern, large commercial buildings are generally well-equipped to respond to incentives to change their demand because they have energy management systems. However, in residences and small commercial buildings, there is little supporting installed technology to communicate to the equipment, though products are emerging.

Increasing demand participation in electric system operations centers on incentive offerings.

A primary issue to increasing demand participation in electric system operations centers on incentive offerings. Smart grid related technology, such as advanced meters, has enabled dynamic pricing programs implemented across the U.S. [Metric 1]. Generally, these tariffs take the following forms:

- Time of use (TOU). Under TOU, prices are differentiated based solely on a peak versus off-peak period designation, with prices set higher during peak periods. TOU pricing is not dynamic because it does not vary based on real-time conditions. It is included here though because it is viewed as an intermediate step towards a more dynamic real-time pricing (RTP) tariff.

- Critical peak pricing (CPP). Under a CPP tariff, the higher critical peak price is restricted to a small number of hours (e.g., 100 of 8,760) each year, with the peak price being set at a much higher level relative to normal conditions.

- Real-time pricing (RTP). Under RTP, hourly prices vary based on the day-of (real time) or day-ahead cost of power to the utility.

The Federal Energy Regulatory Commission conducted an extensive survey of demand response and advanced metering initiatives in 2008. The FERC survey was distributed to 3,407 organizations in all 50 states. In total, 100 utilities that responded to the survey reported offering some form of a real time tariff to enrolled customers (Table 3.1), as compared to 60 in 2006 (FERC 2008).

FERC also found through its 2008 survey that 315 utilities nationwide offered TOU rates, compared to 366 in 2006. In those participating utilities, approximately 1.3 million residential electricity consumers were signed up for TOU tariffs, representing approximately 1.1% of all U.S. customers (Table 3.1). In 2008, customers were enrolled in CPP tariffs offered by 88 utilities, as compared to 36 in 2006. No studies were found to estimate the total number of customers served by RTP and CPP tariffs.

Table 3.1. Number of Entities Offering and Customers Served by Dynamic Pricing Tariffs

Method of Pricing	Number of Entities	Customer Served	
		Number	Share of Total
Real Time Pricing	100		
Critical Peak Pricing	88		
Time of Use	315	1,270,000	1.1%

The interviews conducted for this report included 21 companies with an annual peak capacity of 150,000-175,000 megawatts. The respondents were asked two questions relevant to dynamic pricing. The first question asked respondents: Do you have dynamic or supply based price plans?

- Seven companies (35 percent) indicated no dynamic price plans were in place.

- Twelve companies (60 percent) indicated they had TOU plans.

- Three companies (15) percent offered CPP plans.

- Seven companies (35 percent) indicated they had both dynamic price plans and the ability to send price signals to customers.

The respondents were also asked whether their utility had automated response to pricing signals for major energy using devices within the premises. Responses were as follows:

- Nine companies (45 percent) indicated there were none.

- Eight companies (40 percent) indicated that automated price signals for major energy using devices were in the development stage.

- Three companies (15 percent) indicated that a small degree of implementation (10-30 percent of the customer base) had occurred.

Smart grid also facilitates the bi-directional flow of energy, enabling customers to generate or store energy and sell it back to the grid during peak periods when prices are highest to the utility. For example, solar panels installed on rooftops by homeowners can safely generate power at a current cost of $10 to $12 per watt. In the future, as electric vehicles (EVs) and plug-in hybrid electric vehicles (PHEVs) penetrate the U.S. light-duty vehicle market, these alternative-fuel vehicles could also advance load shifting through their energy storage capabilities [Metric 8]. Finally, the more than 12 million backup generators operated in the U.S., representing 200 GW of generating capacity, could also be used to help alleviate peak load, provide needed system support during emergencies, and lower the cost of power provided by the utility [Metric 7] (Gilmore and Lake 2007).

Utilities could realize enormous cost savings over the long term.

Utilities that facilitate the integration of these resources and use them effectively could realize enormous cost savings over the long term. Most projections show increasing deployment of these resources, especially in the commercial sector where power quality and reliability are a serious consideration. Smart grid technologies may be required, along with DG-friendly regulatory structures, in order to integrate DG technologies.

Consumer participation of DG can be facilitated with agreed upon policies for interconnection to the grid. A 2008 EPA study found that only 15 states have "favorable" interconnection standards, with 12 states being "neutral." There are five states classified as having unfavorable policies towards distributed generation [Metric 3] (EPA 2008).

3.1.2 Managing Supply and Demand

Measures, such as turning off or adjusting water heaters, dishwashers, and heating and cooling systems, result in load shifting and reduced costs through the smoothing of peak power consumption throughout the day. With appropriate metering capability in place, dynamic pricing signals received by customers can encourage greater demand response.

Traditionally, demand participation has principally taken place through interruptible demand and direct control load management programs implemented and controlled by electricity suppliers. Nationally, demand-response participation is low. Potential load managed is 1.5% based on 2008 projections. [Metric 5]. Figure 3.2 shows the aggregate demand response by type and region in 2006 and 2007. From this graph one sees the general increase in direct load control, with little or no increase in interruptible demand.

Though dynamic-pricing and demand-response programs have historically been responsible for modest levels of load shifting, current research suggests that there is significant potential for the programs to manage supply and demand in the future. For example, a recent study, sponsored by the Electric Power Research Institute (EPRI) and the Edison Electric Institute (EEI), estimated that 37 percent of the growth in electricity sales (419 TWh) between 2008 and 2030 could be offset through energy-efficiency programs and 52 percent of peak demand growth (164 GW of capacity) could be offset by a combination of energy-efficiency and demand-response programs. More specifically, approximately 2,824 MW of peak demand could be offset by 2010 through price-responsive policies, 13,661 MW of peak demand could be offset through price response by 2020, and 24,869 MW could be offset by 2030. The largest share of the price-response benefits are forecast to take place in the residential sector (10,838 MW or 43.6% of the offset in 2030), with the commercial (8,350 MW or 33.6% of the offset in 2030) and industrial sectors (5,681 MW or 22.8% of the offset) trailing behind (Richmond et al. 2008).

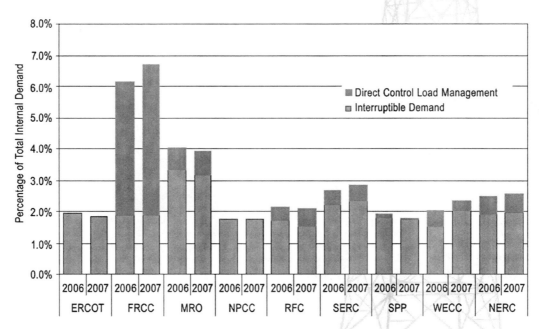

Figure 3.2. Demand Response by NERC Region

3.2 Accommodates All Generation and Storage Options

The ability to accommodate a diverse range of generation types, including centralized and distributed generation as well as diverse storage options, is central to the concept of a smart grid. Through these generation and storage types, a smart grid can better meet consumer load demand, as well as accommodate intermittent renewable-energy technologies. Distributed resources can be used to help alleviate peak load, provide needed system support during emergencies, and lower the cost of power provided by the utility. Accommodating a large number of disparate generation and storage resources requires anticipation of intermittency, unavailability, while balancing costs, reliability, and environmental emissions. Accommodating the diverse nature of these options requires an interconnection process similar to the computer industry's "plug-and-play" environment (DOE/OEDER 2008b).

The primary metrics that measure progress for this characteristic include grid-connected DG and storage, progress in connecting diverse generation types, and a standard distributed-resource-connection policy. There are a number of the other metrics that describe the current ability of a smart grid to accommodate all generation and storage options.

Related Metrics

1, 3*, 6, 7*, 8, 9, 12

Distributed generation and interconnection standards to accommodate generation capacity appear to be moving in positive directions. Distributed-generation (DG) systems are noted for their smaller-scale local power-generation (10 MVA or less) and distribution systems, and generally have low installation and maintenance costs. DG includes power generators such as wind turbines connected at the distribution system level, micro hydro installations, solar panels, diesel, etc. DG also covers energy-storage devices such as batteries and flywheels which could be used to store energy produced or purchased during off-peak hours and then sold or consumed during on-peak hours. Distributed generation, although a small fraction (1%) of total available summer capacity, appears to be increasing rapidly [Metric 7]. In addition, 31 states have interconnection standards in place, with 10 states progressing towards a standard, 1 state with some elements in place, and only 8 states with none. The bad news is that only 15 states had interconnection standards that were deemed to be favorable to distributed generation [Metric 3]. More complete details on the metrics discussed in this section can be found in Annex A.

Microgrids can play a larger role once dynamic pricing and interconnection standards are universally available.

Another measure that impacts this category is dynamic pricing. In this case, time-of-use pricing seems to be gaining momentum, at 1.1% of customers served, and real-time tariffs appear to be slowing increasing. Real-time tariffs would seem to drive the most efficient use of DG, bringing it on-line when prices are high and using more cost-effective central capacity when loads are more manageable [Metric 1] (FERC 2006). The use of smart meters, a driving force behind being able to evaluate grid load and support pricing conditions, has been increasing significantly, almost tripling between 2006 and 2008 to 19 million meters, although the increase from 2007 to 2008 was slower [Metric 12]. Grid responsive load is just beginning to develop with 10% of utilities indicating limited entry into this field, with 45% saying it is in development and 45% having no plans [Metric 9]. The business case for microgrids needs to be made [Metric 6] before commercial capacity will be developed. Currently, universities and petrochemical facilities comprise most of the capacity in microgrids. However, both grid responsive load and microgrids can play a larger role once dynamic pricing and interconnection standards are universally available.

18

Accommodating distributed resources impacts a wide range of stakeholders, including end-users, service providers, regulators, and third-party developers. The interests of these stakeholders will need to be balanced to ensure appropriate evolution of a smart grid while improving the cost efficiency of grid resources (DOE/OEDER 2008b). For example, end-users who implement grid-connected distributed-generation devices will directly affect utilities and service providers whose revenues may decline. Both end-users and service providers must see recovery of investments in distributed resources, smart meters and other smart-grid accessories that allow the grid resources and entities to communicate and respond to changing grid conditions (DOE 2006). Otherwise, end users and service providers will not invest.

The following sections describe distributed generation and storage and interconnection standards in more detail.

3.2.1 Distributed Generation and Storage

Carnegie Mellon's Electricity Industry Center reports that there are now about 12 million backup generators in the United States, representing 200 GW of generating capacity; backup generation is growing at a rate of 5 GW per year [Metric 7] (Gilmore and Lave 2007). Utilities that facilitate the integration of these resources and use them effectively can realize enormous cost savings over the long term. Of the 200 GW of backup generators, currently grid-connected distributed-generation capacity is a small part of total power generation, with combined total grid-connected distributed-generation capacity ranging from 5,423 MW in 2004 to 12,702 MW in 2007 (DOE/EIA 2009a) (see Figure 3.3 and Table 3.2). Available U.S. generating capacity in 2007 comprised 915,292 MW, while summer peak demand reached 782,227 MW and winter peak demand was 637,905 (DOE/EIA 2009b). Thus, while wind and other grid-connected distributed generation increased 134 percent over three years, it still only represented 1.4 percent of grid capacity, 1.6 percent of summer peak and 2.0 percent of winter peak.

Actively-managed fossil-fired, hydrogen, and biofuels distributed generation reached 10,173 MW in 2007, up 112 percent from 2004. This represented approximately 1.1 percent of total generating capacity and 80 percent of total DG. Wind and other renewable-energy sources grew significantly between 2004 and 2007, increasing by 941 percent, yet renewable energy represents only 0.6 percent of total available generating capacity, 0.18 percent of summer peak capacity, and

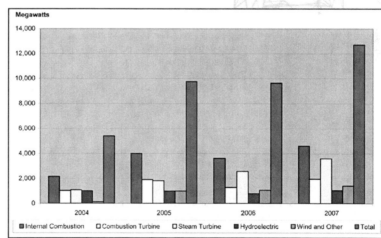

Figure 3.3. Yearly Installed DG Capacity by Technology Type (DOE/EIA 2009b).

0.22 percent of winter peak. Intermittent renewable-energy resources such as wind may not be effective countermeasures for peak-demand reduction, although solar has the potential to be more coincident with summer peak-demand periods. Central wind farms are not included in grid-connected wind DG resources; central wind farms are connected at the transmission level rather than at the distribution level.

Table 3.2.   Capacity of Distributed Generators by Technology Type 2004 to 2006 (count and MW) (DOE/EIA 2009b)

Period	Internal Combustion Capacity	Combustion Turbine Capacity	Steam Turbine Capacity	Hydroelectric Capacity	Wind and Other Capacity	Total Number of	Total Capacity
2004	2,169	1,028	1,086	1,003	137	5,863	5,423
2005(*)	4,024	1,917	1,831	998	994	17,371	9,766
2006	3,625	1,299	2,580	806	1,078	5,044	9,641
2007	4,614	1,964	3,595	1,053	1,427	7,103	12,702

(*) Distributed generator data in 2005 includes a significant number of generators reported by one respondent that may be for residential applications.
Note: Distributed generators are commercial and industrial generators that are connected to the grid. They may be installed at or near a customer's site, or at other locations. They may be owned by either the customers of the distribution utility or by the utility. Other Technology includes generators for which technology is not specified.

Interviews conducted for this report (see Annex B) indicated the following about grid-connected DG:

- Grid connected DG was reported by only 0.9 percent of the customers of the companies interviewed.

- Three entities indicated they have some customers with storage capacity which comprised about 0.3 percent of their total customer base

- Non-dispatchable renewable generation was reported by only 1.4% of total customers.

Battery storage continues to pose cost-effectiveness problems by requiring a large degree of maintenance and adding significantly to the overall costs of building DG systems, and thus increasing the payback period. PHEVs are unlikely to play a significant role as a storage mode or as a distributed generator in the near term due to cost considerations. Some forecasts indicate that it will be at least 2020 before PHEVs hit the market in significant quantities (DOE/EIA 2008b). Microgrids could eventually provide resource capacity to supplement low-cost centralized facility power. Currently the number of microgrids is very small but is expected to grow to 5.5 gigawatts by 2025 [Metric 6].

Forecasts of utility-owned-and-controlled DG capacity indicate DG will reach 5.1 GW by 2010 (DOE/EIA 2007b) (see Figure 3.4). This forecast accounts for only about one-fifth of total DG based on 2006 data, which indicates the DG had more than 9 GW total. The trend however, is positive and significant.

3.2.2 Standard Distributed-Resource Connection Policy

Federal legislation attempting to deal with this issue emerged in progressively stronger language, culminating in the Energy Policy Act of 2005 (EPACT 2005), which requires all

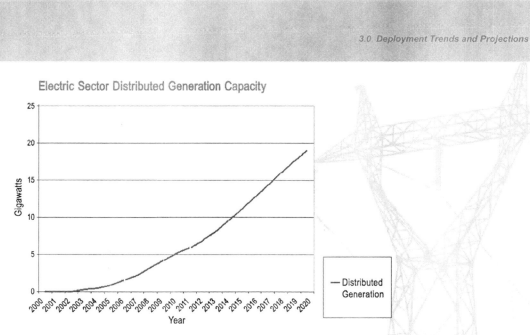

Figure 3.4. Projected DG Capacity in GW (Eynon 2002)

state and non-state utilities to consider adopting interconnection standards based on Standard IEEE 1547. IEEE 1547, which was published in 2003, looks strictly at the technical aspects of distributed generation interconnection, providing a standard that limits the negative impact of these resources on the grid [Metric 3] (Cook and Haynes 2006). In part to address some of the permitting aspects of interconnection, the FERC issued FERC Order 2006, which mandated that all public utilities that own transmission assets provide a standard connection agreement for small generators (under 20 MW) (FERC 2005).

While compliance with the FERC 2006 order is mandatory for public utilities that own transmission assets, other utilities have come under similar legislation at the state level. The progress in developing these laws, however, has been fairly slow. Even states complying with the mandatory FERC order have taken over two years to enact these relatively simple rules. States that have taken aggressive action on distributed generation have tended to do so for other reasons, such as meeting renewable-portfolio standard requirements.

In February 2008, the EPA did a study of the 50 states and the District of Columbia, assessing their standards for interconnection. They found 31 states with standard interconnection rules for distributed resources, and 11 additional states in the process of developing rules (see Figure 3.5). Of these, the EPA found that 55% had standard interconnection forms, 29% had simplified procedures for smaller systems, 35% had a set timeline for application approval, and 45% had larger system-size limits (over 10 kW for residential and over 100 kW for commercial systems) (see Figure 3.6) (EPA 2008b).

By multiplying the percentages above by the number of utilities in each state, it is estimated that roughly 61% of utilities have a standard interconnection policy in place, and that 84% of utilities either have a policy in place or will have one soon based on pending legislation or regulation (DOE/EIA 2002).

21

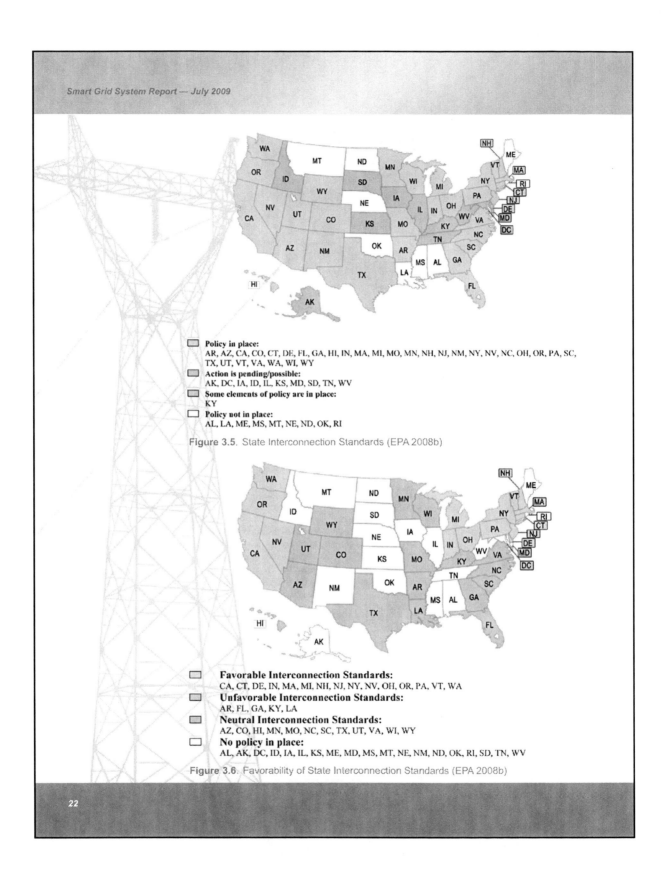

Policy in place:
AR, AZ, CA, CO, CT, DE, FL, GA, HI, IN, MA, MI, MO, MN, NH, NJ, NM, NY, NV, NC, OH, OR, PA, SC, TX, UT, VT, VA, WA, WI, WY
Action is pending/possible:
AK, DC, IA, ID, IL, KS, MD, SD, TN, WV
Some elements of policy are in place:
KY
Policy not in place:
AL, LA, ME, MS, MT, NE, ND, OK, RI

Figure 3.5. State Interconnection Standards (EPA 2008b)

Favorable Interconnection Standards:
CA, CT, DE, IN, MA, MI, NH, NJ, NY, NV, OH, OR, PA, VT, WA
Unfavorable Interconnection Standards:
AR, FL, GA, KY, LA
Neutral Interconnection Standards:
AZ, CO, HI, MN, MO, NC, SC, TX, UT, VA, WI, WY
No policy in place:
AL, AK, DC, ID, IA, IL, KS, ME, MD, MS, MT, NE, NM, ND, OK, RI, SD, TN, WV

Figure 3.6. Favorability of State Interconnection Standards (EPA 2008b)

22

The EPA's study based its criteria for favorability on whether or not standard forms were in place, time frames for application approval, insurance requirements, distributed resource sizes allowable, and interconnection study fees. With these factors considered, only 15 states were classified as having "favorable" interconnection standards, with 27 states either being "favorable" or "neutral." The fact that there are five states with unfavorable policies towards distributed generation is also cause for concern, although it is worth noting that that these states are all in the southeast region of the United States perhaps indicating a regional issue.

There are currently about 10 states with new DG interconnection standards under consideration (AK, DC, IA, ID, IL, KS, MD, SD, TN, and WV). Most projections show increasing deployment of these resources, especially in the commercial sector where power quality and power reliability are of increasing concern.

States and regions may have different regulations for the quality of the power being sold or the way the power is produced. Some states may value DG capacity differently from others and offer different subsidies and/or taxes based on those values. For example, Oregon state law has specific plant site emissions standards for minor sources emitting pollutants such as NO_x, SO_2, CO, particulate matter (PM), etc. whereas Ohio relies on the Best Available Technology (BAT) standard with specific limitations for PM and SO_2 based on location, generator type, and size (EEA 2004). The following are further examples of different policies for interconnection standards:

Only 15 states were classified as having "favorable" interconnection standards.

- California's progressive distributed-generation interconnection policies place no limits on the size of the resource. This is coupled with strong incentives for renewable sources of energy, such as photovoltaic solar panels, primarily for the purpose of promoting cleaner alternative power sources and reducing transmission congestion. California's policies have had a strong impact along the west coast (Shirley 2007).

- New York, which was one of the first states to adopt a standard interconnection policy in 1999, has continued to provide support for distributed generation. Among the driving forces for this have been power outages and transmission congestion, which continue to plague much of the state.

- The Mid-Atlantic Distributed Resources Initiative (MADRI), representing the utility interests of Delaware, the District of Columbia, Maryland, New Jersey, and Pennsylvania, has been a strong driver of interconnection standards for distributed resources and has proposed a model that has been adopted by many states.

- Many states in the Southeast region have been resistant to implementing favorable standards for interconnection (Figure 3.6). This resistance may be due to regional challenges that must be overcome specifically in those states, which would require special assistance.

The electricity industry's ability to accommodate a diverse range of generation types, including centralized and distributed generation as well as diverse storage options, is an important aspect of realizing a smart grid. The business case for real-time pricing needs to be made to end-users and interconnection standards need to be put in place universally before significant progress can be made to accommodate all generation options. Real-time pricing may be making a resurgence after declining from peaks in the early 1990s. Dynamic pricing and favorable interconnection standards are necessary to encourage more grid-connected distribution. Grid-connected DG provided 9,600 MW of generating capacity in 2006. Intermittent renewable energy resource DG needs more cost-effective storage to reach its

23

maximum potential. Once favorable interconnection standards are completed, DG will have more opportunity to expand. Currently, less than one-third of (only 15) states have standards that are favorable to DG and 27 states have interconnection standards that are neutral to favorable.

3.3 Enables New Products, Services, and Markets

Markets that are correctly designed and operated can efficiently reveal benefit-cost tradeoffs to consumers by creating an opportunity for competing services to bid. A smart grid accounts for all of the fundamental dynamics of the value/cost relationship. Some of the independent grid variables that must be explicitly managed are energy, capacity, location, time, rate of change, and quality. Markets can play a major role in the management of those variables. Regulators, owners/operators, and consumers need the flexibility to modify the rules of business to suit operating and market conditions.

> **Related Metrics**
> 1, 3, 4*, 6, 7, 8*, 9, 12, 17, 19*, 20*

While the primary objectives for implementing a smart grid may encompass environmental, energy efficiency, and national security goals, the effort falls short if utilities are unable or unwilling to make an effective business case to regulatory agencies. Smart-grid investments are often capital intensive, expensive, and include multiple jurisdictions within a utility's service area. While smart-grid investments can enable numerous new products (e.g., advanced meters, solar panels, electric vehicles, and smart appliances) and operational efficiencies (e.g., reduced meter reading costs, fewer field visits, enhanced billing accuracy, improved cash flow, and enhanced response to outages), such benefits may be difficult to quantify and to build into business cases given the nascent stages in which these technologies often exist, and the lack of industry standards and best practices for integrating smart-grid technologies.

Because a smart grid holds great potential for enabling new products, services, and markets, public and private interests have aligned in support of smart-grid technologies. For example, the Energy Independence and Security Act (EISA) provided incentives for utilities to undertake smart grid investments in Section 1306, which authorizes the Secretary of Energy to establish the Smart Grid Investment Matching Grant Program. This program was designed to provide reimbursement for up to 20 percent of a utility's investment in smart grid technologies. Section 1306 of EISA also defined what constituted a qualified investment and outlined a process for applying for cost reimbursement. Section 1307 of EISA encouraged states to require utilities prior to investing in non-advanced grid technologies to demonstrate consideration for smart grid investments. Section 1307 also encouraged states to consider regulatory requirements that included the reimbursement of book-value costs for any equipment rendered obsolete through smart grid investment [Metric 4].

Public and private interests have aligned in support of smart-grid technologies.

The private sector is also supporting smart grid investment [Metric 20]. This interest is spurred by several investment drivers:

- Significant increases in fossil fuel prices

- Peak demand growing at a time when energy infrastructure is in need of updating and replacement

- New infrastructure costs to meet load of approximately $500B over the next 20 years

24

- Shrinking capacity margin
- Increasing recognition of clean and efficient technologies.

These drivers suggest that in the future new products, services, and markets will be required to address the growing demand for energy over the long term. As a result, investment in smart-grid technologies has continued to gain traction in recent years. In 2008 alone, numerous significant venture-capital deals were announced:

- Optimal Technologies International, Inc. received $25 million towards the development of software for managing electrical grids.

- SmartSynch, Inc. secured $20 million to develop wirelessly communicating meters.

- Trilliant Incorporated secured $40 million toward the development of intelligent networks powering smart grid related functions.

- Tendril Networks received $12 million to develop smart grid networking products.

- Fat Spaniel Technologies received $18 million toward the development of an energy intelligence platform.

- GridPoint, Inc. received $15 million for their management of distributed storage, renewable generation, and load, bringing the firm's total funding to over $100 million.

- eMeter Corporation secured $12.5 million to support development of advanced metering technologies.

The surge in private-sector interest in smart-grid investment was validated using data from Cleantech Group LLC, which reported venture-capital funding secured by smart-grid startups of $194.1 million in 2007 and $129.3 million during the first two quarters of 2008.[1] In total, the Cleantech Group identified 99 deals during the 2000-2008 timeframe totaling $964.4 million; (the average deal was $9.7 million). Figure 3.7 documents recent trends in venture-capital funding of firms

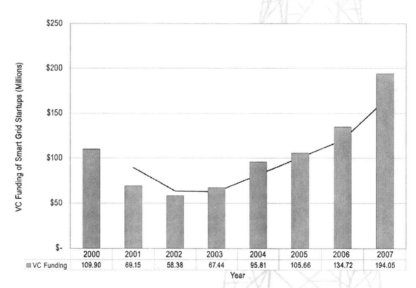

	2000	2001	2002	2003	2004	2005	2006	2007
VC Funding	109.90	69.15	58.38	67.44	95.81	105.66	134.72	194.05

Year

Figure 3.7. Venture-Capital Funding of Smart-Grid Startups (2002-2007)

(1) per email communication with Brian Fan of the Cleantech Group on September 10, 2008.

25

developing smart-grid technologies. As shown, venture-capital funding of startups slumped between 2000 and 2002 but has since rebounded, growing from $58.4 million in 2002 to $194.1 million in 2007. Cleantech Group data suggest that 2008 levels could well exceed those of 2007 as venture-capital funding has topped $129 million in the first two quarters of 2008. Between 2002 and 2007, venture-capital funding of smart-grid startups grew at an average annual rate of 27.2 percent.

3.3.1 Enabling New Products and Services

A smart grid enables new products and services through automation, communication sharing, facilitating and rewarding shifts in customer behavior in response to changing grid and market conditions, and encouraging development of new technologies (e.g., AMI, plug-in hybrid electric vehicles). For example, Carnegie Mellon's Electricity Industry Center reports that there are now approximately 12 million backup generators in the United States, representing 200 GW of generating capacity [Metric 7]. This backup generating capacity is growing at a rate of 5 GW per year (Gilmore and Lave 2007). A smart grid, by enabling the bi-directional flow of energy in combination with programs allowing customers to use these backup generating devices to sell energy back to the grid during high-cost peak periods, could enhance the market for these new products and the services they provide.

Approximately 4.7% of all U.S. customers are currently served by AMI.

A smart grid that incorporates real-time pricing structures and bi-directional information flow through metering and information networks is expected to support the introduction of numerous technologies into the system [Metric 20]. Enabling AMI technology itself represents a major driver in smart-grid investment, as evidenced by several large-scale deployment programs:

- The three largest utilities in California are installing millions of smart meters at homes and businesses and charging customers $4.6 billion for the enhanced service.

- Duke Energy is installing 800,000 smart meters.

- Texas utility Oncor is installing smart meters at a cost of $690 million.

- Pacific Gas and Electric is retrofitting 9 million meters with communications electronics to enable TOU pricing.

- The Los Angeles Department of Water and Power has purchased 9,000 smart meters to enable transmission of real-time data through public wireless networks (Wesoff 2008).

Approximately 4.7% of all U.S. customers are currently served by AMI with states in the Mid-Atlantic and Midwest experiencing the highest rates of usage at around 8-11 percent. Pennsylvania has the highest penetration with 23.9 %.

The smart grid also supports the deployment of new vehicle technologies (e.g., EVs and PHEVs) [Metric 8].[1] Real-time pricing enabled through AMI would allow customers to recharge vehicles at reduced cost during off-peak hours. Bi-directional metering would enable customers to purchase energy at off-peak hours and sell unused, stored energy back to the utility during peak periods at higher rates. These two elements could feasibly enhance the customer's return on investment (ROI) for EV and PHEV technologies, and accelerate market penetration. Note, technical challenges with regard to battery performance due to charge and discharge cycles need further investigation and remediation.

---

(1) The PHEV is a hybrid electric vehicle with batteries that can be recharged when plugged into the electric wall outlet and an internal combustion engine that can be activated when batteries require recharging.

26

The market penetration of EVs and PHEVs demonstrates the potential application of a new technology enabled by the smart grid. Table 3.3 shows that the number of EVs reached 28,891 in 2006, representing roughly .01 percent of all light-duty vehicles in use. Light-duty vehicles include automobiles, vans, pickups, and sport utility vehicles (SUVs) with a gross vehicle weight rating of 8,500 pounds or less.[1] The U.S. DOE does not estimate current PHEV sales. There are several companies that perform aftermarket conversions (e.g., Amberjac Products, Hybrids-PlusTM, Plug-In Conversions Corp.) but there are no original-equipment manufacturers (OEMs) currently marketing PHEVs. Recent announcements by the automotive industry suggest that PHEVs and EVs will be commercially available in the 2010 to 2012 timeframe. PHEV sales are forecast by DOE to reach 237, 212 (1.4% of light-duty vehicle sales) by 2020 and 443,207 (2.2% of light-duty vehicle sales) by 2030.

Table 3.3. EV and PHEV Market Penetration (DOE/EIA 2009c)

Year	EVs On-Road Total in Use	EVs On-Road % of Light-Duty Vehicles	PHEVs On-Road Total in Use	PHEVs On-Road % of Light-Duty Vehicles	EV Sales Total Suites	EV Sales % of Light-Duty Market	PHEV Sales Total Sales	PHEV Sales % of Light-Duty Vehicles
2006	28,891	0.01%	-	0.00%	173	0.00%	-	0.00%
2010	24,247	0.01%	35,526	0.02%	130	0.00%	35,526	0.26%
2015	17,840	0.01%	442,570	0.18%	149	0.00%	139,164	0.86%
2020	11,453	0.00%	1,322,438	0.51%	153	0.00%	237,212	1.43%
2025	6,787	0.00%	2,701,419	0.98%	165	0.00%	350,386	1.95%
2030	4,351	0.00%	4,282,767	1.44%	184	0.00%	443,207	2.21%

Table 3.3 presents a forecast to 2030 of the number of EVs and PHEVs operating on-road based on the Energy Information Administration's (EIA) Annual Energy Outlook 2009, Early Release. As shown, the number of light-duty EVs in use reached nearly 29,000 in 2006 but is forecast to decline to 4,351 by 2030. The decline in EVs in use does not reflect a trend away from alternative vehicle technologies but rather a transition towards more competition among alternative technologies, some of which have not yet entered the marketplace (e.g., PHEVs). The PHEV share of on-road light-duty vehicles is forecast by U.S. DOE to grow slowly through 2030, reaching 4.3 million (DOE/EIA 2008b).

The U.S. DOE forecast presented in the Annual Energy Outlook is conservative compared to a small number of recent forecasts prepared by industry. While most forecasts estimate ultimate hybrid-electric and EV penetration of the light-duty vehicle market in the 8-16 percent range, (Greene et al. 2004), the EPRI and Natural Resources Defense Council (NRDC) were more aggressive, estimating PHEV market penetration rates ranging between 20% and 80% by 2050 (medium PHEV scenario estimate of 62% in 2050). EPRI and NRDC used a consumer choice model to estimate market penetration rates (EPRI/NRDC 2007).

A report recently prepared for the U.S. DOE presented and examined a series of PHEV market-penetration scenarios given varying sets of assumptions governing PHEV market potential. Based on input received from technical experts and industry representatives contacted for the U.S. DOE report and data obtained through a literature review, annual market-penetration rates for PHEVs were forecast from 2013 through 2045 for three

[1] The definition of light-duty vehicles includes motorcycles. Although electric motorcycles are commercially available, plug-in hybrid motorcycles are unlikely to be pursued as a product. Therefore, we omitted motorcycles from this analysis.

27

scenarios. Under the scenario that examines market penetration assuming all the current U.S. DOE goals for PHEV development are achieved (e.g., $4,000 marginal cost of PHEV technology over existing hybrid technology, 40 miles all-electric range, 100 miles-per-gallon equivalent, and that PHEV batteries meet industry standards regarding economic life and safety), PHEV market penetration is forecast to ultimately reach 30 percent of new light-duty vehicle sales, reaching 9.9 percent by 2023 and 27.8 percent by 2035 (Balducci 2008).

The forecasts in Balducci (2008) and EPRI/NRDC (2007) were designed with scenarios based on increasingly aggressive assumptions. These scenarios assume that the PHEV will ultimately become the dominant alternative fuel vehicle. The EPRI/NRDC study was focused on the potential environmental impact of PHEV market penetration. Therefore, aggressive assumptions were required under some of the scenarios to generate a reasonably significant and measurable environmental impact. Neither study presents the scenarios as definitive or assigns probabilities to their outcomes. Rather, the studies are designed to measure the impact, or in the case of Balducci (2008) estimate the penetration rate, given certain sets of assumptions. If the goals outlined in Balducci (2008) are not reached, market penetration rates would certainly be lower than estimated. DOE estimates are generated by the National Energy Modeling System (NEMS), which does not use aggressive assumptions to determine the market potential of PHEVs. Instead, the light-duty alternative fuel vehicle market is forecast by NEMS to be dominated by diesel, flex fuel, and hybrid electric vehicles, not PHEVs.

A smart grid will also enable consumer-oriented "smart" equipment, including communicating thermostats, microwaves, space heaters, refrigerators, clothes washers and dryers, and water heaters [Metric 9]. This equipment will be fitted with signaling software, or more specifically firmware, which enables the device to communicate with other components of a smart grid. These technologies will allow the customer and/or the utility or other authorized third parties to dynamically control the device's energy consumption based on energy prices and grid conditions.

In addition to specific appliances, this category of equipment encompasses other devices including meters, switches, power outlets and various other controllers that could be used to retrofit or otherwise enable existing equipment to respond to smart-grid conditions. For example, a new "smart" water heater may be equipped with a device that coordinates with the facility's energy management system to adjust temperature controls within user-specified limits based on energy prices. These technologies are under development but not yet commercially available on a widespread basis.

There are a number of other technologies that are currently commercially available that take advantage of smart grid features. For example, solar panels can be easily installed on rooftops by homeowners and safely generate power for years. Solar power can generate power at a cost of $10 to $12 per watt [Metric 7c] (Solar Guide 2008). These costs could be much lower in the future.

The new products and services highlighted in this section depend on regulatory recovery for smart grid investments. Historically, utilities have been rewarded for investment in capital projects and energy throughput. That is, expanded peak demand has driven the need for additional capital projects, which increase the rate base. As energy sales grow, revenues increase. Both factors run counter to encouraging smart grid investments. Thus, regulatory frameworks can discourage energy efficiency, demand reduction, demand response, distributed generation, and asset optimization (Anders 2007).

New products and services depend on regulatory recovery for smart grid investments.

Electricity service providers participating in interviews conducted for this report indicated that, on average, they are recovering 8.1 percent of their investment through rate structures but predict that regulatory recovery rates will expand significantly in the coming years, ultimately reaching 90 percent. In addition, there are opportunities for expanded smart grid investment when sales are decoupled from revenues. When states decouple sales from revenues, a significant disincentive for utility investment in energy efficiency measures, including those that may be enabled by smart grid, is removed. In addition, consumer concerns that efficiency measures should reduce electric bills also needs to be addressed. This may be done in part by energy efficiency programs.[1] There are currently 10 states with energy efficiency programs where decoupling is not used, 11 states with energy efficiency programs where decoupling was proposed but not adopted, three states plus the District of Columbia with energy efficiency programs where decoupling is being investigated, nine states with energy efficiency programs where decoupling has been approved for at least one utility, and one state with no energy efficiency program where at least one utility has been approved for decoupling [Metric 4].

3.3.2 Enabling New Markets

A smart grid enables a more efficient allocation of resources through the use of information systems enabling communication between the grid and "smart" appliances, distributed generation units, and other consumer-oriented devices. Further, a smart grid rewards customers who engage in load-shifting behavior through the use of advanced meters, communication of real-time prices, and other incentive structures. In doing so, a smart grid establishes markets to manage these resources and reduce costs to consumers and utilities.

Advanced metering technology is a key facilitator of new markets through its ability to record energy usage at finer time intervals [Metric 12]. Bi-directional communication with the meters and with customers enables an exchange of information to support transactions with customers who can alter their consumption and may even generate excess energy to sell energy back to the grid. Smart-grid technology increases the accuracy of pricing policies, demand forecasts, and responses to grid disturbances and outages. The exchange of real-time prices and market data allows utility customers unprecedented access to information that, when acted upon, may impact energy costs and promote electric system savings. As noted previously, AMI penetration has reached 4.7 percent of total electric meters. In some areas (e.g., Midwestern states) AMI penetration rates have reached 8-11 percent. Pennsylvania has the highest penetration with 23.9 % (FERC 2008).

A smart grid, with "smart" meters, appliances, and real-time information exchange between customers and service providers uses dynamic pricing programs to encourage energy efficiency and load shifting. The most prevalent pricing strategies include time of use (prices are differentiated based solely on a peak versus off-peak period designation), critical peak price (higher critical peak price is restricted to a small number of hours each year), and real time pricing (hourly prices vary based on the day-of (real-time) or day ahead cost of power to the utility [Metric 1]. These pricing strategies incentivize investment in a broad spectrum of energy efficiency programs and equipment by offering customers the opportunity to shift load and reduce marginal energy cost.

A smart grid establishes markets to manage resources and reduce costs to consumers and utilities.

(1) A policy and program framework to encourage greater end-use energy efficiency can be found in (EPA 2008c).

29

Demand-response equipment also enables the design and function of new markets. Demand response attempts to capture why consumers want electric power, when they want it, and how much they are willing to pay for this consumption. Demand response is typically viewed from a system operations point of view as a form of additional capacity and is discussed in terms of MW. Thus, a 2008 FERC survey estimated that the potential reduction due to such demand-response programs is approximately 41,000 MW per year, an increase of 9 percent and 5.8 percent of U.S. peak demands (FERC 2008).

Microgrids also represent a new smart grid-enabled market area [Metric 6]. A microgrid is a distribution system with distributed energy sources, storage devices and controllable loads, which may generally operate connected to the main power grid but is capable of operating as an island. From the grid's perspective, the primary advantage of a microgrid is that it can operate as a single collective load within the power system. Customers benefit from the quality of power produced and the enhanced reliability over relying solely on the grid for power. In the U.S. Department of Energy's vision of the electric power infrastructure (Grid 2030), microgrids are one of the three technical cornerstones. Microgrids are envisioned as local power networks that use distributed energy resources and manage local energy supply and demand. In 2006, less than 0.1% of the nation's generation capacity was met by microgrids, indicating that this is a nascent aspect of smart grid deployment.

The ability to better manage where power is going, how it is being used, and when it is being used also enables markets for premium power [Metric 17]. That is, managing load served by service type, such as firm service or interruptible service and their corresponding tariffs will enable utility and government agencies to differentiate between consumer types, enable demand curve estimation, and identify energy consumption schedules.

The cost of connecting and configuring smart devices and systems into the electric grid remains an obstacle to the high volume penetration levels anticipated. For automation components to connect and work, alignment is needed in communications networks, information understanding, business processing, and business and regulatory policy (see Figure 3.8). This alignment results in interoperability and it is aided by integration methods, tools, as well as adherence to standards and agreements that cover all these aspects [Metric 19]. Given the fast pace of change in technology, the evolving and competitive approaches to conducting business, and the local, state, and federal aspects of policymaking, interoperability cannot rest on a fixed set of standards, but requires a flexible framework, much like contract law. Progress is being made at technical levels with

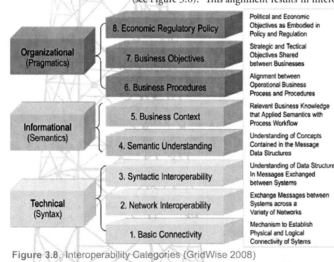

Figure 3.8. Interoperability Categories (GridWise 2008)

openly available communications architectures and standards, while work is just beginning to understand the information and business processes involved in application areas such as demand response.

Determining a quantifiable measurement of progress to improving interoperability for a smart grid is difficult; however, significant progress has been made in educating stakeholders on the nature of the issues and their importance. In the Energy Independence and Security Act of 2007, NIST was given the directive to develop an interoperability framework of protocols and standards to support a smart grid. Stakeholders are assembling to contribute and align their ideas for such a framework. As interoperability improvement is akin to software quality improvement, a more quantifiable measurement based upon a Capability Maturity Model (SEI 2008) is proposed for future reports. With such a model, assessments can be made by interviewing a representative sample of smart grid projects.

3.4 Provides Power Quality for the Range of Needs

Not all commercial enterprises, and certainly not all residential customers, need the same quality of power. Examples of power quality (PQ) issues include voltage sag, flicker, and momentary interrupts. Some customers have critical computer systems and complex processes that require high PQ, while others, such as most residential customers would not appreciate paying for better PQ. A smart grid supplies varying grades of power and supports variable pricing accordingly. The cost of premium PQ features can be included in the electrical service contract. Advanced control methods monitor essential components; enabling rapid diagnosis and precise solutions to PQ events, such as arise from lightning, switching surges, line faults and harmonic sources. A smart grid also helps buffer the electrical system from irregularities caused by consumer electronic loads.

When consumers consider PQ, they are typically concerned with the ability of the electrical grid to provide a continuous flow of energy with a quality to power all their electrical requirements. Not all customers, however, have the same energy needs. Residential customers tend to be affected more by sustained interruptions while commercial and industrial customers are troubled most by sags and momentary interruptions. With greater flexibility to locally target power quality resources, the ability to offer several pricing levels for varying grades of power can be considered. For those customers who are deemed power sensitive, the extra cost of premium power would be a worthwhile investment when compared with the lost revenue due to a loss of power.

A smart grid enables enhanced PQ through a number of specific technologies and approaches, including:

- PQ meters
- System-wide PQ monitoring
- Smart appliances
- Premium-power programs
- Demand-response programs
- Storage devices (e.g., batteries, flywheels, superconducting magnetic energy storage)

Not all customers need the same quality of power.

31

- Power electronic devices with the capacity to correct waveform deformities

- Monitoring systems used to identify system health and correct impending failures

- New distributed-generation devices with the ability to provide premium power to sensitive loads (NETL 2007)

- Active control of voltage regulators, capacitor banks, and inverter-based distributed generation and storage to manage voltage and VARs

- Remote fault isolation

- Dynamic feeder reconfiguration

- Microgrids

- Distribution state estimation

Together, these technologies and other elements of a smart grid could offer tremendous benefits to energy consumers through cost avoidance and associated productivity gains. While PQ is generally viewed in terms of both disruptions and disturbances, this section focuses entirely on PQ issues relating to power disturbances. See Section 3.6 for a discussion of power disruptions.

Related Metrics

5, 6*, 7, 9, 11, 17*

3.4.1 The Cost of Poor Power Quality

Power quality incidents in the past were often rather difficult to observe and diagnose due to their short interruption periods. The increase in power-sensitive and digital loads has forced us to more narrowly define PQ. For example, ten years ago a voltage sag might be classified as a drop by 40% or more for 60 cycles, but now it may be a drop by 15% for five cycles (Kueck et al. 2004).

There have been several PQ studies completed in the U.S. over the past 30 years [Metric 17]. The two most widely cited studies were the 1969-1972 Allen-Segall (IBM) study and the 1977-1979 Goldstein-Speranza (AT&T study). A third more recent and considerably larger study was conducted by the National Power Laboratory (NPL) in the earlier 1990's. The consistent conclusion among all three aforementioned PQ studies was that disturbances are a practical reality, and there is a need for different grades of power to protect sensitive loads (Dorr 1991).

Comparing the studies and assessing trends, however, is more difficult, as each study uses different definitions, parameters, and instrumentation. NPL filtered data to compare it with that of IBM and then the AT&T surveys in their PQ paper to examine trends in disturbances and outages. When the data examined by NPL were compared to both the IBM and AT&T studies, the NPL research team found a decrease in total disturbances per month but an increase in outages and sag disturbances. Thus, the data suggest the electrical grid has improved in terms of its ability to provide clean power free of disturbances but has become less capable over time to meet the growing demand placed on it and provide an uninterrupted power supply to electricity consumers.

Loss and fluctuations in power cause users to lose valuable time and money each year.

A loss of power or a fluctuation in power causes commercial and industrial users to lose valuable time and money each year. Cost estimates of power interruptions and outages vary. A 2002 study prepared by Primen concluded that power quality disturbances alone cost the US economy between $15-$24 billion annually (McNulty and Howe 2002). In 2001 EPRI

32

estimated power interruption and power quality cost at $119 billion a year (EPRI 2001), and a more recent 2004 study from Lawrence Berkeley National Laboratory (LBNL) estimated the cost at $80 billion a year (Hamachi LaCommare and Eto 2004).

3.4.2 Smart-Grid Solutions to Power Quality Issues

Interviews were conducted for this report with 21 companies meeting an annual peak demand of 150,000-175,000 megawatts and 0.8-1.2 billion megawatt hours of generation served. The companies were asked to estimate the percentage of customer complaints related to PQ issues (excluding outages). The utilities indicted that 3.1 percent of all customer complaints were related to PQ issues.

A smart grid can address PQ issues at various stages in the electricity delivery system. For example, smart-grid technologies address transmission congestion issues through demand response and controllable load. Smart-grid-enabled distributed controls and diagnostic tools within the transmission and distribution system help dynamically balance electricity supply and demand, thereby helping the system to respond to imbalances and limiting their propagation when they occur [Metric 11]. This reduces the occurrence of outages and power disturbances attributed to grid overload.

Smart-grid-enabled distributed controls and diagnostic tools help the system dynamically respond to power imbalances.

There are a number of technologies that serve to automate the transmission and distribution system and are enabled by a smart grid, including: Supervisory Control and Data Acquisition (SCADA) technologies, remote sensors and monitors, switches and controllers with embedded intelligence, and digital relays [Metric 11]. Nationwide data has shown that transmission automation has penetrated the market, while distribution automation is primarily led by substation automation, with feeder automation still lagging. Recent research shows that while 84% of utilities had substation automation and integration plans underway in 2005 and about 70% had deployed SCADA systems to substations, the penetration of feeder automation is still limited to approximately 20% (ELP 2008; McDonnell 2008).

Microgrids are also serving to enhance PQ at specific sites [Metric 6]. Technologic, regulatory, economic, and environmental incentives are changing the landscape of electricity production and transmission in the United States. Distributed production using smaller generating systems, such as small-scale combined heat and power (CHP), small-scale renewable energy sources (RES) and other DERs can have energy efficiency, and therefore, environmental advantages over large, central generation. The growing availability of new technologies in the areas of power electronics, control, and communications supports efforts in this area. These new technologies enable small power generators, typically located at user sites where the energy (both electric and thermal) they generate is used, to provide sources of reliable, quality power, which can be organized and operated as microgrids.

A microgrid is defined as a distribution system with distributed energy sources, storage devices, and controllable loads, that may generally operate connected to the main power grid but is capable of operating as an island. Currently, approximately 20 microgrids can be found at universities, petrochemical facilities and U.S. defense facilities. According to RDC (2005), the microgrids provided 785 MW of capacity in 2005. They noted additional microgrids that were in planning at the time as well as demonstration microgrids. RDC also noted that by examining the Energy Information Administration's database they could determine approximately 375 potential sites for microgrids if they weren't already microgrids. Outside of the petrochemical microgrids, there are no commercial microgrids in the United States

33

(PSPN 2008). Given EIA's net summer capacity of 906,155 MW and assuming no devolution of microgrid capacity from 2005, the percentage of capacity met by microgrids is about 0.09% in 2006.

Table 3.4. Capacity of Microgrids in 2005 (MW) (RDC 2005)

	University	Petrochemical	DoD
Capacity (MW)	322	455	8

Navigant Consulting, in their base case scenario, projected 550 microgrids installed and producing approximately 5.5 GW by 2020 (Navigant 2005) or about 0.5% of projected capacity (DOE/EIA 2009a). Navigant (2005) predicts a range of 1 13 GW depending on assumptions about pushes for more central power, requirements and demand for reliability from customers and whether there is a environmental requirement for carbon management.

Grid-connected distributed generation (DG) and storage technologies can enhance PQ due to their smaller scale, localized support for power generation and distribution systems, and potential ability to respond to power disruptions and disturbances (e.g., islanded operation). These technologies include power generators, such as wind turbines connected at the distribution system level, micro hydro installations, solar panels, and gas microturbines. These distributed generators produce power for onsite or adjacent consumption and could sell surplus power back into the grid under an established fee-in tariff. These technologies also include energy storage devices such as batteries and flywheels, which could be used to store energy produced or purchased during off-peak hours and then sold or consumed on-peak. While these technologies have considerable potential for enhancing PQ, distributed generation capacity is currently a small part of total power generation, with combined total distributed generation capacity reaching 12,702 megawatts in 2007 [Metric 7] (DOE/EIA 2007).

The ability to track where power is going, what is being done with it, and when it is being used is paramount to addressing PQ issues. Further, the tracking of load served by service type, such as firm service or interruptible service, and their corresponding tariffs (fixed or marginal-cost based) will enable utility and government agencies to discriminate between consumer types, enable demand-curve estimation, and identify energy-consumption schedules.

According to estimates published in the 2008 Annual Energy Outlook, residential and commercial energy sales are expected to outpace industrial energy sales (DOE/EIA 2008a). With both residential and commercial energy demands approaching approximately double their 1995 values by 2030, the ability to disaggregate and track not only who is consuming the most energy, but how it is being consumed, will become an increasingly more valuable asset of a smart grid as utility and government agencies strive to further increase energy efficiencies, manage ever-increasing loads, and provide high PQ.

Load management involves demand-response equipment that can respond to load conditions [Metric 5]. There are a number of organizations (e.g., Electric Reliability Council of Texas, Public Utility Commission of Texas) that act to balance and curtail loads to avoid and manage power disruptions and disturbances. Nationally, however, demand response is low. Table 3.5 shows the number of entities with demand response programs.

34

Table 3.5. Entities Offering Load-Management and Demand-Response Programs (FERC 2008)

Type of Program	Number of Entities
Direct Load Control	209
Interruptible/Curtailable	248
Emergency Demand Response Program	136
Capacity Market Program	81
Demand Bidding/Buyback	57
Ancillary Services	80

Grid-responsive demand-side equipment includes "smart" appliances (e.g., communicating thermostats, microwaves, space heaters, hot water heaters, refrigerators) and other devices, including switches, power-outlets, and various other controllers that could be used to retrofit or otherwise enable existing equipment to respond to smart grid conditions. This type of equipment enhances power quality by enabling customers, utilities, and/or third parties to dynamically control energy consumption based on energy prices and grid conditions. A recent smart grid experiment conducted by the U.S. Department of Energy, which tested thermostats, washers and dryers, and water heaters fitted with "smart" grid-responsive equipment, found these "smart" devices reduced load fluctuations, decreased peak loads, and significantly reduced energy costs. However, only approximately 8% of U.S. energy customers have any form of time-based or incentive-based price structure that would enable customers to reap the benefits associated with load shifting behavior [Metric 9] (FERC 2008). Only 5 percent were reported as having a time-based rate in the 2006 FERC Survey (FERC 2006a).

3.5 Optimizes Asset Utilization and Operating Efficiency

Related Metrics
2, 3, 5, 7, 11*, 13, 14*, 15*, 16*

One of the key features of a smart grid is its lower costs of operations, maintenance, and expansion compared with those of traditional forms of operation. A smart grid is able to optimize operating efficiency and utilization of assets by employing advanced information and communication technologies; this allows better monitoring of equipment maintenance, minimizes operation costs, and "replaces iron with bits" (DOE/OEDER 2008b) by reducing the need for increased generation and infrastructure through demand-response measures and other technologies.

This section looks at asset utilization and operating efficiency of the bulk generation, transmission and distribution delivery infrastructure, and the distributed energy resources in the electric system. It concludes with an overall view of system efficiency.

3.5.1 Bulk Generation

The United States crept closer to its generation capacity limits for at least the ten years preceding 1998-2000, according to NERC, but reversed that trend during the next 5 years and returned to more conservative capacity factors [Metric 14]. Figure 3.9 shows measured and predicted winter and summer peak generation capacity factors from 1999 and projected

35

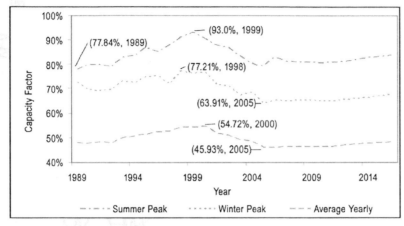

Figure 3.9. Measured and Predicted Peak Summer, Peak Winter, and Yearly Average Generation Capacity Factors in the U.S. (NERC 2008)

to 2014. It indicates that, after a recent decline, the generation capacity factors are predicted to increase slightly in the next 8 years. The large differential between available capacities and average capacity is built to accommodate a few hours of peak demand during winter and summer regionally.

For bulk generation, efficiencies for coal, petroleum, and gas remain almost constant for the last 20 years; there is no new breakthrough in sight (see Figure 3.10). The combination of coal, petroleum, and natural gas makes up about 80% of the nation's electric power-generation base [Metric 15].

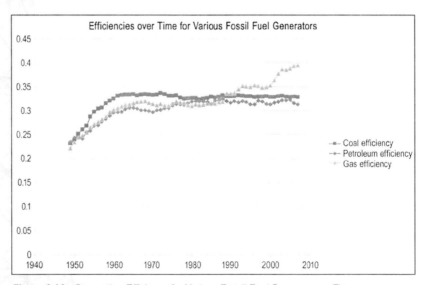

Figure 3.10. Generation Efficiency for Various Fossil Fuel Sources over Time (DOE/EIA 2007a)

Table 3.6 shows the 2006 and projected 2008 peak demand and generation capacities. The grid currently runs with a generation capacity factor of about 46%.

Chapter 7: Smart Grid System Report, U.S. Department of Energy

Table 3.6. Measured and Projected Peak Demands and Generation Capacities for Recent Years in the U.S., and Calculated Capacity Factors (NERC 2008)

Measurement	2006 Measured	2008 Projected
Summer peak demand (MW)	789,475	801,209
Summer generation capacity (MW)	954,697	991,402
Capacity factor, peak summer (%)	82.69	80.82
Winter peak demand (MW)	640,981	663,105
Winter generation capacity (MW)	983,371	1,018,124
Capacity factor, peak winter (%)	65.18	65.13
Yearly energy consumed by load (GWhr)	3,911,914	4,089,327
Capacity factor, average (%)[1]	46.08	46.46

(1) The average of the NERC (2006) summer and winter capacities was used for this calculation.

3.5.2 Delivery Infrastructure

T&D automation devices communicate real-time information about the grid and their own operation and then make decisions to bring energy consumption and/or performance in line with their operator's preferences. These smart devices, which exchange information with other substation devices or area control centers, can increase asset utilization and smart-grid reliability as well as reduce operating expenses by increasing device and system responsiveness to grid events. T&D automation devices can aid in reducing the differential between average load and peak load. Recent research found that about 60% of the control centers in North America have linkages with other utilities [Metric 2] (Newton-Evans 2008).

Data from utilities across the nation show a clear trend of increasing T&D automation and increasing investment in these systems. Key drivers for the increase in investment include operational efficiency and reliability improvements to drive cost down and overall reliability up. The lower cost of automation with respect to T&D equipment (transformers, conductors, etc.) is also making the value proposition easier to justify. With higher levels of automation in all aspects of the T&D operation, operational changes can be introduced to operate the system closer to capacity and stability constraints [Metric 11].

Results of interviews undertaken for this report (see Annex B) indicate that:

- 28% of the total substations owned were automated
- 46% of the total substations owned had outage detection
- 46% of total customers had circuits with outage detection
- 81% of total relays were electromechanical relays
- 20% of total relays were microprocessor relays (presumed rounding error)

Other nationwide data has shown that transmission automation has already penetrated the market highly, while distribution automation is primarily led by substation automation, with feeder equipment automation still lagging. Recent research shows that while 84% of utilities had substation automation and integration plans underway in 2005, and about 70% of utilities had deployed SCADA systems to substations, the penetration of feeder automation is still limited to about 20% (ELP 2008; McDonnell 2008). Because feeder automation lags other automation efforts so significantly, this should be an area addressed directly in future work.

37

A significant component of the measurement, analysis, and control of the T&D infrastructure relates to control centers at the transmission and distribution levels of the system (SCADA, energy management systems – EMS, and distribution management systems - DMS). According to a recent survey by Newton-Evans Research, almost all utilities with over 25,000 customers have SCADA/EMS systems in place, while only about 17% of utilities have DMS systems (Newton-Evans 2008). One smart grid trend is to integrate other functions with these centers. For example, about 30% of the SCADA/EMS systems are linked to Distribution Automation/DMS. Figure 3.11 shows the projected integration of EMS/SCADA/DMS systems to a variety of other data systems.

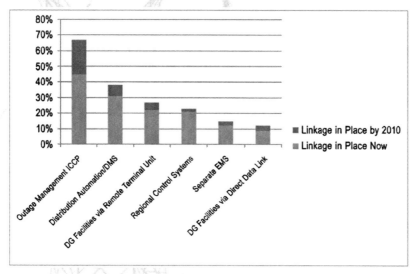

Figure 3.11. Current/Future Plans for Connecting EMS/SCADA/DMS Systems to Other Data Systems (Newton-Evans 2008)

The investment in T&D automation can be estimated either from total industrial output of specific automation products to US markets or from the receiving demand side (utility company) as purchases. Market statistics for T&D investment already exist and could be readily utilized. Newton-Evans Research provides market-volume estimates on automation products aggregated to categories such as shown in Figure 3.12. The figure shows that significant increases in T&D automation are expected between 2007 and 2010. For example, spending on distribution automation is expected to almost triple by 2010 to nearly $180 million. Protective relays are expected to increase 25% to $235 million, feed-switch investment by 225% to $65 million, control-center upgrades by 29% to $155 million; and substation investment by 35% to $540 million (Newton-Evans 2008).

Data sharing from the field and between control centers and reliability coordination centers improves the true operational view of the system. Without an accurate view, operating procedures are developed with engineering buffers that allow for inaccuracy or unpredictable situations. The level of situation awareness is also being raised by advanced measurements, such as synchro-phasors, that are beginning to be shared across large regions of the country. The North American Synchro-Phasor Initiative (NASPI) reports that in 2008, 175 phasor measurement units are operating in North America [Metrics 2, 13]. In addition, control centers have more data to gather from the field with the growth in T&D automation. This further reduces the number of system disruptions, the downtime from a disruption, and the total impact of such an occurrence [Metric 11]. Distribution automation investment is expected to triple to almost $180 million in 2010, while transmission automation investment is expected to increase by 35% to $540 million.

Dynamic line ratings [Metric 16] can also help increase grid utilization by allowing the delivery infrastructure to operate closer to its true limits. Concern rises as long-term growth of transmission capacity is dramatically short of keeping up with growth of peak demand. Forecasts predict there will be a less than 1% increase in total miles of transmission cables and GW-miles between 2002 and 2012 (Hirst 2004). Dynamic line ratings have the potential to provide an additional 10-15% transmission capacity 95% of the time and fully 20-25% more transmission capacity 85% of the time (Seppa 1997). Currently, only a small fraction of the nation's transmission lines are monitored to support dynamic line ratings. The interviews of electricity service providers conducted as part of this report (see Annex B) reveal that, on average, only 0.5% of respondents' transmission lines were dynamically rated, and that number dropped to 0.3% when weighted by the number of customers served by each respondent.

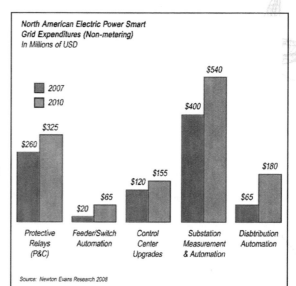

Figure 3.12. North American Electric Power T&D Automation Expenditures (in Millions of USD) (Newton-Evans 2008; Ockwell 2008)

3.5.3 Distributed Energy Resources

Smart grid applications, such as demand response [Metric 5] and grid-connected distributed generation (DG) [Metric 7], should also improve grid operating efficiency by controlling load and adding localized resources when required. In order for this to occur, favorable DG interconnection standards are needed [Metric 3]. Currently the amount of load-managed distributed generation has been declining since 1995 and is currently just above 1% of net summer capacity. Grid-connected DG, on the other hand, increased 134% between 2004 and 2007. This still represents only 1.6% of summer peak capacity. Once favorable interconnection standards are approved by all states, the amount of DG should become a more significant portion of grid capacity. Currently only 15 states have favorable standards although significantly more have interconnection standards.

At present about 10 states are considering new DER interconnection standards, and it is estimated that 85% of utilities will have a policy in place in the near future [Metric 3]. Only 15 states have what are considered favorable interconnection standards (EPA 2008b).

The demand for smart technologies will only increase as grid demand increases. In fact, current trends suggest a significant increase in load in the

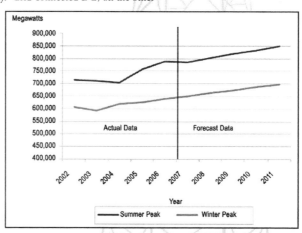

Figure 3.13. Peak Demand for the Contiguous U.S. (DOE/EIA 2007c)

Smart Grid System Report — July 2009

near future. For example as shown in Figure 3.13, the Energy Information Administration predicts that the winter peak demand for energy will increase to almost 700,000 megawatts by 2011 (DOE/EIA 2007c).

3.5.4 Overall System Efficiency

Currently, gross annual measures of grid operating efficiency have been steady or improving slightly as the amount of energy lost in generation dropped 0.6 percentage points to 67.7% in 2007 and transmission and distribution losses improved very slightly by 0.05 percentage points to 9.44% of net generation [Metric 15]. Presently, load is growing at almost double the rate of growth in transmission capacity; however, most regions have very limited plans to expand generation and transmission facilities. Using traditional planning and operations practices, the current delivery infrastructure is not capable of bringing renewable-energy generation online at a capacity that is consistent with the amount of construction.

Figure 3.14 shows overall grid operating efficiency. In this figure T&D losses are shown to be approximately 1.34 quadrillion BTU. Compared with net generation of electricity at 14.19 quadrillion BTU, T&D losses are about 9.4 %. This is a very slight improvement over 2004.

DG represents one of the most promising technologies in this regard [Metric 7]. The Electric Power Research Institute, for example, forecasts that 25% of new electric power generation by the year 2010 will be distributed generation (Dugan et al. 2001). Currently only 1.2 percent of net summer capacity is met by grid-connected DG.

With these considerations in mind, the benefits of a smart grid become clear; advanced sensors and control technologies will enable more efficient management and delivery of existing capacity, and will provide a strong framework for infrastructure support and the

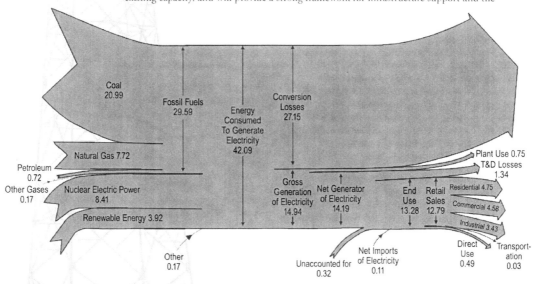

Figure 3.14. Electricity Flow Diagram 2007 (Quadrillion Btu) (DOE/EIA 2007a)

40

development of additional, distributed generators. This will be especially true in the case of renewable energy sources such as solar, wind, and small hydro generation, as current regional initiatives support the focus and development of these technologies.

3.6 Operates Resiliently to Disturbances, Attacks, and Natural Disasters

"Resiliency" refers to the ability of a system to react to events such that problematic consequences are isolated with minimal impact to the remaining system, and the overall system is restored to normal operation as soon as practical. These self-healing actions result in reduced interruption of service to consumers and help service providers more effectively manage the delivery infrastructure. Resiliency includes protection against all hazards, whether accidental or malicious, and needs to span natural disasters, deliberate attack, equipment failures, and human error. A smart grid inherently addresses security from the outset as a requirement for all the elements, and ensures an integrated and balanced approach across the system.

Resiliency is embedded in operational culture: policy, procedures, and vigilance.

From the point of view of the Nation's national security, this characteristic is arguably the most important. Resiliency in the face of adverse conditions or aggression, particularly high-consequence events, underlies all aspects of a smart grid and cuts across the other characteristics. Resiliency is embedded in operational culture: policy, procedures, and vigilance. It is embodied through effective risk management, with thorough understanding and management of threats, vulnerabilities, and consequences.

Given the great numbers of automation components interacting with a smart grid, an important operational paradigm going forward is distributed decision making. That is, equipment and smart-grid subsystems need to share actionable information so that local decision making not only serves local self-interest, but collaboratively supports the overall health of the system. As individual components of the system fail, including processing and communications components, the remaining connected components have the ability to adapt and reconfigure themselves to best achieve their objectives much like a society of devices. Though hierarchical command-and-control approaches will continue to occupy important roles in system design, distributed decision-making approaches are becoming more prevalent.

The strength of our electricity system does not lie in its ability to optimally reach a predefined mission or objective, but in the fact that its business and infrastructure components can adapt and evolve to meet the changing needs of an unpredictable future. As in nature, disturbances may impact portions of the ecosystem to varying degrees, and in the case of natural disasters, render regions incapacitated; however, the remainder of the system reacts to contain the damage, and amass a reconstruction effort once the event has past.

Operational resiliency has three basic descriptive properties (Caralli et al. 2006):

1. ability to change (adapt, expand, conform, contort) when a force is enacted,

2. ability to perform adequately or minimally while the force is in effect,

3. ability to return to a predefined, expected normal state whenever the force relents or is rendered ineffective.

41

Related Metrics

1, 2*, 5, 6, 7, 8, 9*, 10*, 11, 12, 13*, 16, 18*

A majority of the metrics identified for measuring smart-grid advancement contribute in some way to resiliency.

3.6.1 Area, Regional, National Coordination

At the transmission-system level, area control centers and regional reliability coordination centers have been exchanging system status information for many years. These systems are continually being upgraded to share more information including SCADA data, state-estimation results, and market data. The communication links between these systems now cover the country with increasing exchange of information between electric utility companies. There is also an increased level of data exchange between transmission and distribution levels within the system.

According to a recent survey by Newton-Evans Research (Newton-Evans 2008), one smart grid trend is to share this information between other reliability and control centers. For example, the research found that about 60% of the control centers in North America have linkages with other utilities. Figure 3.15 shows the projected integration of EMS/SCADA systems to a variety of other area and regional control systems as well as operations planning and DMS [Metric 2].

An interview of electricity service providers conducted for this report finds that 40% of the companies interviewed have new information flowing across functions and systems due to recent project implementations.

A transformational aspect of a smart grid is its ability to incorporate distributed energy resources, particularly demand-side resources, into system operations. The ability to send area, regional, and national signals to these resources, which enables distributed decision making, supports adaptation of these resources to impending threats, disturbances, and

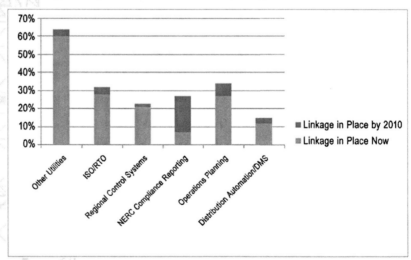

Figure 3.15. Current/Future Plans for Connecting EMS/SCADA/DMS Systems to Other Data Systems (Newton-Evans 2008)

42

attacks. In particular, critical-peak and real-time pricing programs provide a mechanism for system operators and reliability coordinators to access these resources to enhance operational resiliency. In a 2006 survey, FERC found that approximately 1.1% of the total customer base was served by time-of-use price offerings [Metric 1], nearly all of which were residential customers.

Finally, situational awareness of grid behavior is being transformed by wide-area-measurement networks. Initiatives in the western interconnection have been underway for many years and have contributed to reviews of major outages and questioned system dynamics models for planning and operations. Only recently have time-synchronized, high quality measurements (from phasor measurement units – PMUs) worked their way into operating rooms of reliability coordinators and balancing authorities [Metric 13]. The North American Synchro-Phasor Initiative (NASPI), led by NERC and supported by DOE, is advancing the coordination of the deployment of PMUs and the networking of their measurements for wide-area situation awareness and other applications. Currently there are approximately 165 PMUs installed. In the eastern interconnection, there are 104 PMUs with 89 networked and 61 PMUs in the western interconnection (Dagle 2008). One trade source indicates there were 150 PMUs installed in early 2008 within the eastern and western interconnections (Galvan et al. 2008) up from the 100 PMUs indicated in 2006 (DOE 2006a).

Figure 3.16 shows, as of 2007, the existing and planned PMU deployment locations in North America. There are many PMUs installed that are not networked across organizations not shown on the map, with many more projected in the future.

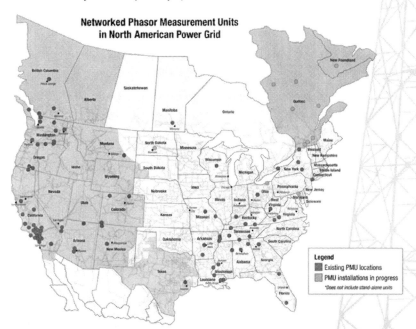

Figure 3.16. Networked Phasor Measurement Units in the North American Power Grid (PNNL/EIOC 2008)

43

3.6.2 DER Response

A smart grid provides the flexibility to adapt to a changing mix of demand-side resources, including changeable load, dispatchable distributed generation and storage, as well as variable-output local generation such as wind and solar. In the event of a disturbance, attack, or natural disaster, these resources can help alleviate constraints or support electrically energized islands that can mitigate the impact to events, and improve response times for post-disturbance reconstruction.

According to a 2008 FERC Survey (FERC 2008), only about 8% of customers have time-based rates or are involved in some form of incentive-based program [Metric 5]. Similarly, the number of entities offering such programs is low, with direct load control (DLC) and interruptible/curtailable tariffs listed as the most common incentive-based demand response programs. It also indicates increases in direct controlled-load management in nearly every region of the country. Interruptible loads, primarily industrial loads, are more mixed with several regions indicating decreases.

Grid-connected distributed generation and storage increased from 5,423 MW in 2004 to 12,702 MW in 2007 (DOE/EIA 2007d) [Metric 7]. While grid-connected distributed generation increased 134 percent over two years, it still only represented 1.4 percent of grid capacity, 1.6 percent of summer peak and 2.0 percent of winter capacity. Growth projections indicate a doubling of distributed generation capacity in five years (Eynon 2002).

Only about 8% of customers are involved in some form of incentive-based program.

Other distributed energy resources are just now emerging on the scene. These include microgrids [Metric 6] that are designed to operate in islanded and grid-connected modes, distributed storage, electric vehicles [Metric 8], and grid-responsive appliances in homes and other facilities [Metric 9]. The ability of a microgrid to run autonomously as an island (off of the main grid) provides these communities with greater reliability to withstand disturbances that may affect the greater electric system, while still being able to use grid resources should internal equipment fail or need maintenance. Today the amount of these resources is extremely small. While they will be good indicators for smart-grid progress, they are expected to have little impact on overall operational resiliency in the near future.

Grid-responsive equipment [Metric 9] has the potential to significantly enhance the resiliency of the overall system. Communicating thermostats, responsive appliances, responsive space conditioning equipment, etc. can quickly respond to frequency deviations or voltage changes. This can enhance the system reserve capacity that provides the necessary margin to respond to contingencies, and it can do it by measuring local system conditions or responding to communicated information. Currently there is significant interest in this field. Businesses such as LG Electronics and Westinghouse are designing and producing more "web-enabled" household appliances. Research and development in these fields will poise producers to easily transition into "smart" devices. However, incorporating electronics into increasing numbers of appliances, as well as developing and maintaining software for these appliances, will require a new look at the products' life-cycle costs. Manufacturers and grid entities have not yet settled on standards that would give manufacturers the confidence necessary to fully integrate and launch grid-responsive equipment.

3.6.3 Delivery Infrastructure

For many years, electric-service providers have realized the benefits of adding sensing, intelligence, and communications to equipment in the transmission and distribution infrastructure. Smart-grid initiatives are encouraging faster deployment of this capability with ever-greater functionality. Transmission and distribution substation automation projects and efforts to deploy advanced measurement equipment for applications such as wide-area situational awareness and dynamic line ratings, are helping to improve the ability to respond resiliently and adapt to system events.

Smart-grid-enabled distributed controls and diagnostic tools within the delivery system will help dynamically balance electricity supply and demand, thereby helping the system respond to imbalances and limiting their propagation when they occur. This could reduce the occurrence of outages and power disturbances attributed to grid overload as well as reduce planned rolling brownouts. These technologies could also quickly diagnose outages due to physical damage of the transmission and distribution facilities and direct crews to repair them quickly (Baer et al. 2004).

The national averages for outage disruptions (SAIDI, SAIFI, and MAIFI) [Metric 10] were estimated in a 2004 LBNL study at 106 minutes, 1.2 interruptions per year, and 4.3 minutes respectively. Recent trends in the reliability indicators SAIDI, SAIFI, and CAIDI are shown in Figure 3.17. An IEEE 2005 benchmarking study analyzed data from 55 companies between 2000 and 2005. Results showed an eight percent increase in CAIDI, a 21% increase in SAIDI and a 13% increase in SAIFI.

The relatively worsening trend in these indices suggests that a lack of investment in the delivery infrastructure is having an impact. The North American Electric Reliability Council's 2007 Long Term Reliability Assessment (NERC 2006) predicts capacity margins declining in the coming years, suggesting that the reliability indices can be expected to continue to increase given current operating practices. While it is difficult to attribute the ability of smart grid implementation to slow any degradation or enhance the increase in reliability, smart grid related resources in terms of substation automation equipment, sensing and management should play a significant role.

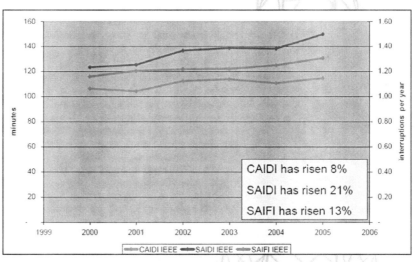

Figure 3.17. Trends for 55 Utilities Providing Data Between 2000-2005 (IEEE 2006)

Data from utilities across the nation show a clear trend of increasing T&D automation and increasing investment in these systems [Metric 11]. Increases in investment in T&D automation systems should lead to greater operational resiliency in the delivery infrastructure. Recent research shows that while 84% of utilities had substation automation and integration plans underway in 2005, and about 70% of utilities had deployed SCADA systems to substations, the penetration of automation at the distribution-feeder level is still limited to about 20% (ELP 2008; McDonnell 2008). This suggests a large area for growth in smart-grid deployment.

Another important trend within the electricity-delivery infrastructure is the rollout of advanced meters [Metric 12]. The capabilities of these systems promise to dramatically increase the accuracy of data to pricing policies, demand forecasts, and consumer applications, as well as increase the ability of the grid to respond to emergency occurrences such as blackouts and brownouts. Because AMI can play an enabling role in load participation in system operations it can have a significant influence on operational resiliency. Currently, AMI composes about 4.7% of total U.S. electric meters (FERC 2006a). Activity in the use of advanced metering has been increasing rapidly, growing nearly 700% from 2006 to 2008. While it is difficult to assess precisely which functions these AMI deployments support, the penetration rates indicate that this enabling technology is being positioned to support greater participation by distributed energy resources to the benefit of operational resiliency.

An increase in the penetration of dynamic line ratings and the associated measurement equipment will also contribute to understanding the status of the deployment of a smart grid [Metric 16]. The capacity of transmission equipment is not static, but can change significantly according to several variables, most notably conductor sag caused by thermal properties. Sensors for measuring the impact on sag are appearing more frequently, particularly in pilot programs for critical corridors. Though the number of miles of transmission with dynamic ratings is anticipated to increase, it is so small now as to be negligible on an interconnection basis.

3.6.4 Secure Information Networks

A vigilant security culture needs to permeate the stakeholder base to continually assess evolving cyber threats, risks, and response.

Economic forces and technology development are making the power system more dependent on information systems and external communications networks. The interconnected nature of the communications systems that support regional and interregional grid control, and the need to continue supporting older legacy systems in parallel with newer generations of control systems, further compound these security challenges. Additionally, with the advent of inexpensive microcontrollers and smart-grid implementation, there is a growing trend for increased intelligence and capabilities in field equipment installed in substations, within the distribution network, and even at the customer's premises. This increased control capability, while vastly increasing the flexibility and functionality to achieve better economies, also introduces cyber-vulnerabilities that have not previously existed and presents a significantly larger number of targets.

An understanding of component and associated system vulnerabilities will be necessary to quantify cyber-security issues inherent in smart grid deployments, particularly when these systems can be used to control or influence the behavior of the system. Assessments will be needed, both in controlled laboratory and test-bed environments, and in actual deployed field conditions, to explore and understand the implications of various cyber-attack scenarios, the resilience of existing security measures, and the robustness of proposed countermeasures.

46

Vendor and operator adoption of these countermeasures will be critical in broadly influencing the installed base of future deployments. The asset owners remain responsible for their legacy systems as smart grid technologies are deployed. A security culture that is vigilant to continually assess evolving threats and risks, then balance those with countermeasures needs to permeate the stakeholder base.

The interviews with service providers in Annex B of this report offer a sampling of data with regard to industry compliance with NERC cyber-security standards, including percentage of utilities that have conducted assessments at various frequencies for NERC Critical Infrastructure Protection (CIP) Standards 002 through 009 (see Table 3.7). The interviews indicate 5% of the utility respondents have never conducted an assessment. It's not clear whether this is because these utilities are not large enough to have an impact on the bulk electric power system or because they are still in the process of phasing in their compliance. As the timeline for mandatory compliance of all entities associated with the bulk electric system becomes fully implemented, and NERC establishes procedures for more formally tracking compliance with these standards, it will become increasingly easier to gather data for this metric and assess it for trends.

Table 3.7. Summary of the NERC Critical Infrastructure Protection Standards

NERC Standard	Subject Area
CIP-001-1	Sabotage Reporting
CIP-002-1	Critical Cyber Asset Identification
CIP-003-1	Security Management Controls
CIP-004-1	Personnel & Training
CIP-005-1	Electronic Security Perimeter(s)
CIP-006-1	Physical Security of Critical Cyber Assets
CIP-007-1	Systems Security Management
CIP-008-1	Incident Reporting and Response Planning
CIP-009-1	Recovery Plans for Critical Cyber Assets

Additionally, the interviews of 21 electricity service providers (Annex B) included a question about specific security measures that utilities are implementing. The sample results are shown in Table 3.8. While this information can be valuable for trending as a preliminary view, the interview questions need to be focused to better reveal the security culture instituted as more smart-grid capabilities are integrated by system operators, customers, and oversight organizations.

Table 3.8. Sample Security Question from Service Provider Interviews

Have you deployed the following security features? (Select all that apply)	Affirmative Responses
a. Intrusion detection	65.0%
b. Key management systems	50.0%
c. Encrypted communications	70.0%
d. Firewalls	95.0%
e. Others (Please describe)	30.0%

47

4.0 Challenges to Deployment

Among the significant challenges facing development of a smart grid are the cost of implementing a smart grid, with estimates for just the electric utility advanced metering capability ranging up to $27 billion, and the regulations that allow recovery of such investments. For perspective, the Brattle Group estimates that it may take as much as $1.5 trillion to update the grid by 2030 (Chupka et al. 2008). Ensuring interoperability of smart-grid standards is another hurdle state and federal regulators will need to overcome. Major technical barriers include developing economical storage systems; these storage systems can help solve other technical challenges, such as integrating distributed renewable-energy sources with the grid, addressing power-quality problems that would otherwise exacerbate the situation, and enhancing asset utilization. Without a smart grid, high penetrations of variable renewable resources (e.g., wind or solar) may become increasingly difficult and expensive to manage over time as they penetrate to high levels due to the greater need to coordinate these resources with dispatchable generation (e.g., natural gas combined cycle) and demand.

There are a variety of technical challenges facing a smart grid.

Another challenge facing a smart grid is the uncertainty of the path that its development will take over time with changing technology, changing energy mixes, and changing energy policy. Trying to legislate or regulate the development of a smart grid or its related technologies can severely diminish the benefits of the virtual, flexible, and transparent energy market it strives to provide. Conversely, with the entire nation's energy grid potentially at risk, some may see the introduction of a smart grid in the United States as too important to allow laissez-faire evolution. Thus, the challenge of development becomes an issue of providing flexible regulation that leverages desired and developing technology through goal-directed and business-case-supported policy that promotes a positive economic outcome. These and other challenges are discussed in the following sections.

4.1 Technical Challenges

There are a variety of technical challenges facing a smart grid, some of the greatest being developing, implementing, and deploying the array of different technologies required to enable both sides of the meter to communicate in a cost-effective way. One of the most important developments facing a smart grid is AMI technology. These devices help coordinate consumer equipment, as well as receive market signals and adjust household consumption based on a combination of this data and consumer preferences. However, alternatives to such AMI systems do exist. For example, market information such as prices and grid conditions can be decoupled from communication of energy consumption. Thus, the meter can be separate while pricing signals and the like can be transmitted via other public communication mechanisms such as phone, internet, cable, and wireless radio. A decoupled situation can make sense for commercial buildings and industrial uses where energy savings can be significant, while a more traditional bundled AMI package may be more desirable for residential consumers due to its "all-in-one" and "plug-and-play" aspects. Implementing price- and consumption-bundled AMI technology has been estimated to cost as much as $27 billion (Kuhn 2008) and will require very aggressive deployment to meet desired market penetration levels in the near future. Failure to successfully deploy technology that captures bi-directional power flow rather than net consumed energy, as well as dynamic pricing support, such as AMI technology or others, will keep the two sides of the market from properly communicating, and a smart grid will not function as desired regardless of other

49

Smart Grid System Report — July 2009

successful technical deployments, such as distributed generation, demand-response measures, or automated distribution schemes. Without real-time demand-response signals being promptly communicated and quickly addressed by consumers, the power system will not be flexible enough to provide the market transparency or the price signals required for a functioning energy market (FERC 2006a). Further, AMI billing techniques and the machines themselves may require regional customization reducing potential economies of scale in production and deployment. Regional customization may be required because of differences in consumer preferences, aggressiveness of service providers, state and local regulations, and the speed with which smart grid structures and technology change over time. Not all regions are likely to respond identically and may have different needs.

Another significant technical consideration is the impact of high levels of new technology penetration on existing grid infrastructure. Implementing new improvements into the grid, including smart-grid technologies, is pivotal to increasing efficient operations, as the operating efficiency gains from familiar technologies have begun to plateau (DOE/EIA 2007a). In addition, a NERC survey recently ranked the number one challenge to grid reliability as "aging infrastructure and limited new construction." How this aging infrastructure will function when combined with new "smart" technology remains to be seen, particularly with regard to solar, wind, and other forms of distributed generation (NERC 2007). Adding large amounts of variable and distributed generation, for example, requires a fundamental reworking of how the delivery system is managed, power quality is monitored, faults are detected, and maintenance is handled (Pai 2002). This problem is compounded when PHEVs and EVs are considered, potentially making each vehicle its own DG resource and requiring supporting infrastructure to draw, generate, and price power transactions.

However, these technologies themselves face several technical challenges. Cost-effective battery technology continues to be a challenge for PHEVs and EVs and local wind and solar resources. Issues such as discharge, battery life, size and weight are all serious considerations. Additionally, incorporating battery power storage into current automobile frames will require manufacturing adjustments; including systems to monitor the status of the battery (including battery charge and temperature) as well as structural design changes to accommodate the battery itself.

> There may be smaller sub-markets that would be better served if differentiated power quality standards existed.

A smart grid is needed at the distribution level to manage voltage levels, reactive power, potential reverse power flows, and power conditioning, all critical to running grid-connected DG systems, particularly with high penetrations of solar and wind power and PHEVs. Advanced voltage regulation, fault-detection, and system-protection practices need to be rethought as an increasing number of DG resources become available. This may require new equipment to identify and isolate DG resources in the event of a fault occurrence (Driesen and Belmans 2006). Another consideration for power-generation systems, distributed or otherwise, is power quality. Customers and the utilities that serve them lack standards for classifying varying qualities of power. Because customers have different power quality requirements (e.g., willingness to accept outages of varying durations, and load sensitivity to power harmonics) and with the increasing availability of DG resources to produce power locally, there may be smaller sub-markets for power that would be better served if such differentiated power standards existed.

Designing and retrofitting household appliances, such as washers, dryers, and water heaters with technology to communicate and respond to market signals and user preferences via home automation technology will be a significant challenge. Substantial investment will be required to implement user-friendly communication equipment which ensures that data storage and transmissions are tamper proof, reliable, and do not corrupt or break down over the lifetime of an appliance. Devices that communicate wirelessly with their facility energy-management systems must broadcast powerful-enough signals, or other technical barriers to effective communication must be resolved. For example, a washer/dryer located in a house's basement attempting to communicate with an energy-management system on the far side of the building will require a stronger signal than a closer device on the ground floor. Therefore communication equipment may need a flexible and dynamic range of broadcast strengths.

Finally, aggregating and sharing system data involves its own concerns; for example, providing infrastructure to communicate wide-area measurement data across the grid requires agreement by the stakeholders on the information network architecture, the supported functions, data exchange interface definitions, and legal conditions for granting use of the data.

4.2 Business and Financial Challenges

The business case for a smart grid needs to be firmly established for deployment decisions to progress. In many situations, individual applications may not be cost effective in isolation, but where common hardware and information network infrastructure can be leveraged to accomplish a number of objectives, the value proposition can become compelling. The business challenge is to prove that out with field deployments. Smart grid investments often require large upfront costs relative to their benefits. However, future benefits may come at small incremental costs. Utilities and regulators may need to look at full system life cycle costs and benefits in order to fully justify added investments. Some of the benefits may come in the form of societal benefits which will need to be clearly understood and evaluated. Payback periods may be longer than stakeholders would like. The service providers, regulators, and ultimately ratepayers are going to have to believe it before such substantial investments are made.

Utilities and regulators may need to look at full system life cycle costs and benefits in order to fully justify investments.

Since the technology and value propositions are emerging, utility companies may be reluctant to expend the significant amount of capital required to move toward a smart grid, especially because expected cost-recovery timelines are only theoretical and have no precedent. Currently, regulated utilities and their flat-rate customers have no risk or reward signal. Regulation makes it difficult for them to raise rates and recover costs, and makes them reluctant to change. Moreover, transmission-planning difficulties may or may not offset revenue losses incurred from reduced transmission; with uncertainty about market penetration of DG these effects can be difficult to model. Without effective cost recovery mechanisms in place, increased market penetration of DG will translate into lost demand for utilities. The uncertainty about market penetration is increased when utilities start to consider the time and cost of training a new smart-grid-skilled labor force (NERC 2007). Thus, utilities seeking to balance costs and operating efficiency will seek to increase asset utilization through the implementation of demand response measures and AMI technology, as opposed to expensive infrastructure upgrades. Further, as more and more devices become "web enabled" and move toward becoming fully "smart" devices, the inclusion of electronics in these devices, as well as the development and maintenance of this hardware and its respective software, will require manufacturers to reevaluate these devices' life-cycle costs. A smart grid will require service providers to operate in new ways and be willing to take reasonable risks for reasonable rewards. Regulators will need to design rules such that customers who do not change are not worse off, but that businesses can pursue advantageous arrangements between participating suppliers and consumers.

Aside from making a strong analytical business case with existing distribution models, the first few successful deployments of these new "smart" technologies will be pivotal to ensuring deep market penetration. Not all of these technologies are necessarily complementary. For example, when metering residential customers, drive-by and walk-by meters (AMR) are considered a competing technology and currently are out-shipping AMI products. Other than the more-convenient data gathering over traditional meters, AMR meters offer very few to none of the benefits and functions necessary to enable residential customers to meaningfully participate in a smart grid. However, implementing smart-grid technologies is daunting; the cost to implement AMI technology alone has been forecast between $19 and $27 billion (Kuhn 2008). Customers desire good value for the investments reflected in their power bills and they may want more options to manage their energy usage and bills, especially during a rate increase.

> Data from wide-area measurement systems could have eliminated the $4.5 billion in losses as a result of the 2003 blackout of the northeastern U.S. and Canada.

While utilities must be able to recover their investment costs, the potential savings from some of these technologies is considerable. For example, use of data from wide-area measurement systems (WAMS), including synchro-phasor measurements, could have mitigated or even avoided the estimated $4.5 billion in losses suffered by over 50 million people in the 2003 blackout of the northeastern U.S. and Canada (DOE 2004). To fully realize these benefits, high levels of market penetration must be encouraged; to accomplish this, new technologies will need simple, streamlined user interfaces, "plug-and-play" setups, and cost models that accurately forecast a reasonable payback period for newly developed and installed technologies for both utility companies and consumers, followed by reports on actual and successful deployments. Prior to successful deployments, important questions remain, including identifying winners and losers with bulk system reliability, evaluating those losses and gains, and how reasonable investments are recouped.

As consumer participation increases, a higher level of distributed-generation resources are expected to become available (Eynon 2002). The costs of making these DG resources dispatchable are estimated to be high and vary significantly between utilities. Storing energy generated by DG resources will continue to be a problem until a cost-effective, low-maintenance solution is introduced. Trends suggest this might be done with highly efficient batteries or by pre-heating and cooling buildings. Until then however, viable payback strategies, such as storing generated power during off-peak hours and selling it back into the grid during high-price on-peak hours, will not be feasible. The lack of cost-effective, low-maintenance batteries is a particular hindrance for renewable energies such as solar and wind generation, because their generation varies over time and may not match demand patterns.

Lastly, consumer concerns about hybrid electric vehicles including price, insufficient power, and dependability will need to be addressed by PHEV and EV manufacturers. The cost to convert a hybrid vehicle to a PHEV is currently considered prohibitive; it can vary between six and eight thousand dollars and consumers may consider the payback period too long. Because of these concerns, PHEVs will be unlikely to penetrate all markets, leaving heavy-duty and long-range vehicles, such as semi-trucks, and high-performance vehicles such as sports cars requiring contemporary infrastructure, such as gas stations, while PHEVs and EVs require new supporting infrastructure, such as charge stations. Economies of scale for these services may or may not exist.

52

5.0 Recommendations for Future Reports

The transition toward a smart grid has made advances within the electric transmission and distribution infrastructure as information technology and communications has advanced; however, recent visions of an electric system that not only services, but integrates and interacts with its use, generation, and storage in factories, businesses, and homes is driving new business and policy models and the technology deployments to support them. Winning directions are only emerging making calibration of potentially useful metrics small or otherwise difficult to measure.

As the first in a series of biennial smart-grid status reports, information gathered for this report should form a framework and measurement baseline for future reports. The metrics identified are indicators of smart grid deployment progress that facilitate discussion regarding the main characteristics of a smart grid, but they are not comprehensive measures of all smart grid concerns. Because of this, they deserve to be reviewed for continued relevance and appropriate emphasis of major smart grid attributes. For example, a desirable metric considered for reporting in this report is smart grid cost savings. The report attempts to address cost savings through capacity factors [Metric 14] and generation, transmission and distribution efficiencies [Metric 15], subject to non-deteriorating trends in transmission and distribution system reliability [Metric 10] and power quality [Metric 17]. However, future reports should consider improvements to this approach. As smart grid business cases are developed and gain acceptance, a new cost savings or value added metric may become apparent. Also, this report describes the flexibility of the smart grid to support renewable and non-renewable generation sources [Metric 7] while emphasizing the controllable versus variable aspects of distributed generation. Future reports may wish to better distinguish progress on renewable generation as well as the environmental impacts of the electric system.

In addition, the status of smart grid deployment should project as balanced a view as possible across the diverse stakeholder perspectives related to the electric system. Workshops, interviews, and research into smart grid related literature needs to reflect a complete cross-section of the stakeholders. Future reports should review the stakeholder landscape to ensure coverage of these perspectives. In particular, the smart grid environmental aspects and the electricity consumer perspectives are important areas that arguably deserve greater attention.

Given the time period for developing the report, investigation was restricted to existing literature research and interviews with 21 electricity-service providers, representing a cross-section of organizations by type, size, and location (see Table B.1 in Annex B of this report). Further research is needed to better gage the metrics and gain insights into deployment directions, as well as engage the other stakeholder groups. A more extensive interview process can facilitate gathering this information. Coordination with other smart grid information collection activities whose products can be used in the creation of this report should also be supported. For example, the Department of Energy is collaborating with other organizations to create a clearinghouse of smart grid related information that should be useful for this report.

This report should form a framework and measurement baseline for future reports.

53

In addition, future reports require the development of assessment models that support those metrics that are difficult to measure, particularly regarding progress on cyber-security and automation-system interoperability related to open architecture and standards. Other models and tools to measure smart grid progress may also be useful.

Besides reviewing the progress of measurements to the metrics identified in this report, future reports should consider addressing the following potential improvements:

- Track significant smart grid demonstration and deployment projects.

- Review progress toward resolving smart grid challenges, identify new challenges, and describe places where opportunities to advance smart grid concepts are taking place.

- Track the evolvement of legislative and regulatory decisions and structures to describe how government agencies are embracing smart grid objectives and are working on paths that enable and support smart grid advancement.

- The sixth characteristic in the table is a merger of the Smart Grid Implementation Workshop characteristics a) Addresses and Responds to System Disturbances in a Self-Healing Manner and b) Operates Resiliently Against Physical and Cyber Attacks and Natural Disasters. Though this report found that the same metrics substantially contribute to both of these concerns, future reports may find it advantageous to keep these characteristics separate.

- Support a glossary of terms related to smart grid deployment status.

Further recommendations specific to each metric can be found in Annex A, which presents the detailed results of investigation into the metrics. The end of each metric description includes a subsection on metric recommendations. Future reports should include a review of these recommendations in addition to those summarized above.

A final consideration for future reports on the status of smart grid deployments is perhaps more of a warning; attempts to be comprehensive about all things related to a smart grid can overwhelm the investigation effort and threaten to create so much material that the report compromises its ability to convey the major aspects of smart grid progress. Care should be taken to avoid the tendency to proliferate the number of metrics. In deciding if a new metric is merited, consideration should be given to how it fits with the other metrics, if a previous metric can be retired, and the strength of a metric's contribution to explaining the smart grid progress regarding the identified characteristics.

6.0 References

Anders, S. 2007. *Implementing the Smart Grid: A Tactical Approach for Electric Utilities.* Energy Policy Initiatives Center presentation, University of San Diego School of Law, October 15, 2007. Del Mar, California.

Baer WS, B Fulton, and S Mahnovski. 2004. *Estimating the Benefits of the GridWise Initiative: Phase I Report.* TRI-160-PNNL, prepared by Rand Science and Technology for the Pacific Northwest National Laboratory, Richland, Washington. Accessed in October 2008 at http://www.rand.org/pubs/technical_reports/2005/RAND_TR160.pdf

Balducci P. 2008. *Plug-In Hybrid Electric Vehicle Market Penetration Scenarios.* PNNL-17441, Pacific Northwest National Laboratory, Portland, Oregon.

Caralli RA, JF Stevens, CM Wallen, and WR Wilson. 2006. *Sustaining Operational Resiliency: A Process Improvement Approach to Security Management.* CMU/SEI-2006-TN-009, Software Engineering Institute, Carnegie Mellon University, Pittsburgh, Pennsylvania.

Chupka M, R Earle, P Fox-Penner, and R Hledik. 2008. *Transforming America's Power Industry: The Investment Challenge 2010-2030.* Prepared by The Brattle Group for The Edison Foundation, Washington, D.C. Accessed in January 2009 at http://www.eei.org/ourissues/finance/Documents/Transforming_Americas_Power_Industry.pdf

Cook C and R Haynes. 2006. "Analysis of U.S. Interconnection and Net Metering Policy." In *Proceedings of the ASME International Solar Energy Conference - Solar Engineering,* July 9-13, Denver, Colorado. Accessed in October 2008, at http://www.ncsc.ncsu.edu/research/documents/policy_papers/ASES2006_Haynes_Cook_.pdf

Dagle J. 2008. *Number of PMUs.* Email communication on September 20, 2008, Pacific Northwest National Laboratory, Richland, Washington.

DOE - U.S. Department of Energy. 2004. *Final Report on the August 14, 2003 Blackout in the United States and Canada: Causes and Recommendations.* U.S.-Canada Power System Outage Task Force, U.S. Department of Energy, Washington, D.C.

DOE - U.S. Department of Energy. 2006a. *Five-Year Program Plan for Fiscal Years 2008 to 2012 for Electric Transmission and Distribution Programs.* Report to Congress Pursuant to Section 925 of Energy Policy Act of 2005, U.S. Department of Energy, Washington, D.C. Accessed in October 2008 at http://www.oe.energy.gov/DocumentsandMedia/Section_925_Final.pdf

DOE - U.S. Department of Energy. 2006b. *Modern Grid v1.0: A Systems View of the Modern Grid, Appendix A5: Accommodate All Generation and Storage Options.* Prepared by National Energy Technology Laboratory, Pittsburgh, Pennsylvania. Accessed in October 2008 at http://www.masstech.org/dg/benefits/2006_DER_ModernGrid_a5_v1.pdf

DOE/EIA – U.S. Department of Energy, Energy Information Administration. 2002. *Contact Information for Electric Utilities by State.* Accessed in October 2008 at http://www.eia.doe.gov/cneaf/electricity/utility/utiltabs.html (last updated July 16, 2003).

DOE/EIA - Energy Information Administration. 2007a. *Annual Energy Review 2007*. DOE/EIA-0384(2007), Energy Information Administration, Washington, D.C. Accessed in November 2008, at http://www.eia.doe.gov/aer/elect.html

DOE/EIA - U.S. Department of Energy, Energy Information Administration. 2007b. *Energy Market Impacts of a Clean Energy Portfolio Standard – Follow-Up*. SR-OIAF/2007-02, Energy Information Administration, Washington, D.C. Accessed in October 2008, at http://www.eia.doe.gov/oiaf/servicerpt/portfolio/pdf/sroiaf(2007)02.pdf

DOE/EIA – U.S. Department of Energy, Energy Information Administration. 2007c. "Noncoincident Peak Load, Actual and Projected by North American Electric Reliability Council Region." *Electric Power Annual With Data for 2007*, Energy Information Administration, Washington, D.C. Accessed in October 2008, at http://www.eia.doe.gov/cneaf/electricity/epa/epat3p1.html

DOE/EIA – U.S. Department of Energy, Energy Information Administration. 2007d. *Total Capacity of Dispersed and Distributed Generators by Technology Type*. Accessed in November 2008 at http://www.eia.doe.gov/cneaf/electricity/epa/epat2p7c.html

DOE/EIA - U.S. Department of Energy, Energy Information Administration. 2008. *Annual Energy Outlook 2008*. Energy Information Administration, Washington, D.C. Available at http://www.eia.doe.gov/oiaf/archive/aeo08/index.html

DOE/EIA - U.S. Department of Energy, Energy Information Administration. 2008. "Supplemental Table 47: Light-Duty Vehicle Sales by Technology Type." In *Annual Energy Outlook 2008*. Energy Information Administration, Washington, D.C. Accessed in October 2008 at http://www.eia.doe.gov/oiaf/aeo/supplement/pdf/suptab_47.pdf

DOE/EIA - U.S. Department of Energy, Energy Information Administration. 2009a. *Capacity of Distributed Generators by Technology Type, 2004 and 2007*. Excerpt from the Electric Power Annual (2007), Energy Information Administration, Washington, D.C. Accessed in January 2009 at http://www.eia.doe.gov/cneaf/electricity/epa/epaxlfile2_7b.pdf

DOE/EIA - U.S. Department of Energy, Energy Information Administration. 2009b. *Electric Power Industry 2007: Year in Review*. Energy Information Administration, Washington, D.C. Accessed in January 2009 at http://www.eia.doe.gov/cneaf/electricity/epa/epa_sum.html

DOE/EIA - U.S. Department of Energy, Energy Information Administration. 2009c. "Supplemental Tables 57 and 58." In *Annual Energy Outlook 2009, Early Release*. Energy Information Administration, Washington, D.C. Accessed in January 2009 at http://www.eia.doe.gov/oiaf/aeo/supplement/sup_tran.xls

DOE/OEDER - U.S. Department of Energy, Office of Electricity Delivery and Energy Reliability. 2008a. *The Smart Grid: An Introduction*. Prepared by Litos Strategic Communication, East Providence, Rhode Island. Accessed in November 2008 at http://www.oe.energy.gov/1165.htm

DOE/OEDER - U.S. Department of Energy, Office of Electricity Delivery and Energy Reliability. 2008b. "Metrics for Measuring Progress Toward Implementation of the Smart Grid: Results of the Breakout Session Discussions," *Smart Grid Implementation Workshop*,

56

6.0 References

June 19-20, 2008. Prepared by Energetics, Inc., Columbia, Maryland. Accessed in October 2008 at http://www.oe.energy.gov/DocumentsandMedia/Smart_Grid_Workshop_Report_Final_Draft_08_12_08.pdf

DOE/OEDER - U.S. Department of Energy, Office of Electricity Delivery and Energy Reliability. 2008c. *Advanced Metering Infrastructure.* Prepared by National Energy Technology Laboratory, Pittsburgh, Pennsylvania. Accessed in October 2008 at http://www.netl.doe.gov/moderngrid/docs/AMI%20White%20paper%20final%20021108.pdf

Dorr DS. 1991. "AC Power Quality Studies: IBM, AT&T and NPL." *13th International Telecommunications Energy Conference,* Kyoto, Japan November 5-8, pp. 552-559.

Driesen J and R Belmans. 2006. "Distributed Generation: Challenges and Possible Solutions." In *2006 IEEE Power Engineering Society General Meeting,* Institute of Electrical and Electronics Engineers, Piscataway, New Jersey.

Dugan RC, TE McDermott, DT Rizy, and SJ Steffel. 2001. "Interconnecting Single-Phase Backup Generation to the Utility Distribution System." *IEEE Transmission and Distribution Conference and Exposition,* Oak Ridge National Laboratory, Oak Ridge, Tennessee. Accessed in October 2008, at http://www.ornl.gov/~webworks/cppr/y2001/rpt/112434.pdf

EEA - Energy and Environmental Analysis, Inc. 2004. *Economic Incentives for Distributed Generation.* Energy and Environmental Analysis, Inc., Arlington, Virginia. Accessed in November 2008, at http://www.eea-inc.com/rrdb/DGRegProject/Incentives.html

EAC - Electricity Advisory Committee. 2008. *Smart Grid: Enabler of the New Energy Economy.* Prepared by Energetics, Inc., Columbia, Maryland. Accessed in January 2009 at http://www.oe.energy.gov/DocumentsandMedia/final-smart-grid-report.pdf

ELP - Electric Light and Power. 2008. *T&D Automation News.* PennWell Corporation, Tulsa, Oklahoma. http://uaelp.pennnet.com/resources/transmission%20and%20distribution

EPA - U.S. Environmental Protection Agency. 2008a. *eGRID FAQ: What do the eGRID and NERC Region Maps Look Like?* Accessed in November 2008 at http://www.epa.gov/cleanenergy/energy-resources/egrid/faq.html#egrid6

EPA - U.S. Environmental Protection Agency. 2008b. *Interconnection Standards.* Combined Heat and Power Partnership, U.S. Environmental Protection Agency, Washington, D.C. Accessed in November 2008 at http://www.epa.gov/chp/state-policy/interconnection.html (last updated July 3, 2008).

EPA- U.S. Environmental Protection Agency. 2008c. National Action Plan for Energy Efficiency *National Action Plan for Energy Efficiency Vision for 2025: A Framework for Change.* http://www.epa.gov/eeactionplan.

EPRI/NRDC - Electric Power Research Institute and National Resources Defense Council. 2007. *Environmental Assessment of Plug-In Hybrid Electric Vehicles – Volume 1: Nationwide Greenhouse Gas Emissions.* EPRI-1015325. Final Report, Electric Power Research Institute, Palo Alto, California. Accessed in November 2008 at http://mydocs.epri.com/docs/public/000000000001015325.pdf

57

241

Smart Grid System Report — July 2009

Eynon RT. 2002. *The Role of Distributed Generation in U.S. Energy Markets.* Energy Information Administration, Washington, D.C. Accessed in October 2008, at http://www.eia.doe.gov/oiaf/speeches/dist_generation.html

FERC - Federal Energy Regulatory Commission. 2005. *Standardization of Small Generator Interconnection Agreements and Procedures.* Docket No. RM02-12-001; Order No. 2006-A. Federal Energy Regulatory Commission, Washington, D.C. Accessed in October 2008, at http://www.caiso.com/14ea/14ead07a4660.pdf

FERC - Federal Energy Regulatory Commission. 2006a. *Assessment of Demand Response and Advanced Metering Staff Report.* Docket Number AD06-2-000, FERC, Washington, D.C. Accessed in November 2008 at http://www.ferc.gov/legal/staff-reports/demand-response.pdf

FERC - Federal Energy Regulatory Commission. 2006b. *The 2006 Assessment of Demand Response and Advanced Metering.* Staff report. Docket Number AD-06-2-000, Federal Energy Regulatory Commission, Washington, D.C. Accessed in November 2008 at http://www.ferc.gov/legal/staff-reports/09-07-demand-response.pdf

FERC - Federal Energy Regulatory Commission. 2007. *The 2007 Assessment of Demand Response & Advanced Metering.* Staff Report. Federal Energy Regulatory Commission, Washington, D.C. Accessed in October 2008 at http://www.ferc.gov/legal/staff-reports/09-07-demand-response.pdf

FERC - Federal Energy Regulatory Commission. 2008. *The 2008 Assessment of Demand Response and Advanced Metering.* Staff report. Federal Energy Regulatory Commission, Washington, D.C. Accessed in December 2008 at http://www.ferc.gov/legal/staff-reports/12-08-demand-response.pdf

Galvan F, L Beard, J Minnicucci, and P Overholt. 2008. "Phasors Monitor Grid Conditions." *Transmission and Distribution World.* Accessed in October 2008, at http://tdworld.com/overhead_transmission/phasors_monitor_grid_conditions/

Gilmore E and L Lave. 2007. "Increasing Backup Generation Capacity and System Reliability by Selling Electricity during Periods of Peak Demand." Presented at *26th UAAEE/IAEE North American Conference, September 16-19, 2007.* Carnegie Mellon Electricity Industry Center, Pittsburgh, Pennsylvania. Accessed in October 2008 at http://www.usaee.org/usaee2007/submissions/Presentations/Elisabeth%20Gilmore.pdf

Greene D, K Duleep, and W McManus. 2004. *Future Potential of Hybrid and Diesel Powertrains in the U.S. Light-Duty Vehicle Market.* ORNL/TM-2004/181, Oak Ridge National Laboratory, Oak Ridge, Tennessee. Accessed in October 2008 at http://www.cta.ornl.gov/cta/Publications/Reports/ORNL_TM_2004_181_HybridDiesel.pdf

GridWise Architecture Council. 2008. *Interoperability Context-Setting Framework, v1.1.* Accessed in November 2008 at http://www.gridwiseac.org/pdfs/interopframework_v1_1.pdf

Hamachi LaCommare K and J Eto. 2004. *Understanding the Cost of Power Interruptions to U.S. Electricity Consumers.* LBNL-55718, Ernest Orlando Lawrence Berkeley National Laboratory, Berkeley, California. Accessed in November 2008 at http://certs.lbl.gov/pdf/55718.pdf

58

6.0 References

Hammerstrom DJ, R Ambrosio, TA Carlon, DP Chassin, JG DeSteese, RT Guttromson, OM Jarvegren, R Kajfasz, S Katipamula, P Michie, T Oliver, and RG Pratt. 2007. *Pacific Northwest GridWise Testbed Projects: Part 1. Olympic Peninsula Project.* PNNL-17167, Pacific Northwest National Laboratory, Richland, Washington. Accessed in November 2008 at http://gridwise.pnl.gov

Hirst E. 2004. *U.S. Transmission Capacity: Present Status and Future Prospects.* Edison Electric Institute, Washington, DC. Accessed in November 2008 at http://www.eei.org/industry_issues/energy_infrastructure/transmission/USTransCapacity10-18-04.pdf

IEEE - Institute of Electrical and Electronics Engineers, Working Group on Distribution Reliability. 2006. *IEEE Benchmarking 2005 Results.* Power Engineering Society, Institute of Electrical and Electronics Engineers, Piscataway, New Jersey. Accessed November 7, 2008 at http://grouper.ieee.org/groups/td/dist/sd/doc/2006-07-BenchmarkingUpdate.pdf

Kueck JD, BJ Kirby, PN Overholt, and LC Markel. 2004. *Measurement Practices for Reliability and Power Quality: A Toolkit of Reliability Measurement Practices.* ORNL/TM-2004/91, Oak Ridge National Laboratory, Oak Ridge, Tennessee. Accessed in November 2008 at http://www.ornl.gov/sci/engineering_science_technology/eere_research_reports/power_systems/reliability_and_power_quality/ornl_tm_2004_91/ornl_tm_2004_91.pdf

Kuhn TR. 2008. *Legislative Proposals to Reduce Greenhouse Gas Emissions: An Overview.* Testimony before the United States House of Representatives Subcommittee on Energy and Air Quality, June 19, 2008. Accessed in November 2008 at http://energycommerce.house.gov/Press_110/110st177.shtml

McDonnell D. 2008. *Beyond the Buzz: The Potential of Grid Efficiency.* Global Smart Energy, Redmond, Washington. Accessed in November 2008 at http://www.smartgridnews.com/artman/publish/industry/Beyond_the_Buzz_The_Potential_of_Grid_Efficiency_180_printer.html

McNulty S and B Howe. 2002. *Power Quality Problems and Renewable Energy Solutions.* Prepared for the Massachusetts Renewable Energy Trust, Madison, Wisconsin. Available at http://www.mtpc.org/rebates/public_policy/DG/resources/2002-09_MA_PQ-Report_Primen-MTC.pdf

Navigant Consulting. 2005. *Base Case Scenarios.* Burlington, Massachusetts. Available at http://www.navigantconsulting.com

NERC - North American Electric Reliability Council. 2006. *2006 Long-Term Reliability Assessment: The Reliability of the Bulk Power Systems in North America.* North American Electric Reliability Council, Princeton, New Jersey. Available at http:// www.nerc.com/files/LTRA2006.pdf

NERC - North American Electric Reliability Corporation. 2007. *Results of the 2007 Survey of Reliability Issues.* North American Electric Reliability Corporation, Princeton, New Jersey. Accessed in October 2008, at http://www.nerc.com/files/Reliability_Issue_Survey_Final_Report_Rev.1.pdf

59

NERC - North American Electric Reliability Corporation. 2008. *Electricity Supply & Demand (ES&D): Frequently Requested Reports. 2007 Reports (with 2006 actuals).* North American Electric Reliability Corporation, Princeton, New Jersey. Accessed in October 2008 at http://www.nerc.com/page.php?cid=4|38|41 (last updated December 12, 2007).

NETL - National Energy Technology Laboratory. 2007. *A System View of the Modern Grid. Vol 2.* National Energy Technology Laboratory, Pittsburgh, Pennsylvania. Accessed in October 2008, at http://www.netl.doe.gov/moderngrid/docs/ASystemsViewoftheModernGrid_Final_v2_0.pdf

NETL - National Energy Technology Laboratory. 2008. *The Modern Grid Strategy: Characteristics of the Modern Grid.* National Energy Technology Laboratory, Pittsburgh, Pennsylvania. Accessed in October 2008 at http://www.netl.doe.gov/moderngrid/opportunity/vision_characteristics.html

Newton-Evans Research Company. 2008. *Market Trends Digest.* Newton-Evans Research Company, Ellicott City, Maryland. Accessed in November 2008 at http://www.newton-evans.com/mtdigest/mtd3q08.pdf

Ockwell G. 2008. "The Smart Grid Reaches Main Street USA." *Utility Automation & Engineering T&D.* 13(5), PennWell Corporation, Tulsa, Oklahoma. Accessed in October 2008 at http://uaelp.pennnet.com/display_article/328726/22/ARTCL/none/none/1/Th

Pai MA. 2002. *Challenges in System Integration of Distributed Generation with the Grid.* University of Illinois, Urbana, Illinois. http://www.nfcrc.uci.edu/2/UfFC/PowerElectronics/PDFs/04_Pai_Challenge_Part1.pdf

PNNL/EIOC - Electricity Infrastructure Operations Center. 2008. *North American SynchroPhasor Initiative.* Pacific Northwest National Laboratory, Richland, Washington. Accessed in November 2008 at http://eioc.pnl.gov/research/synchrophasor.stm (last updated July 2008).

PSPN - Penn State Policy Notes. 2008. *Reducing Demand, Promoting Efficiency Key to Defusing Electric Rate Increases.* Center for Public Policy Research in Environment, Energy and Community Well-Being, University Park, Pennsylvania. Accessed in November 2008 at http://www.ssri.psu.edu/policy/GeneralPolicyBrief_0415.pdf

RDC – Resource Dynamics Corporation. 2005. *Characterization of Microgrids in the United States: Final Whitepaper.* Prepared for Sandia National Laboratory by Resource Dynamics Corporation, Vienna, Virginia. Accessed in November 2008 at http://www.electricdistribution.ctc.com/pdfs/RDC_Microgrid_Whitepaper_1-7-05.pdf

Rohmund I, G Wikler, A Furuqui, O Siddiqui, and R Tempchin. 2008. "Assessment of Achievable Potential for Energy Efficiency and Demand Response in the U.S. (2010-2030)." In *ACEEE Summer Study on Energy Efficiency in Buildings.* American Council for an Energy-Efficient Economy, Washington D.C.

SEI - Software Engineering Institute. 2008. *The Capability Maturity Model for Software.* Carnegie Mellon University, Pittsburgh, Pennsylvania. Accessed in October 2008, at http://www.sei.cmu.edu/cmm/

6.0 References

Seppa TO. 1997. "Real Time Rating Systems." Presented at the *EPRI Workshop on Real Time Monitoring and Rating of Transmission and Substation Circuits: A Technology Increasing Grid Asset Utilization*, San Diego, California, February 26-28, 1997.

Shirley W. 2007. "Survey of Interconnection Rules." *Workshop on Interconnection of Distributed Generation.* Prepared for the Florida Public Service Commission, Tallahassee, Florida. Accessed in October 2008 at http://www.epa.gov/chp/documents/survey_interconnection_rules.pdf

Solar Guide. 2008. *Solar Cost FAQ.* Moxy Media, Guelph, Ontario, Canada. Accessed in November 2008 at http://www.thesolarguide.com/solar-power-uses/cost-faq.aspx

Wesoff E. 2008. *The Not so Smart Grid: Utilities and Consumers in the 21st Century.* Greentech Media, Cambridge, Massachusetts.

61

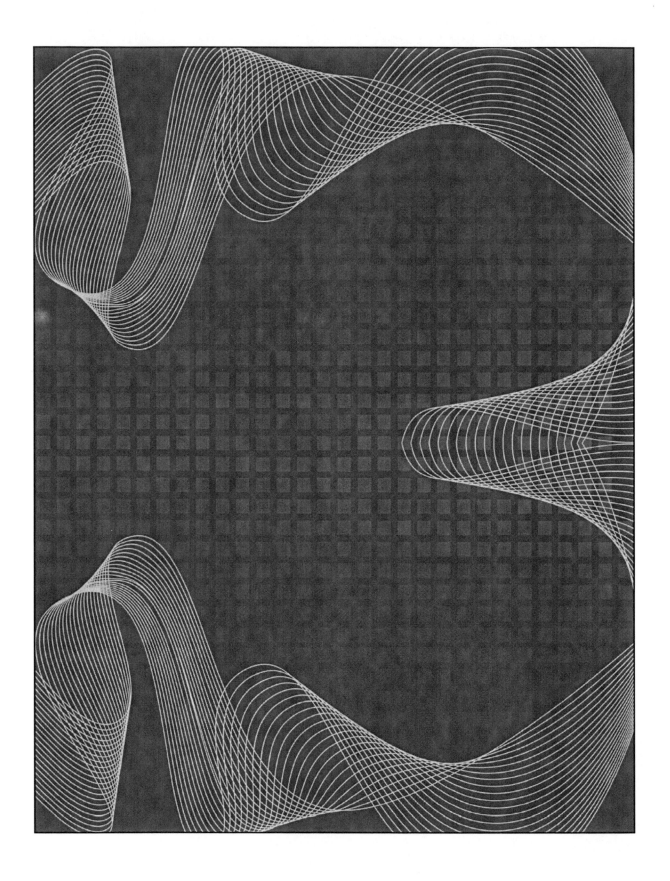

Testimony of Chairman Jon Wellinghoff
Federal Energy Regulatory Commission
Before the Energy and Environment Subcommittee
Of the Committee on Energy and Commerce
United States House of Representatives
Hearing on the Future of the Grid: Proposals for Reforming National
Transmission Policy
June 12, 2009

Mr. Chairman, and members of the Subcommittee:

My name is Jon Wellinghoff, and I am the Chairman of the Federal Energy Regulatory Commission (Commission). Thank you for the opportunity to appear before you today to discuss the critical topic of the development of our Nation's electric transmission grid.

Transmission facilities are critical to meeting the goal of reducing reliance on carbon-emitting sources of electric energy and bringing new sources of renewable energy to market. A reliable and robust transmission grid is essential to allow regions, states, and utilities to access least-cost resource options to meet state and national environmental, economic and security goals. To meet the challenges of building needed new transmission facilities, we must address not only the role of federal siting authority but also the closely-related issues of transmission planning and cost allocation. In doing so, we must focus on maintaining the reliability of the electric system. The time has come to develop a regulatory framework that will allow us to successfully meet these challenges. I commend you, Mr. Chairman, and the Subcommittee for your decision to hold a hearing on these important issues.

Introduction

President Obama has stated that the country that harnesses the power of clean, renewable energy will lead the 21st century. The President also stated that we will need to build power lines that can carry new clean energy to cities and towns across this country. He also said we should be "starting to build a new smart grid that will save us money, protect our power sources from blackout or attack, and deliver clean, alternative forms of energy to every corner of our nation."

A majority of states have adopted renewable portfolio standards that require utilities to acquire renewable generation capacity, some of which are quite aggressive. For example, the Connecticut standard requires that 27% of the energy consumed in the state be generated using renewable resources by 2020.

1

Both houses of Congress are considering a federal renewable energy standard as well.

Clean power is essential to meeting energy goals such as promoting fuel diversity, reducing greenhouse gas emissions, strengthening our national security, enhancing competition, ensuring reliability, and revitalizing our economy. The need for additional federal authority to achieve these goals is clear. Historically, the Nation's electric utilities transported fuels to generate electricity to plants located near load centers. Many of today's clean energy resources are located far from consumers and existing transmission facilities and those resources cannot be moved. Moreover, they are not evenly dispersed throughout the country. Delivering the power generated by these resources to consumers will require the planning, siting and construction of interstate and inter-regional transmission facilities. Only Congress, exercising its authority to regulate commerce among the states, can address this problem.

The requirement for greater fuel diversity, whether as a result of federal or state goals, cannot be accomplished unless we ensure that the renewable, and sometimes variable, generation resources that we will rely upon to meet these goals can be reliably integrated into the power grid and ultimately deliverable to consumers. Renewable energy resources, particularly those whose operation follow a natural but variable cycle, must be integrated into the transmission system in a manner consistent with reliable operation of the grid. We know that the grid can accommodate some level of renewable generation, but we also know that, with the current configuration of the grid and the variability of some forms of renewable generation, it cannot accommodate 100%. Compounding the challenges of integrating renewable generation, we also know that the grid is aging, was designed for more traditional types of generation, and is characterized by decreasing reserve margins. These conditions mean that smaller disturbances on the grid cause larger fluctuations and increase the risk of outages.

Because of these factors, Commission staff is conducting a study to determine the appropriate metrics for use in assessing the reliability impact of integrating large amounts of variable renewable power generation onto the existing power grid. That study, which is being undertaken by Lawrence Berkeley National Laboratory and overseen by Commission staff, is due to be completed by November 2009. When the study is complete, it will help answer the question of how variable resources can be reliably integrated onto the existing grid, which will help inform policy makers about the current limitations of the grid and identify what new resources and transmission facilities will be necessary to reliably accommodate future renewable resources and those currently under development.

2

I believe that, if the Nation is to meet its goals, there must be a mechanism that, after the states have had an opportunity, allows a transmission developer to invoke federal authority to site the transmission facilities necessary to interconnect renewable power to the electric transmission grid and move that power to consumers. We need a national policy commitment to develop the transmission infrastructure to bring renewable energy from remote areas where it is produced most efficiently into our metropolitan areas where most of this Nation's power is consumed. This transmission infrastructure is likely to be comprised of extra-high voltage facilities, related feeder lines that will interconnect remote renewable energy resources to the transmission grid, and supporting upgrades to the existing grid (hereinafter, "transmission infrastructure"). Without this national commitment, we will not be able to take advantage of our capacity to develop clean power.

We must develop a structured regulatory framework that will enable the United States to build the transmission infrastructure necessary to deliver our Nation's high quality, location-constrained renewable resources to load centers. That framework must adequately address transmission siting and the related issues of transmission planning and cost allocation. And above all, we must ensure that we preserve the reliability of the electric grid so that consumers and businesses continue to receive the highest level of service, protecting the safety of our citizens, the security of our Nation, and the health of our economy.

There is a real opportunity to make the United States a world leader in developing the clean energy industries of the future. Without a coherent drive for a smart grid that is designed and built (or rebuilt) to achieve our national energy and environmental goals in a timely fashion, the jobs and sustainable economic development options from those potential new industries could very well manifest in Europe or Asia rather than here.

Though the focus of this hearing is on ensuring that the development of the interstate transmission grid allows our country to meet national and state goals, we should not lose sight of the critical role of local renewable energy, distributed resources, and demand response. We must focus on ensuring that we remove barriers to entry for local renewable and distributed resources. Developing and reliably delivering these local resources is important as we expand our capacity to generate clean power, but that effort must be made in concert with and not separate from developing the transmission infrastructure that I describe above. An optimal blending of both resources will be necessary to achieve our Nation's energy goals. That optimization process will require a collaborative effort between the states and the Federal government with an expanded Federal role.

The Need for an Expanded Federal Role

3

The electric grid is actually a combination of individual systems, separated into three electrical interconnections. Though there has been some expansion of regional and inter-regional transmission facilities over the last 15 years, that expansion is not sufficient to address the need I have identified to develop our transmission infrastructure to allow us to meet state and national goals. In the Energy Policy Act of 2005, Congress directed the Commission to develop incentive-based rate treatments for certain new transmission facilities, and while this effort has been effective in encouraging developers to come forward with new transmission projects, it does not ensure that the projects will be constructed and placed in operation. Without new siting authority, the Commission's ability to address these challenges is limited. For this reason, I recommend that the Congress enact legislation that will enable transmission developers to invoke federal authority in appropriate circumstances to site the transmission facilities necessary to interconnect renewable power to the electric transmission grid and move that power to consumers. Such legislation should also address cost allocation and planning of such facilities. Each of these issues is a crucial aspect of developing an effective power grid that can spur the production and reliable movement to market of renewable energy.

Siting. States should continue to have the opportunity to site transmission facilities, but transmission developers should have recourse to a federal siting authority under appropriate circumstances. With additional authority, the Commission could play an important role in this grid-building effort as it has the institutional structure, capacity, and experience to make important contributions. The Commission is well-versed in reviewing and authorizing critical energy infrastructure projects, and in establishing a regulatory regime that encourages the development of appropriate energy projects, while at the same time protecting the interests of consumers and safeguarding the environment.

Since 1920, the Commission has been charged with licensing and overseeing the operation of the Nation's non-federal hydropower projects. Using existing authority under Part I of the Federal Power Act, the Commission has sited thousands of miles of electric transmission lines related to these projects that have delivered this power to the Nation's consumers. Likewise, under the Natural Gas Act, the Commission has authorized the construction of natural gas pipelines for over 65 years. Under the Commission's oversight, the country has developed a robust, comprehensive pipeline grid that moves natural gas supplies from distant producing areas to consuming regions. Based on its decades of experience in siting natural gas pipelines and in siting hydropower projects and associated transmission lines, the Commission has developed comprehensive, efficient processes that provide for public notice and extensive public participation, including participation by affected states.

4

The Commission's existing transmission siting authority is limited. The Energy Policy Act of 2005 gave the Commission authority to site and permit interstate electric transmission facilities to relieve congestion under limited circumstances and only within geographic areas designated by the Secretary of Energy as national interest electric transmission corridors. However, the United States Court of Appeals for the Fourth Circuit has recently held that the limited authority granted by Congress to the Commission to review and site facilities needed to transmit electric energy in interstate commerce is not available in situations where a state agency has timely denied an application for a proposed project, regardless of how important the project may be in relieving congestion on the interstate grid. The court's ruling is a significant constraint on the Commission's already-limited ability to site appropriate projects to transmit electricity in interstate commerce. To date, no applicant has sought Commission authority to site transmission facilities under this law.

Congress should consider the question of how best to exercise its authority over interstate commerce to ensure that the necessary transmission facilities are built in a timely manner to deliver location-constrained renewable power to customers. Federal siting authority would be helpful even if limited only to transmission facilities needed to reliably meet renewable energy goals and only in those cases where the states have had an opportunity to address a proposal in the first instance. It is clear, however, that without some broader federal siting authority, it is unlikely that the Nation will be able to achieve its renewable energy goals.

Planning. Effective regional and inter-regional transmission planning will improve reliability, reduce congestion, increase the deliverability of existing power supplies, and identify investments necessary to integrate significant potential sources of renewable energy that are constrained by a lack of adequate transmission capacity or facilities. Increasingly, such planning must look beyond the needs of a single utility or even a single state to examine the grid requirements of the entire region. The Commission has recognized the need for improvements in transmission planning. To improve the coordination of transmission planning among utilities, it required all public utility transmission providers to establish and participate in open and transparent regional transmission planning processes (Order No. 890, February 2007). The Order No. 890 regional planning processes are in their second year, and the Commission is reviewing how well those planning efforts are working, is monitoring implementation, and will be looking for ways to improve the regional planning processes.

Meeting our national energy goals will require building on such regional planning initiatives and expanding their scope. I urge the Congress not to be distracted by the false choice between so called "bottom-up" and "top-down" planning models.

5

It is indisputable that local and sub-regional planning and coordination must continue, addressing such issues as smaller upgrades that must proceed in a timely way, without awaiting regional or inter-regional review. But to achieve greater benefits and efficiencies, we must also create a structure that includes coordination on an inter-regional basis, which will facilitate, for example, the development of facilities to transport electric energy from areas rich in renewable energy resources to load centers or the deployment of key smart grid equipment and systems. The American Recovery and Reinvestment Act of 2009 includes funding of an initial analysis to implement this approach through the appropriation of $80 million to the Department of Energy to conduct, in consultation with the Commission, a thorough resource assessment for each interconnection to facilitate regional transmission planning. Going forward, Congress could help by clarifying the Commission's authority to ensure that state and regional planning is consistent with national energy goals. I recommend, however, that any new transmission planning requirements be harmonized with, rather than supplant, planning efforts already taking place at the regional, state and local levels.

Cost Allocation. Renewable energy resources such as wind, solar, and geothermal are usually found in large quantities at dispersed locations remote from load centers. For this reason, there are often high costs associated with developing transmission facilities needed to deliver power from such resources. If the resource developer or the host utility is compelled to bear all of the cost of these transmission facilities, they may not be developed.

Under Federal Power Act sections 205 and 206, the Commission ensures that public utilities' (investor-owned utilities) rates, terms and conditions of transmission service in interstate commerce are just, reasonable, and not unduly discriminatory or preferential. This responsibility includes setting rates for recovering the costs of new transmission facilities built by public utilities. At present, the Commission has greater ability to assign such costs over broad geographic areas where there is a regional transmission organization (RTO) or independent system operator (ISO).

If Congress determines that there are broad public interest benefits in developing the transmission infrastructure necessary to accommodate the Nation's renewable energy potential, and therefore that in some cases it may be appropriate for the costs of transmission facilities needed to meet our renewable energy potential to be fairly spread to a broad group of energy users (for example, across a region or multiple regions), then Congress should consider clarifying the Commission's authority to allocate such transmission costs to all load-serving entities within an interconnection or part of an interconnection where it is appropriate to do so. Of course, the Commission would need to ensure, as it does today, that the costs are allocated fairly to the appropriate entities and that regions work together to

6

develop cost allocation mechanisms that garner broad support. However, I urge the Subcommittee to avoid including unduly restrictive language on cost allocation in any new legislation, particularly language that would impose a requirement to calculate the precise monetary benefits expected to accrue from a new transmission facility. Rather, Congress should maintain the Commission's flexibility to address cost allocation for each facility under the facts and circumstances presented.

It is important to acknowledge that appropriately allocating the costs of transmission facilities to connect remote resources will not disrupt the implementation of state resource policies or disadvantage local renewable or demand resources. Rather, a fair cost allocation will eliminate a barrier to the development of new, clean resources and thus will facilitate competition, which should ensure that utilities may access least-cost resource options to meet state and national environmental, economic and security goals. Development of the necessary transmission infrastructure will enable those resources options to reach load centers, and, as discussed below, ensure that they may do so without jeopardizing the reliability of the system. The issue is not how to choose between nearby renewable resources and more distant renewable resources: we need both. The issue is ensuring that costs are allocated fairly, sending the right economic signals without unduly impeding development of location-constrained resources.

Conclusion

In summary, to achieve the Nation's renewable energy goals, Congress and federal and state regulators, including the Commission, must address in a timely manner the issues of transmission planning, transmission siting and transmission cost allocation. Congressional action on all three of these related areas, particularly siting and cost allocation authority for transmission infrastructure needed to deliver high quality, location-constrained renewable energy, would provide greater ability to achieve these important goals. I recognize that the concepts we are discussing today can seem threatening or overreaching to some and that the Commission's actions have not always been perceived as benevolent. I recognize that we need to retain state and local expertise and authorities that are critical to everyday grid operations and regulation, but we also need to expand regional and national cooperation. We are not seeking to usurp local prerogatives but to make sure the Nation's electricity grid is prepared to meet the challenges and realize the opportunities of the 21st century. There are elements of the various bills under development in the Senate and the House that address the matters I have discussed, and I would be happy to answer follow-up questions in writing about the specific provisions in those bills.

7

Thank you for the opportunity to appear before you today to provide my insight as you consider legislation to provide a regulatory framework for tackling the challenging energy issues that we face. I stand ready to work with Congress, state and federal regulators, industry, and other stakeholders on these important issues. I would be happy to answer any questions you may have.

8

254

Testimony of Commissioner Suedeen G. Kelly
Federal Energy Regulatory Commission
Before the Committee on Energy and Natural Resources
United States Senate
March 3, 2009

Introduction and Summary

Mr. Chairman and members of the Committee, thank you for the opportunity to speak here today. My name is Suedeen Kelly, and I am a Commissioner on the Federal Energy Regulatory Commission (FERC or Commission). My testimony addresses the efforts to develop and implement a range of technologies collectively known as the "Smart Grid."

Our nation's electric grid generally depends on decades-old technology, and has not incorporated new digital technologies extensively. Digital technologies have transformed other industries such as telecommunications. A similar change has not yet happened for the electric grid. As detailed below, a Smart Grid can provide a range of benefits to the electric industry and its customers, enhancing its efficiency and enabling its technological advancement while ensuring its reliability and security.

Smart Grid efforts involve a broad range of government agencies, at both the Federal and state levels. The Federal agencies include primarily the Department of Energy (DOE), the National Institute of Standards and Technology (NIST) and FERC. DOE's tasks include awarding grants for Smart Grid projects and developing a Smart Grid information clearinghouse. NIST has primary responsibility for coordinating development of an "interoperability framework" allowing Smart Grid technologies to

- 2 -

communicate and work together. FERC is then responsible for promulgating

interoperability standards, once FERC is satisfied that NIST's work has led to sufficient

consensus.

Development of the interoperability framework is a challenging task. Recent

funding for NIST's efforts will help, but cooperation and coordination among

government agencies and industry participants is just as important. DOE, NIST and

FERC have been working with each other and with other Federal agencies to ensure

progress, and those efforts will continue. FERC also has been coordinating with state

regulators, to address common issues and concerns.

FERC can use its existing authority to facilitate implementation of Smart Grid.

For example, FERC can provide rate incentives for appropriate Smart Grid projects, and

can provide guidance on cost recovery for such projects.

A critical issue as Smart Grid is deployed is the need to ensure grid reliability and

cyber security. The significant benefits of Smart Grid technologies must be achieved

without taking reliability and security risks that could be exploited to cause great harm to

our Nation's citizens and economy.

Finally, if the intent of Congress is for the Smart Grid standards to be mandatory

beyond the scope of the Federal Power Act, additional legislation should be considered.

EISA

Section 1301 of the Energy Independence and Security Act of 2007 (EISA) states

that "it is the policy of the United States to support the modernization of the Nation's

electricity transmission and distribution system to maintain a reliable and secure

- 3 -

electricity infrastructure that can meet future demand growth and to achieve" a number of benefits. Section 1301 specifies benefits such as: increased use of digital technology to improve the grid's reliability, security, and efficiency; "dynamic optimization of grid operations and resources, with full cyber-security;" facilitation of distributed generation, demand response, and energy efficiency resources; and integration of "smart" appliances and consumer devices, as well as advanced electricity storage and peak-shaving technologies (including plug-in hybrid electric vehicles).

Section 1305(a) of EISA gives NIST "primary responsibility to coordinate the development of a framework that includes protocols and model standards for information management to achieve interoperability of smart grid devices and systems." NIST is required to solicit input from a range of others, including the GridWise Architecture Council and the National Electrical Manufacturers Association, as well as two international bodies, the Institute of Electrical and Electronics Engineers and the North American Electric Reliability Corporation (NERC). Many of the organizations working with NIST on this issue develop industry standards through extensive processes aimed at achieving consensus.

Although EISA does not define interoperability, definitions put forth by others often include many of the same elements. These include: (1) exchange of meaningful, actionable information between two or more systems across organizational boundaries; (2) a shared meaning of the exchanged information; (3) an agreed expectation for the response to the information exchange; and (4) requisite quality of service in information

- 4 -

exchange: reliability, accuracy, security. (See GridWise Architecture Council, "Interoperability Path Forward Whitepaper," www.gridwiseac.org)

Pursuant to EISA section 1305, once FERC is satisfied that NIST's work has led to "sufficient consensus" on interoperability standards, FERC must then "institute a rulemaking proceeding to adopt such standards and protocols as may be necessary to insure smart-grid functionality and interoperability in interstate transmission of electric power, and regional and wholesale electricity markets." Section 1305 does not specify any other prerequisites to Commission action, such as a filing by NIST with the Commission or unanimous support for individual standards or a comprehensive set of standards.

FERC's role under EISA section 1305 is consistent with its responsibility under section 1223 of the Energy Policy Act of 2005. Section 1223 directs FERC to encourage the deployment of advanced transmission technologies, and expressly includes technologies such as energy storage devices, controllable load, distributed generation, enhanced power device monitoring and direct system state sensors.

Smart Grid Task Force

As required by EISA section 1303, DOE has established the Smart Grid Task Force. The Task Force includes representatives from DOE, FERC, NIST, the Environmental Protection Agency and the Departments of Homeland Security, Agriculture and Defense. The Task Force seeks to ensure awareness, coordination and integration of Federal Government activities related to Smart Grid technologies, practices, and services. The Task Force meets on a regular basis, and has helped inform

- 5 -

the participating agencies on the Smart Grid efforts of other participants as well as the efforts outside the Federal Government.

<u>Smart Grid Collaborative</u>

A year ago, FERC and NARUC began the Smart Grid Collaborative. I and Commissioner Frederick F. Butler of the New Jersey Board of Public Utilities co-chair the collaborative. The collaborative was timely because state regulators were increasingly being asked to approve pilot or demonstration projects or in some cases widespread deployment in their states of advanced metering systems, one key component of a comprehensive Smart Grid system.

The Collaborative began by convening joint meetings to hear from a range of experts about the new technologies. A host of issues were explored. Key among them were the issues of interoperability, the types of technologies and communications protocols used in Smart Grid applications, the sequence and timing of Smart Grid deployments, and the type of rate structures that accompanied Smart Grid projects.

Through these meetings, Collaborative members learned of a range of Smart Grid projects already in place around the country. The Smart Grid programs in existence were varied in that they used a mix of differing technologies, communications protocols and rate designs. Collaborative members began discussing whether a Smart Grid information clearinghouse could be developed that would then allow an analysis of best practices. This information could help regulators make better decisions on proposed Smart Grid projects in their jurisdictions. As discussed below, recent legislation requires DOE to establish such a clearinghouse.

- 6 -

The Collaborative members have begun to look beyond the information clearinghouse to who could best analyze this information to identify best practices from Smart Grid applications. The Collaborative has met with staff from DOE to discuss possible funding for a project under the auspices of the Collaborative that could act as an analytical tool to evaluate Smart Grid pilot programs, using the information developed by the clearinghouse. This issue is still being explored.

The Stimulus Bill

The American Recovery and Reinvestment Act of 2009 (the "Stimulus Bill") appropriated $4.5 billion to DOE for "Electricity Delivery and Energy Reliability." The authorized purposes for these funds include, inter alia, implementation of programs authorized under Title XIII of EISA, which addresses Smart Grid. Smart Grid grants would provide funding for up to 50 percent of a project's documented costs. In many cases, state and/or Federal regulators could be asked to approve funding for the balance of project costs. The Secretary of Energy is required to develop procedures or criteria under which applicants can receive such grants. The Stimulus Bill also states that $10 million of the $4.5 billion is "to implement [EISA] section 1305," the provision giving NIST primary responsibility to coordinate the development of the interoperability framework.

The Stimulus Bill also directs the Secretary of Energy to establish a Smart Grid information clearinghouse. As a condition of receiving Smart Grid grants, recipients must provide such information to the clearinghouse as the Secretary requires.

- 7 -

As an additional condition, recipients must show that their projects use "open protocols and standards (including Internet-based protocols and standards) if available and appropriate." These open protocols and standards, sometimes also referred to as "open architecture," will facilitate interoperability by allowing multiple vendors to design and build many types of equipment and systems for the Smart Grid environment. As the GridWise Architecture Council stated, "An open architecture encourages multi-vendor competition because every vendor has the opportunity to build interchangeable hardware or software that works with other elements within the system." (See "Introduction to Interoperability and Decision-Maker's Checklist," page 4, www.gridwiseac.org.)

The Collaborative has begun discussing additional criteria that regulators would like to see applied to projects seeking Smart Grid grants. The Collaborative members are focusing on criteria that could help them fulfill their legal responsibilities as to Smart Grid projects they would be asked to approve. For example, cost-effectiveness could be a key criterion and could inform regulatory decisions on rate recovery issues. Upgradeability could be another criterion. Once the Collaborative reaches consensus on the criteria, the Collaborative intends to ask the Secretary of Energy to consider its recommended criteria.

Initial Deployments Are Still In Progress

Initial efforts to use Smart Grid technologies are still being implemented and analyzed. Even comprehensive pilot projects such as Xcel's project in Boulder, Colorado (which includes smart meters, in-home programmable control devices, smart substations

- 8 -

and integration of distributed generation), are in the early stages of development and data gathering. Thus, it is too early to assess the "lessons learned" from such efforts.

A particularly interesting project, however, is under development by Pepco Holdings, Inc. (PHI). At the transmission level, Smart Grid can be equated with widespread deployment of advanced sensors and controls and the high-speed communications and IT infrastructure needed to fully use the additional data and control options to improve the electric system's reliability and efficiency. PHI's proposal follows this model. In a filing with FERC seeking approval of incentive rates, PHI committed to promote interoperability through insistence "upon open architecture, open protocols and 'interoperability'" when dealing with potential vendors, and to adhere to "available standards which have been finalized, proven, and have achieved some levels of broad industry acceptance" as much as possible for its Smart Grid deployments. Furthermore, PHI committed to "provide a method of upgrading systems and firmware remotely (through the data network as opposed to local/site upgrades) and ensure that unforeseen problems or changes can be quickly and easily made by PHI engineers and system operators on short notice." Adherence to such principles, along with adequate consideration of cyber security concerns, is essential at this early stage of Smart Grid development. The Commission granted incentive rates for this project, and construction is expected to start in 2009.

Next Steps

As Congress recognized in enacting EISA, the development of an interoperability framework can accelerate the deployment of Smart Grid technologies. The process of

- 9 -

developing such a framework may take significant time. NIST has primary responsibility for this task, and must coordinate the efforts and views of many others. As a non-regulatory agency, NIST is used to serving as a neutral mediator to build consensus towards standards. Achieving consensus among the many, diverse entities involved in Smart Grid may be difficult. Coordinated leadership is needed to help minimize conflicting agendas and unnecessary delay. The Stimulus Bill's funding will help NIST's efforts, but may not guarantee quick achievement of the goals.

In the meantime, the Commission may be able to take steps to help hasten development and implementation of Smart Grid technology. For example, the Commission's day-to-day knowledge of the electric industry may allow it to suggest aspects of the interoperability framework that should be prioritized ahead of others. This prioritization may facilitate progress on the Smart Grid technologies that will provide the largest benefits for a broad group of participants.

An overarching approach for prioritization could focus initially on the fundamental standards needed to enable all of the functions and characteristics envisioned for the Smart Grid. This may include, for example, standards for cyber security, since the electric grid and all devices connected to it must be fully protected. This approach also may include standards that promote common software semantics throughout the industry, which would enable real-time coordination of information from both demand and supply resources.

The next set of targets for prioritization could be standards needed to enable key Smart Grid functionalities identified by relevant authorities including FERC. For

- 10 -

example, challenges associated with integrating variable renewable resources into the generation mix and reliably accommodating any new electric vehicle fleets could be addressed, at least in part, through certain capabilities envisioned for the Smart Grid. Accordingly, priority could be placed on the development of: (1) standards permitting system operators to rely on automated demand response resources to offset an unplanned loss of variable generation such as wind turbines or to shift load into off-peak hours with over-generation situations; (2) standards permitting system operators to rely on emerging electric storage technologies for similar purposes; (3) standards permitting transmission operators to rely on technologies such as phasor measurement units for wide-area system awareness and congestion management; and, (4) standards permitting some appropriate control over the charging of plug-in hybrid electric vehicles, particularly encouraging such charging to occur during off-peak hours.

Even before NIST's work has led to sufficient consensus, the Commission could provide rate incentives to jurisdictional public utilities for early implementation of certain Smart Grid technologies, if adequate steps are taken to ensure reliability and cyber security while minimizing the risk of rapid obsolescence and "stranded costs." The Commission also may be able to use its ratemaking authority, apart from incentives, to encourage expansion of Smart Grid technologies. Providing clear guidance on the types of Smart Grid costs recoverable in rates, and on the procedures for seeking rate recovery, may eliminate a major concern for utilities considering such investments.

While FERC, by itself, may be able to take steps such as these to foster Smart Grid technologies, achieving the full benefits of a Smart Grid will require coordination among

- 11 -

a broad group of entities, particularly DOE, NIST, FERC and state regulators. For
example, DOE's authority to support up to 50 percent of the cost of a Smart Grid project
may elicit little interest from utilities if they are uncertain of their ability to recover the
rest of their costs. Similarly, Congress itself recognized, in EISA section 1305(a)(1), the
need for NIST to seek input from FERC, the Smart Grid Task Force established by DOE
and "other relevant Federal and state agencies." Also, the concurrent jurisdiction of
FERC and state commissions over many utilities will require regulators to adopt
complementary policies or risk sending conflicting regulatory "signals." More
fundamentally, a Smart Grid will require substantial coordination between wholesale and
retail markets and between the Federal and state rules governing those markets.
Similarly, Smart Grid standards may require changes to business practice standards
already used in the industry, such as those developed through NAESB, and the industry
and government agencies should support the work needed to evaluate and develop those
changes.

Concerns about access to, and security of, Smart Grid control systems and/or data
also must be resolved. For example, data on how and when individual customers use
electricity could be valuable to various commercial entities, but customers may have
privacy concerns about unauthorized dissemination or marketing of this data. Similarly,
generation owners and operators may be concerned about cyber access to control systems
that operate their facilities. Access to information enabling the identification of critical
energy infrastructure must also be limited. Issues about who owns Smart Grid-generated
data and the security of some of its products are unresolved.

- 12 -

An additional issue involves enforcement of Smart Grid standards promulgated by the Commission under EISA section 1305. This section, which is a stand-alone provision instead of an amendment to the Federal Power Act (FPA), requires FERC to promulgate standards, but does not provide that the standards are mandatory or provide any authority and procedures for enforcing such standards. If FERC were to seek to use the full scope of its existing FPA authority to require compliance with Smart Grid standards, this authority applies only to certain entities (i.e., public utilities under its ratemaking authority in Sections 205 and 206, or users, owners and operators of the bulk power system under its reliability authority in Section 215). FERC also has asserted jurisdiction in certain circumstances over demand response programs involving both wholesale and eligible retail customers. However, FERC's authority under the FPA excludes local distribution facilities unless specifically provided, its authority under sections 205 and 206 applies only to public utilities, and its section 215 authority does not authorize it to mandate standards but rather only to refer a matter to NERC's standard-setting process. If the intent of Congress is for the Smart Grid standards to be mandatory beyond the scope of the Federal Power Act, additional legislation should be considered.

Finally, in developing and implementing Smart Grid technologies, the electric industry and vendors must meet the critical need, recognized by Congress in EISA section 1301, for grid reliability and "full cyber-security." An entity subject to FERC-approved reliability standards under FPA section 215 must maintain compliance with those standards during and after the installation of Smart Grid technologies. Also, the interoperability framework and the technology itself must leave no gaps in physical

- 13 -

security or cyber security. Reliability and security must be built into Smart Grid devices, and not added later, to avoid making the grid more vulnerable and to avoid costly replacement of equipment that cannot be upgraded. The significant benefits of Smart Grid technologies must be achieved without taking reliability and security risks that could be exploited to cause great harm to our Nation's citizens and economy.

Conclusion

A properly-coordinated and timely deployment of Smart Grid can provide many positive benefits to the Nation's electric industry and its customers, if we are careful to maintain and enhance grid security and reliability at the same time. Indeed, I would expect Smart Grid to evolve in many unanticipated but beneficial ways. Well-designed standards and protocols are needed to make Smart Grid a reality. They will eliminate concerns about technology obsolescence, allow system upgrades through software applications, and ultimately permit plug-and-play devices, regardless of vendor. FERC is committed to working closely with DOE, NIST and others to facilitate rapid deployment of innovative, secure Smart Grid technologies.

Thank you again for the opportunity to testify today. I would be happy to answer any questions you may have.

**United States House of Representatives
Before the Select Committee on Energy Independence and
Global Warming**

Prepared Statement of James J. Hoecker
Counsel to WIRES (Working group on Investment in Reliable and Economic electric Systems)

**Hearing on "Get Smart on the Smart Grid: How Technology Can
Revolutionize Efficiency and Renewable Solutions"**
February 25, 2009

"We will build the roads and bridges, the electric grids and digital
lines that feed our commerce and bind us together."

President Barack Obama
January 20, 2009

1

Chairman Markey, Ranking Member Sensenbrenner, and Honorable Members of the

Committee, my name is James J. "Jim" Hoecker. Thank you for the opportunity to testify this

morning on the future of smart grid technology deployment within the electric transmission

system, and the grid's contribution to our dynamic clean energy future. I am especially

honored to have the opportunity to appear before this Committee.

I. Introduction

Today I appear before you as Counsel to WIRES, the **W**orking group on **I**nvestment in

Reliable and **E**conomic electric **S**ystems. WIRES is a new national coalition of both publicly-

owned, investor-owned, and cooperatively-owned transmission providers, customers, and

services companies. To my knowledge, WIRES is the only private sector group exclusively

dedicated to promoting investment in the electric transmission system and educating

policymakers and the public on the benefits derived from an upgraded and strengthened grid.

WIRES' most recent work on transmission, including studies on cost allocation and integrating

"location-constrained" resources like wind and solar power into the grid, can be found on it

website (www.wiresgroup.com).

WIRES was formed to highlight the need for electric transmission investment and to explore

ways to facilitate it. I am pleased to say that a range of business and special interests are taking

a fresh look at the grid. Policy makers are coming to recognize that, properly planned, sited,

and animated by digital technologies, transmission is a network industry with diverse benefits

2

and beneficiaries and not simply an adjunct to other utility functions. During the time that I was Chairman of the Federal Energy Regulatory Commission, the focus of regulators was largely on enabling a competitive electric generation market. Those wholesale power markets will grow and endure and deliver benefits to consumers, but they are complex and comprised of thousands of transactions. Federal policymakers and the industry are now rediscovering electric transmission infrastructure in light of the need to utilize those markets to deliver reliable, low carbon energy from entirely new resources to load-serving entities.

The need for a more integrated and extensive transmission network is real. When the individual utility transmission systems were achieving a higher degree of integration a half century ago, we had no plasma TV's or energy-hungry computers; no one seriously conceived of the possibility that automobiles would be plugged into the electric system; large-scale regional bulk power markets were only a blip on the horizon; few people were concerned about the consequences of greenhouse gases in the atmosphere; and extensive deployment of "location-constrained" wind, solar, biomass, or geothermal technologies for electric generation – not to mention low-carbon forms of coal generation – was a fantasy. Today an American consumer uses 13 times the electricity he or she did a half century ago and there are twice as many of us. In most instances, we are asking the transmission system, and indeed the electricity system generally, to perform tasks for which it was not designed. The imperative we face is therefore to both upgrade and expand the system and to make it more interactive and "smarter" – *i.e., more digital and less electro-mechanical.*

3

My testimony today seeks to connect these objectives. As a representative of WIRES, I will focus principally on the challenges facing transmission providers and customers that seek to enlarge the capabilities of the transmission system as a network of wires and the related technologies and equipment that animate it. These challenges must be addressed if the U.S. is to have a chance at changing the energy economy and scaling back its emissions of deleterious greenhouse gases. Climate change is a global problem which demands a range of solutions, among which energy efficiency and demand response are among the most important in our estimation. However, because low-carbon alternative energy resources that utilize some of the most innovative technologies developed in the past quarter century are far from major load centers, transmission is an indispensible enabler of many of the new technological applications now being touted as the engines of energy independence and reduced emissions. In other words, Mr. Chairman, when we speak of the "smart grid," let's not overlook the "grid" itself.[1] The need to invest in smart grid technologies and to strengthen the grid generally are intertwined objectives. WIRES looks forward to working with you, the Committee, and technology companies to create a modern 21st Century electric system. I have attached to this testimony an outline of a legislative proposal that addresses the planning, siting, and cost allocation and recovery issues I discuss below. WIRES is engaged with many groups in an effort to find the best approach to solve the challenges facing the grid.

---

[1] Of course, the distinction is difficult to draw because the terms "grid" and "smart grid" are so often used interchangeably. There is no standard definition of smart grid. I believe it entails two-way communications technologies that provide customers with real-time information and tools that allow them to be responsive to system conditions, help ensure efficient use of the electric grid, and enhance system reliability. The wires network – both transmission and distribution -- is the platform upon which digital technologies will operate to empower customers to manage their carbon footprints and utilize system assets more efficiently.

4

II. The Benefits of Transmission

Electric transmission has several important benefits. The grid's benefits and the benefits of energy efficiency and distributed generation are not mutually exclusive. At one level, high voltage transmission provides network reliability benefits, including coordinating the operation of power production facilities to permit them to reinforce one another, providing a high degree of flexibility to accommodate changing conditions as they occur, and the sharing of generation reserves among interconnected systems across whole regions.

In addition, transmission systems allow electricity to be transported in large quantities from one production area to another. Power can be delivered to industrial, commercial, and residential customers from generators located at a great distance from those loads. This magnifies consumer access to less expensive, more diverse, or environmentally more benign resources. Transmission, assisted by modern communications technologies, enable buyers and sellers of power to engage in trading of electricity, providing opportunities to reduce the cost of power overall. The electric transmission system provides the greatest hedge against extreme conditions and events that could result in large economic dislocations and threats to the public health. Power from readily available resources can be transmitted to the broadest regional markets to maximize the economic and environmental benefits of those resources.

The benefits derived from the grid may be in direct proportion to the technological advances that will accompany its expansion. Investment in technologies that enhance system reliability, reduce line loss, increase transfer capability may be made without expanding the grid's footprint. Techniques that permit the aggregation of variable resources and transmission of remote renewable resources over greater distances are on the horizon. Control technologies that

5

enable the grid to be "self-healing" by detecting frequency fluctuations and re-routing power to avoid interruption will produce a high-quality electrical economy. Those technologies can also increase the efficiency and transfer capability of existing transmission assets, thereby avoiding the need to develop new corridors for transmission facilities in many cases.

Educated estimates of the size of the investment that must be made to ensure that these benefits continue to flow in the face of the demands to be placed on the grid between now and 2030 range in the neighborhood of $300 billion. After a period of declining investment, U.S. companies will have spent about $30 billion on transmission in the period 2006-2009, at a rate roughly double the annual expenditures at the beginning of the century. However, as of mid-2008, only 668 miles of high voltage transmission has been built across state lines since 2000. Remarkably, the staggering expenditure on transmission will remain the smallest component of the investment we must make in the electricity system.

The most important potential benefit of transmission along these lines comes from the historic task undertaken by this Committee as part of a shift in public policy – its potential contribution to addressing climate change. The quest to curb greenhouse gas emissions will not -- indeed cannot -- succeed without squarely coming to terms with the need for greater transmission investment. The reasons for this are clear:

- Transmission is the principal means by which electricity from new clean energy resources such as wind, solar, geothermal, and biomass can be made available to the majority of American consumers. This is equally true for other low-carbon resources such as nuclear power and potential low-carbon coal generation. All of these resources

6

are "location constrained" by their very nature and existing transmission infrastructure is inadequate to serve both the growth in traditional demand and development of these new generation resources.

- By both expanding the high voltage "backbone" network and ensuring that it becomes a "smart grid", we can empower consumers to control their own carbon footprint, enable companies to make optimal use of existing assets, and turn the grid into a driver of energy efficiency and demand response.

- Transmission ensures fuel diversity and provides the needed market access for new technologies like carbon capture and sequestration, wind power, and solar generation. Deployment of new transportation technologies like plug-in hybrid vehicles will necessitate a more uniformly strong transmission system to deliver power on demand.

This climate change challenge can be met. It will require leadership from Congress and the States, industry, and regulators. As the National Clean Energy Project Summit here in Washington amply demonstrated this week, transmission expansion is becoming a national priority.

III. Challenges to Transmission Development

The existing electric transmission system today faces well-recognized challenges, however. New competitive bulk power markets test the limits of the grid's capabilities. Transmission is

7

persistently constrained and congestion costs have risen. As investment in transmission declined for a quarter century, electricity demand grew by 34% between 1992 and 2007. Most importantly, the regulatory path for facilities that could link major renewable and low-carbon resources to consumers many hundreds of miles away is a long and winding road. Barriers to transmission upgrades and expansions often delay or even deter the development of facilities truly needed for a low-carbon energy environment. The National Renewable Energy Laboratory, in a recent report entitled _20% Wind Energy By 2030_ (May 2008), has identified some of these barriers:

A. **Transmission Planning**. Upgrades and expansions of the transmission system serve numerous purposes. They meet the needs of the next increment of generation, sustain the reliability of the electrical system as a whole, and serve an evolving need for a flexible low-carbon energy mix over the long term. Planning these enhancements, and execution of such plans, must be regional and national while accommodating local concerns. It should also anticipate the development of broad areas or "zones" of location-constrained renewable energy resources.

Sound transmission planning (to analyze benefits and costs and the distribution of benefits for the purpose of allocating costs) should incorporate a number of features. Yet, there is no generally accepted planning regime for these interstate facilities. WIRES believes the following:

o Transmission planning and analysis should be done on a regional level – tending toward larger regions as a general rule. While the overall planning

8

process must encompass a large region, the planning studies cannot lose sight of the impacts on sub-regions.

o Transmission planning and analysis should include all of the demand loads (existing and anticipated) and all of the supply resources (existing and anticipated) located within the geographic region for which planning is taking place.

o Transmission planning should occur in a process that is open, transparent, and inclusive, and conducted by a credible entity without particular attachment to specific interests or market outcomes in the region. In other words, it should be compliant with the planning principles of FERC's Order No. 890.

B. Allocation of Costs. Public policy must provide a clear and consistent guide to who pays for additions to the electron superhighway, *i.e.*, the high voltage grid that has such broad regional benefits. While cost allocation may vary regionally, WIRES, as well as the NREL study, believe it should be founded on fixed, clear, and equitable principles, particularly where multi-state facilities are concerned. No generic principles guide the allocation of costs of transmission, which produces great difficulty when the facilities at issue cross multiple jurisdictions with varying regulatory criteria. Where transmission investment was once only a candidate for system-specific rate base, today such costs can be allocated to users of regionally-interconnected systems. They can be very diverse. In both organized markets (i.e., markets run by regional transmission organizations ("RTOs")) and non-RTO bilateral markets, the disputes over cost allocation and cost recovery, and the procedural delays occasioned by these disputes, can be prolonged and counter-productive.

9

There are numerous ways to allocate costs. At one end of a spectrum of approaches is so-called participant funding which seeks to allocate costs of a transmission upgrade or expansion to immediate "cost causers" such as interconnecting generators, even if facilities may have regional reliability or economic benefits. At the other end of the spectrum is the "socialization" of costs, meaning a broad allocation of all project costs to the perceived beneficiaries of the project across the market or region served. Different perceptions of the equities and the reliability or economic benefits of a grid expansion have often chilled transmission investment. The debate over cost allocation remains largely unresolved and many of our members identify cost allocation as the greatest deterrent to transmission development.

In 2007, WIRES commissioned an independent study of how best to allocate the costs of transmission. Entitled A National Perspective On Allocating the Costs of New Transmission: Practice and Principles, it is available on the WIRES website. It does not advocate "one size fits all," but instead a principled approach to determining what is the just and reasonable way to assign cost responsibility.

    C. Cost Recovery. As a general rule, when state-regulated investor-owned companies invest in transmission assets, that investment typically goes into state-jurisdictional rate base subject to retail regulation. Retail customers are then asked to pay for those facilities in their rates even if the benefits of the facilities are traceable to beneficiaries beyond the utility's service territory. These rates can overlap with federal transmission rates established to recover costs from third parties that utilize the lines in an open access environment. This dual-pricing

10

system complicates the allocation of costs and makes cost responsibility subject to various interests that have different public policy agendas.

The NREL study argues that this effectively dilutes incentives for development provided by the FERC under the 2005 Energy Policy Act and other laws and creates substantial regulatory uncertainty.

    D. **<u>Facilities Siting</u>**. Laws governing the siting of transmission date from an era when utilities were generally not interconnected and the modern network of interstate lines and multi-state interconnections did not exist. According to NREL, the need to connect location-constrained generation resources to growing load centers over long distances, in part to implement climate change laws and renewable portfolio standards, requires a new regulatory approach.

Facilities siting is an intractable problem that often leaves all parties dissatisfied and the long-term interests of electricity consumers ignored. Congress sought a balanced approach to siting transmission facilities when it adopted Section 216 of the Federal Power Act in 2005. That provision allows FERC to site transmission as a "back-stop" to state procedures, and grant any necessary federal rights of eminent domain, only (1) if the facilities are located within broadly-defined corridors designated by DOE as experiencing significant market inefficiency, high prices, and threats to reliability that should be resolved through enhancement of the transmission system; (2) after states have had the opportunity to consider a project under their traditional authority to site facilities (or lack of such authority) and have failed to act in a

11

timely manner; and (3) pursuant to its own subsequent review, including environmental analysis under the National Environmental Policy Act and applicable laws, to ascertain what the public interest requires. FERC's effort to expand its ability to utilize the backstop authority in cases where a state provided a reasoned denial of a project application was recently reversed by a Court of Appeals.

The DOE carried out its responsibilities by designating two National Interest Electric Transmission Corridors ("NIETC"). The NIETC process did not site facilities or determine the outcome of transmission siting or a planning processes, or take property or preempt or undermine protection of environmentally or culturally sensitive areas or assets. DOE was hyper-conscientious not to pick winners and losers or specify a required route for any line. Yet, the statutory process resulted in a perfect storm of controversy, delay, and inaction. To this date, FERC has not been formally called upon to exercise its authority under section 216 of the Federal Power Act. The NIETC process was never intended to be a planning device. And it has marginal value as a goad to state action.

While an arguably valid attempt to address the obvious mismatch between the interstate operation of the grid at the high voltages and the exclusive authority of states to determine if such lines are needed and can be constructed, the NIETC process has failed to resolve the delays that inevitably accompany the transmission siting process. Indeed the lead-time for planning and constructing transmission – which is already substantial -- promises to remain so.

The NIETC process may also fail to achieve its goals for two additional but related reasons. First, transmitting large amounts of remotely located renewable generation to load will unquestionably entail entirely new high-voltage network additions that will cross multiple

12

jurisdictions in many circumstances. The need to take advantage of these domestic, "location-constrained" renewable and clean-coal resources will be central to any climate change and energy independence goals. Development of these generating facilities await some indication that transmission capacity will be available to them. Yet, DOE's focus in implementing corridors focuses on transmission constraints and congestion that already exist. Second, upgrades or expansions to the grid may also be necessary to ensure electric reliability for our digital society, promote energy security, or meet economic development and demographic trends. Section 1221 of EPAct, which adopted section 216 of the FPA, permits DOE to take these forward-looking factors into account when designating corridors but it has largely chosen not to do so. I am unsure whether this reflects a reading of the law or a practical decision about the difficulties of formulating future plans for integrating alternative energy resources.

In the final analysis, delay in selecting and building the right transmission in the right place to serve the right generation resources cannot be good for consumers.

IV. Conclusion

WIRES does not argue that transmission is a singular solution to the challenges facing the electric industry. On the other hand, WIRES is persuaded that the high voltage network provides benefits that are unattainable in other ways. It will, however, require modernization and technical innovation. If we are to fulfill our national ambition of a more secure, environmentally sustainable, and efficient power system, we need a workable regulatory process that ensures that transmission can be built on a timely basis, based on collaboration with stakeholders and a clear regulatory path to completion. That regulatory regime must be regional in nature. Under current circumstances, such a regime will require federal leadership.

13

WIRES has proposed a pragmatic redesign of federal regulation of the grid to address each of the challenges I outlined above. It is available on www.wiresgroup.com and I have attached it to this testimony.

"Smart grid" technologies may help reduce the difficulties of siting by obviating the development of new rights of way in many instances. Those smart grid investments will nevertheless encounter the same cost allocation and cost recovery problems that transmission already faces. Finally, if the vision of a clean energy economy with substantial contributions from renewable resources and electric vehicles is to be realized, it will be realized in part by a vibrant and liquid interstate bulk power market based on a platform of adequate transmission capacity.

Thank you once again for inviting me to make this presentation. WIRES looks forward to working with you, Mr. Chairman, and the Committee to attract investment to the transmission system. I will be pleased to take your questions.

14

The NETL Modern Grid Initiative

A VISION FOR
THE MODERN GRID

Conducted by the National Energy Technology Laboratory
for the U.S. Department of Energy
Office of Electricity Delivery and Energy Reliability
March 2007

Office of Electricity
Delivery and Energy
Reliability

V1.0

TABLE OF CONTENTS

WHY WE NEED A VISION

Before we can begin to modernize today's grid, we first need a clear vision of the power system required for the future. Given that vision, we can create the alignment necessary to inspire passion, investment, and progress toward an advanced US grid for the 21st century.

A modernized grid is a necessary enabler for a successful society in the future. Modernizing today's grid will require a unified effort by all stakeholders rallying around a common vision. Throughout the 20th century, the U.S. electric power delivery infrastructure served our nation well, providing adequate, affordable energy to homes, businesses and factories. This once state-of-the-art system brought a level of prosperity to the United States unmatched by any other nation in the world. But a 21st-century U.S. economy cannot be built on a 20th-century electric grid.

Many agree there is an urgent need for major improvements in the nation's power delivery system and that advances in key technology areas can make these improvements possible. But more is needed. The Modern Grid vision will set the foundation for a transition that will focus on meeting the six key goals discussed below:

The grid must be more reliable. A reliable grid provides power dependably, when and where its users need it and of the quality they value. It provides ample warning of growing problems and withstands most disturbances without failing. It takes corrective action before most users are affected.

The grid must be more secure. A secure grid withstands physical and cyber attacks without suffering massive blackouts or exorbitant recovery costs. It is also less vulnerable to natural disasters and recovers more quickly.

The grid must be more economic. An economic grid operates under the basic laws of supply and demand, resulting in fair prices and adequate supplies.

The grid must be more efficient. An efficient grid takes advantage of investments that lead to cost control, reduced transmission and distribution electrical losses, more efficient power production and improved asset utilization. Methods to control the flow of power to reduce transmission congestion and allow access to low cost generating sources including renewables will be available.

The grid must be more environmentally friendly. An environmentally friendly grid reduces environmental impacts through initiatives in generation, transmission, distribution, storage and consumption. Access to sources of renewable energy will be expanded. Where possible, future designs for Modern Grid assets will occupy less land reducing the physical impact on the landscape.

The grid must be safer. A safe grid does no harm to the public or to grid workers and is sensitive to users who depend on it as a medical necessity.

The nation's grid should be modernized not by randomly gathering a group of interesting technologies and calling it modern, but rather by first building a vision and the framework that enables that vision. The systems view taken by the NETL MGI team provides such a comprehensive perspective.

This document describes our vision for the Modern Grid. The vision discussed in this document modifies the traditional approach that was used in developing today's grid. In addition to continuing the traditional approach based on large, remote, central generating stations (coal, nuclear, hydroelectric, natural gas, renewables, etc.) providing energy to consumers through extensive transmission systems, this vision recognizes the major benefits the distribution system and end user involvement can provide. By blending the traditional centralized model with one that embraces distributed resources, demand response, advanced operational tools and networked distribution systems, we can enjoy the benefits of both and minimize the negative aspects of each. The application of modern computing, communications and materials sciences will enable this transformation.

A Call to Action is included at the end of this document to solicit your input on further developing the vision and other aspects of the Modern Grid Initiative.

THE VISION

Vision statements are often used to help define the future state for government and business. A vision statement for the Modern Grid could be stated as:

"To revolutionize the electric system by integrating 21st century technology to achieve seamless generation, delivery, and end-use that benefits the nation."

But a vision statement such as the one above does not adequately convey the level of detail necessary to support a clear understanding for stakeholders.

Additional detail that supports this vision statement is needed to provide a platform for debate and an opportunity to gain consensus among all stakeholders. Beginning with the goals described in the previous section and the broad vision statement presented above, we must move to the next level of detail in defining the overall vision. This will provide the level of understanding needed to stimulate debate and ultimately enable stakeholder alignment.

But what are the defining characteristics of this vision? These characteristics should describe the features of the grid in terms of its functionality rather than in terms of specific technologies that may ultimately be needed. Reaching a vision that includes the seven characteristics described below will enable the Modern Grid to achieve its goals.

First, it will heal itself. The modernized grid will perform continuous self-assessments to detect, analyze, respond to, and as needed, restore grid components or network sections. It will handle problems too large or too fast-moving for human intervention. Acting as the grid's "immune system", self-healing will help maintain grid reliability, security, affordability, power quality and efficiency.

The self-healing grid will minimize disruption of service, employing modern technologies that can acquire data, execute decision-support algorithms, avert or limit interruptions, dynamically control the flow of power, and restore service quickly. Probabilistic risk assessments based on real-time measurements will identify the equipment, power plants and lines most likely to fail. Real-time contingency analyses will determine overall grid health, trigger early warnings of trends that could result in grid failure, and identify the need for immediate investigation and action.

Communications with local and remote devices will help analyze faults, low voltage, poor power quality, overloads and other undesirable system conditions. Then appropriate control actions will be taken, automatically or manually as the need determines, based on these analyses.

Second, it will motivate consumers to be an active grid participant and will include them in grid operations. The active participation of consumers in electricity markets brings tangible benefits to both the grid and the environment, while reducing the cost of delivered electricity.

In the modernized grid, well-informed consumers will modify consumption based on the balancing of their demands and the electric system's capability to meet those demands. Demand for new cost-saving and energy-saving products will benefit both the consumer and the power system.

Demand-response (DR) programs will satisfy a basic consumer need: greater choice in energy purchases. The ability to reduce or shift peak demand allows utilities to minimize capital expenditures and operating expenses while also providing substantial environmental benefits by reducing line losses and the operation of inefficient peaking power plants. Over time, DR will also encourage consumers to replace inefficient end use devices such as incandescent lighting. In addition, emerging products like the plug-in hybrid vehicle will result in substantially improved load factors while also providing huge environmental benefits.

Third, the Modern Grid will resist attack. Security requires a system-wide solution that will reduce physical and cyber vulnerabilities and recovers rapidly from disruptions. The Modern Grid will demonstrate resilience to attack, even from those who are determined and well equipped. Both its design and its operation will discourage attacks, minimize their consequences and speed service restoration.

It will also withstand simultaneous attacks against several parts of the electric system and the possibility of multiple, coordinated attacks over a span of time. Modern grid security protocols will contain elements of deterrence, prevention, detection, response, and mitigation to minimize impact on the grid and the economy. A less susceptible and more resilient grid will make it a less desirable target of terrorists.

Fourth, the Modern Grid will provide the level of power quality desired by 21st century users. New power quality standards will balance load sensitivity with delivered power quality at a reasonable price. The modernized grid will supply varying grades of power quality at different pricing levels.

Additionally, PQ events that originate in the transmission and distribution elements of the electrical power system will be minimized and irregularities caused by certain consumer loads will be buffered to prevent impacting the electrical system and other consumers. The digital, high-tech economy has raised the bar for quality beyond the capabilities of today's grid.

Fifth, the Modern Grid will accommodate all generation and storage options. It will seamlessly integrate many types of electrical generation and storage systems with a simplified interconnection process analogous to "plug-and-play".

Improved interconnection standards will enable a wide variety of generation and storage options. Various capacities from small to large will be interconnected at essentially all voltage levels and will include distributed energy resources such as photovoltaic, wind, advanced batteries, plug-in hybrid vehicles and fuel cells. It will be easier and more profitable for commercial users to install their own generation (including highly efficient combined heat and power installations) and electric storage facilities.

Large central power plants including environmentally-friendly sources such as wind and solar farms and advanced nuclear plants will continue to play a major role in the Modern Grid. Enhanced transmission systems will accommodate their typically remote location. This wide variety of generation and storage options will enable the United State's dependence on foreign energy sources to be reduced.

Sixth, the Modern Grid will enable markets to flourish. Open-access markets expose and shed inefficiencies. The Modern Grid will enable more market participation through increased transmission paths, aggregated demand response initiatives and the placement of energy resources including storage within a more reliable distribution system that is closer to the consumer.

Parameters such as energy, capacity, rate of change of capacity, congestion, and resiliency may be most efficiently managed through the supply and demand interactions of markets. By reducing congestion, the modernized grid expands markets; it brings together more buyers and sellers. Consumer response to price increases felt through real time pricing will mitigate demand, driving lower-cost solutions and spurring new technology development. New, clean energy related products will also be offered as market options.

Finally, the Modern Grid will optimize its assets and operate more efficiently. Asset management and operation of the grid will be fine-tuned to deliver the desired functionality at a minimum cost. This does not imply that assets will be driven to their limits continuously but rather that they will be managed to efficiently deliver what is needed when it is needed.

Improved load factors and lower system losses are cornerstone aspects of optimizing assets. Additionally, advanced information technologies will provide a vast amount of data and information that will be integrated with existing enterprise-wide systems, significantly enhancing their ability to optimize operations and maintenance processes. This same information will provide designers and engineers with better tools for creating optimal designs. Planners who make recommendations for capital projects to increase system capacity will have the data they need to improve their processes. As a result, O&M and capital expenses will be more effectively managed.

The seven characteristics described above represent unique yet interdependent features that refine the Vision of the Modern Grid. Table 1 below summarizes these seven points and contrasts them between today's grid and the Vision of the Modern Grid.

Today's Grid	Principal Characteristic	Modern Grid
Responds to prevent further damage. Focus is on protection of assets following system faults.	**Self-heals**	Automatically detects and responds to actual and emerging transmission and distribution problems. Focus is on prevention. Minimizes consumer impact.
Consumers are uninformed and non-participative with the power system.	**Motivates & includes the consumer**	Informed, involved and active consumers. Broad penetration of Demand Response.
Vulnerable to malicious acts of terror and natural disasters.	**Resists attack**	Resilient to attack and natural disasters with rapid restoration capabilities.
Focused on outages rather than power quality problems. Slow response in resolving PQ issues.	**Provides power quality for 21st century needs**	Quality of power meets industry standards and consumer needs. PQ issues identified and resolved prior to manifestation. Various levels of PQ at various prices.
Relatively small number of large generating plants. Numerous obstacles exist for interconnecting DER.	**Accommodates all generation and storage options**	Very large numbers of diverse distributed generation and storage devices deployed to complement the large generating plants. "Plug-and-play" convenience. Significantly more focus on and access to renewables.
Limited wholesale markets still working to find the best operating models. Not well integrated with each other. Transmission congestion separates buyers and sellers.	**Enables markets**	Mature wholesale market operations in place; well integrated nationwide and integrated with reliability coordinators. Retail markets flourishing where appropriate. Minimal transmission congestion and constraints.
Minimal integration of limited operational data with Asset Management processes and technologies. Siloed business processes. Time based maintenance.	**Optimizes assets and operates efficiently**	Greatly expanded sensing and measurement of grid conditions. Grid technologies deeply integrated with asset management processes to most effectively manage assets and costs. Condition based maintenance.

Table 1: Comparison between Today's Grid and the Modern Grid

The Modern Grid is expected to perform consistent with this vision in all of its various operating modes. These performance modes include:

- **Emergency response** – A modernized grid provides advanced analysis to predict problems before they occur and to assess problems as they develop. This allows steps to be taken to minimize impacts and to respond more effectively.
- **Restoration** – It can take days or weeks to return today's grid to full operation after an emergency. A modernized grid can be restored faster and at lower cost as better information, control and communications tools become available to assist operators and field personnel.
- **Routine operations** – With a modernized grid, operators can understand the state and trajectory of the grid, provide recommendations for secure operation, and allow appropriate controls to be initiated. They will depend on the help of advanced visualization and control tools, fast simulations and decision support capabilities. Some operations will be fully automated when decisions need to be made faster than is possible by operators.
- **Optimization** –A modernized grid provides advanced tools to understand conditions, evaluate options and exert a wide range of control actions to optimize grid performance from reliability, environmental, efficiency and economic perspectives. New peak-shaving and load factor-improving strategies are employed.
- **System planning** – Grid planners must analyze projected growth in supply and demand to guide their decisions about what to build, when to build and where to build. Modern Grid data mining and modeling will provide much more accurate information to answer those questions.

The details described above expand on the vision statement presented earlier and establish a foundation from which debate among all stakeholders can begin. The ultimate objective is to reach a national consensus for the vision of the Modern Grid.

Summary

These seven characteristics describe a vision for the Modern Grid that is generally more resilient and distributed, more intelligent, more controllable and better protected than today's grid.

Advancements in large, centralized generating stations and higher capacity, more controllable transmission lines will continue to be needed and will complement the benefits of shifting to a more distributed grid model. This vision will enable the Modern Grid to benefit from a unique and more synergistic utilization of the transmission and distribution systems and active involvement by end users to meet the 21st century needs of consumers and society. Significant opportunities exist to apply modern communications, computing technologies and advancements in materials to achieve this Modern Grid vision.

Much work remains to be done to achieve this vision. The integration of existing technologies, the development of new ones and integrated testing to show their benefits are all needed. Regulatory and legislative reform to modify regulations and statutes that are inconsistent with this vision is also needed. New standards must be developed and some existing standards will require changes. Various process issues must be resolved. We also need metrics to provide the milestones for measuring our progress towards this vision. And perhaps, most important of all, the totality of societal benefits must be included in the calculus of Modern Grid investments to provide the financial incentive needed to move us forward.

A clear understanding and consensus for this vision among all stakeholders will generate a huge force for change. Only through their aligned efforts can this vision for the Modern Grid become a reality. It is a big job, but we can do it by working together. The work is already underway at MGI and its partner organizations. But we need your active support in making grid modernization an essential part of our national energy policy. Your active participation is essential as we lay out the framework for a modernized grid that can enable our nation's future growth and preserve our global competitiveness and way of life.

CALL TO ACTION

Creating the Modern Grid will require a monumental effort by all stakeholders. With a clear vision, we can generate the alignment needed to inspire passion, investment, and movement toward that vision. Your input is needed, along with your acceptance, which will ultimately lead to a national consensus for the Modern Grid vision.

We want your thoughts. Visit our website at www.TheModernGrid.org to find out how you can become involved or to speak directly with a team member.

For more information

This document is part of a collection of documents prepared by The NETL Modern Grid Initiative team. All are available for free download from the Modern Grid Web site.

The NETL Modern Grid Initiative

Website: www.netl.doe.gov/moderngrid

Email: moderngrid@netl.doe.gov

(304) 599-4273 x101

Appendix B1:
A Systems View of the Modern Grid

INTEGRATED
COMMUNICATIONS

Conducted by the National Energy Technology Laboratory
for the U.S. Department of Energy
Office of Electricity Delivery and Energy Reliability
February 2007

Office of Electricity
Delivery and Energy
Reliability

v2.0

TABLE OF CONTENTS

EXECUTIVE SUMMARY

The United States urgently needs a fully modern power grid if we are to meet our country's requirements for power that is reliable, secure, efficient, economic, and environmentally responsible.

To achieve the modern grid, a wide range of technologies must be put into operation. These technologies can be grouped into five key technology areas, as seen in Figure 1:

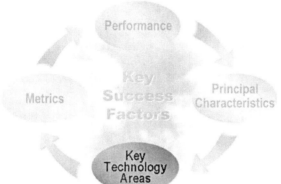

- **Integrated Communications**
- Sensing and Measurement
- Advanced Components
- Advanced Control Methods
- Improved Interfaces and Decision Support

Figure 1: The Modern Grid Systems View provides an "ecosystem" perspective that considers all aspects and all stakeholders.

Of these five key technology areas, the implementation of integrated communications is a foundational need, required by the other key technologies and essential to the modern power grid. Due to its dependency on data acquisition, protection, and control, the modern grid cannot exist without an effective integrated communications infrastructure. Establishing these communications must be of highest priority since it is the first step in building the modern grid.

Integrated communications will create a dynamic, interactive "mega infrastructure" for real-time information and power exchange, allowing users to interact with various intelligent electronic devices in an integrated system sensitive to the various speed requirements (including near real-time) of the interconnected applications.

As a first order of business, there is a need to specify the technical requirements for the system (e.g., speed, redundancy, reliability). Various utility applications have different demands, and these must be fully defined up front.

Second, standards development must be seriously addressed and encouraged. Although communications media technologies are being developed very rapidly, their widespread deployment will be seriously delayed unless the development of universal standards is accelerated.

This paper covers the following four important topics:
- Current state of integrated communications
- Future state of integrated communications
- Benefits of implementation
- Barriers to deployment

Although it can be read on its own, this paper supports and supplements "A Systems View of the Modern Grid," an overview prepared by the Modern Grid Initiative team.

CURRENT STATE

Before we see what the modern grid will look like with integrated communications in place, we will first consider the present state of communications in our nation's power grid.

The communications systems utilized in the power industry today are too slow and too localized to support the integrated communications needed to enable the modern power grid. An open communications architecture that supports "plug and play" interoperability is needed. Further, universally accepted standards for these communications must be defined and agreed upon in the industry.

COMMUNICATIONS STANDARDS

For communications in the grid to be truly effective, they must exist in a fully integrated system. And to be fully integrated, universal standards must be applied. Although numerous communication standards already exist today, the establishment and adoption of *universal* standards by users, vendors, and operators is lacking but greatly needed. Until these universal standards are set for the various functionalities required by the modern grid, investors will be reluctant to invest, and lack of funding will severely limit attainment of a modern grid.

One exception is in the area of substation automation (SA). The International Electrotechnical Commission (IEC), a recognized authoritative worldwide body responsible for developing consensus in global standards in the electro-technical field, has developed IEC 61850 for SA. This appears to have become the universally adopted standard for SA. Additional IEC standards for advanced meter reading (AMR), demand response (DR), and other modern grid features are expected to be adopted in the future.

However, at present universally adopted standards do not yet exist for most user-side features such as AMR and DR. In his *A Strawman Reference Design for Demand Response Information Exchange* (Draft), Erich Gunther of EnerNex Corporation recommends the formation of "an industry-driven working group to work out the details of the reference design and set up the mechanisms for already existing standards bodies to contribute."

The question of setting standards is expected to be addressed by the Open AMI Technical Subcommittee. Open AMI is a task force working under the UCA International Users Group, a non-profit organization whose members are utilities, vendors, and users of communications for utility automation. One of Open AMI's specific objectives is to "define what open standards means for advanced metering and demand response."

In another example of the search for common standards, the Institute of Electrical and Electronics Engineers (IEEE) is currently working on standards for Broadband over Power Line (BPL) technologies. However, user standards, such as advance metering and DR, have not yet been developed. Standards development, testing, and adoption could take five to ten years to complete.

COMMUNICATIONS MEDIA AND TECHNOLOGIES

A variety of communications media are used in today's electric grid, including copper wiring, optical fiber, power line carrier technologies, and wireless technologies. Using these media, many U.S. facilities have deployed SA, an excellent first step in integrating grid communications. However, SA does not yet fully integrate with the other features that will modernize our power grid.

Limited deployment of distribution automation (DA) has also occurred. Low speed transmission supervisory control and data acquisition (SCADA) and energy management system (EMS) applications have been successfully integrated among regional transmission organizations, generators, and transmission providers. But these applications still lack full utilization of the integrated, high-speed communications system required by the modern grid.

Power line carrier technology has been in use for many years in the utility industry. Recently, BPL carrier technologies have been developed and successfully demonstrated on a pilot basis. Also, wireless technologies are currently being developed and demonstrated, but they are not yet used in the grid communications infrastructure on either the system or the user side.

The current state of communications technologies described in the three tables below are in various stages of availability, deployment, or development.

Broadband over Power Line

Originally focused on Internet access and voice over Internet protocol for consumers, BPL is increasingly being deployed to meet utility needs for distributed energy resources (DER), AMR, DR, and consumer portal applications, as well as DA and video monitoring (primarily for security) applications and other high-speed data needs on the system side.

Broadband over Power Line	
Name	**Description**
Broadband over power line	• Meets some utility needs for AMR, DER, DR, and consumer portal applications, as well as DA, video monitoring, and other high-speed data applications
	• Deployable only over low- and medium-voltage distribution facilities
	• Demonstrated in over 30 pilots and trials
	• Has not penetrated the communications market as the lead candidate for supporting the modern grid's communications infrastructure
	• Deployment and integration with distribution facilities currently limited
	• Numerous vendors are aggressively marketing these products
	• Next-generation systems now under development promise lower cost, improved performance, higher speed, and utility applicability
	• Application at transmission voltages may also be viable
	• Radio frequency interference with ham radio identified in some BPL technologies; however, techniques have been developed and appear effective in eliminating the interference

Table 1: Broadband over power line (BPL) technology

Wireless Technologies

Various wireless technologies are emerging as possible candidates for the communications infrastructure of the modern grid. To date, few of them have made significant market penetration in either system- or user-side applications.

Table 2: Wireless Technologies

Wireless Technologies	
Technology	**Description**
Multiple address system radio	• Consists of a master radio transmitter/receiver and multiple remote transmitters/receivers
	• Master can access multiple units
	• Can be used as a repeater radio to transmit signals over or around obstructions
	• Used widely by utilities for SCADA systems and DA systems
	• Flexible, reliable, and compact

Wireless Technologies	
Technology	**Description**
Paging networks	• Radio systems that deliver short messages to small remote mobile terminals • One-way messaging is cost effective, but two-way is generally cost prohibitive • Some paging standards exist, but many systems remain proprietary
Spread spectrum radio systems	• Used in point-to-multipoint radio systems • Can operate unlicensed in 902-928 MHz band but must continually hop over a range of frequencies • Line of sight is needed for optimal coverage • Often used as last-mile connection to a main communications system
WiFi	• Utilizes IEEE 802.11b and IEEE 802.11g • Data transfer rates range from 5 – 10 Mbps for 802.11b and up to 54 Mbps for 802.11g • Effective for in-office or in-home use • Range is only about 100 meters
WiMax	• Utilizes IEEE 802.16 • Provides longer distance communications (10 – 30 miles) with data transfer rates of 75 Mbps • May be used as the spine of a transmission and distribution communications system that will support WiFi applications for SA or DA • Can communicate out-of-sight using IEEE 802.16e and can communicate with moving vehicles • Communicates point-to-point with different vendors
Next-generation cellular (3G)	• Can be applied as a low-cost solution for SA to control and monitor substation performance when small bursts of information are needed • May not meet the quality needs of online substation control and monitoring • Expected to be cost effective and quickly implemented • Coverage may not be 100% (some dead zones)
Time division multiple access (TDMA) Wireless	• Digital cellular communication technology that allocates unique time slots to each user in each channel • Utilizes IS-136 standard • Two major (competing) systems split the cellular market: TDMA and CDMA (see below); third-generation wireless networks will use CDMA

Wireless Technologies	
Technology	**Description**
Code division multiple access (CDMA) wireless	• Has become the technology of choice for the future generation of wireless systems because network capacity does not directly limit the number of active radios; this is a significant economic advantage over TDMA • Has been widely deployed in the United States • Utilizes the IS-95 standard which is being supplanted by IS-2000 for 3G cellular systems
Very small aperture terminal (VSAT) satellite	• Provides new solutions for remote monitoring and control of transmission and distribution substations • Can provide extensive coverage • Can be tailored to support substation monitoring and provide GPS-based location and synchronization of time (important for successful use of phasor measurement units) • Quickly implemented • High cost, except for remote locations • Functionality effected by severe weather

Table 2: Wireless technologies

Other Technologies

The table below includes other communication technologies that support, or could support, the modern grid.

Table 3: Other Technologies

Other Technologies	
Technology	**Description**
Internet2	• Next-generation high-speed internet backbone • More than 200 universities are working to develop and deploy advanced network applications
Power-line carrier	• Supports advanced metering infrastructure (AMI) deployments and grid control functions, such as load shedding • Communicates over electric power lines • Provides low-cost, reliable, low- to medium-speed, two-way communications between utility and consumer

Other Technologies	
Technology	**Description**
Fiber to the home (FTTH)	• Provides a broadband fiber-optic connection to customer sites
	• Costs of installation and associated electronics prohibitive
	• For decades, has been the "holy Grail" of the telecommunications industry, promising nearly unlimited bandwidth to the home user
	• To be cost-effective, needs passive optical network, which permits a single fiber to be split up to 128 times without active electronic repeaters; general decrease in cost of electronics is also helpful
Hybrid fiber coax (HFC) architecture	• Uses fiber to carry voice, video, and data from the central office (head end) to the optical node serving a neighborhood
	• Cable operators have begun plant upgrades using HFC to provide bi-directional services, such as video-on-demand, high-speed Internet, and voice-over-Internet protocol
Radio frequency identification (RFID)	• Uses radio frequency communication to identify objects
	• Provides an alternative to bar codes
	• Does not require direct contact or line-of-sight scanning
	• Low-frequency systems have short ranges (generally less than six feet); high-frequency systems have ranges of more than 90 feet

Table 3: Other technologies

No limitations are expected in the development of any of the media commonly used today (copper, fiber, power-line carrier, and wireless technologies). Radio frequency interference has been identified in some BPL technologies, but this issue is not expected to have a major impact on the future development and deployment of BPL.

The Common Information Model (CIM) is the industry standard for monitoring and controlling enterprise computing environments. Lessons learned from applying CIM to solve past data exchange issues will be applied to the integrated communications infrastructure of the modern grid. Through CIM techniques, the seamless interchange of all data with all applications and users can be achieved.

FUTURE STATE

An effective, fully integrated communications infrastructure is an essential component of the modern grid

Integrated communications will enable the grid to become a dynamic, interactive medium for real-time information and power exchange. When integrated communications are fully deployed, they will optimize system reliability and asset utilization, enable energy markets, increase the resistance of the grid to attack, and generally improve the value proposition for electricity.

Through advanced information technology, the grid system will be self-healing in the sense that it is constantly self-monitoring and self-correcting to keep high quality, reliable power flowing. It will also sense disturbances and instantaneously counteract them or reconfigure the flow of power to mitigate damage before it can propagate. The integrated communications infrastructure is necessary to enable the various intelligent electronic devices (IEDs), smart meters, control centers, power electronic controllers, protection systems, and users to communicate as a network.

Figure 2 gives one view of the complexity of the integrated communications systems required to support the modern grid.

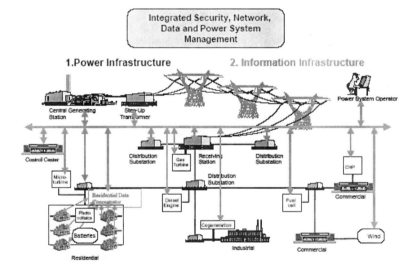

Source: EPRI

Figure 2: Communication environments: Integration of enterprise and power System management. Image courtesy of EPRI.

These integrated systems will provide two fundamental functions that will effectively support modern grid operations:

- **Open communications standards** that have the necessary intelligence to enable information to be recognized and understood by a wide assortment of senders and receivers
- **Appropriate media** that will provide the necessary infrastructure to transmit information accurately, securely, reliably, and at the required speed with the required data throughput

Most importantly, it is these two functions that will instill confidence in investors and motivate them to invest in the other key technology areas required by the modern grid.

High-speed, fully integrated, two-way communications technologies will allow much-needed real-time information and power exchange. Open architecture will create a plug-and-play environment that networks the grid components together for talk and interaction.

Universal standards will provide for all sensors, IEDs, and applications to communicate seamlessly at the speed necessary to support all required functions. These standards, when adopted by all parties, will provide confidence to stakeholders that their investments in integrated communications for the grid will not be stranded.

The integrated communications infrastructure of the modern grid will possess the following characteristics:

- **Universality** – All potential users can be active participants.
- **Integrity** – The infrastructure operates at such a high level of manageability and reliability that it is noticed only if it ceases to function effectively.
- **Ease of use** – Logical, consistent, and intuitive rules and procedures are in place for the user.
- **Cost effectiveness** – The value provided is worth the cost.
- **Standards** – The basic elements of the infrastructure and the ways in which they interrelate are clearly defined and remain stable over time.
- **Openness** – The public part of the infrastructure is available to all people on a nondiscriminatory basis.
- **Security** – The infrastructure is able to withstand security attack, and users have no fear of interference from others.
- **Applicability** – The infrastructure will have sufficient bandwidth to support not only current functions but also those that will be developed in the future.

BENEFITS OF IMPLEMENTATION

One of the main benefits to be gained from implementation of integrated communications will be the grid's ability to self-heal.

The near real-time acquisition and transfer of data will support the grid's ability to detect, analyze, and respond autonomously to adverse trends and conditions. Further, integrated communications will enable the development of new, real-time analytical tools, including wide area measurement technologies that will assist system operators in predicting and preventing events that negatively affect grid reliability and will also aid in the post-mortem analysis of such events.

Another benefit is that the grid will become more reliable when integrated communications are in place because they will make possible a broader application of alternative resources, including renewables that depend on an integrated communications system to become an effective part the grid system. A more effective and reliable dispatch of centralized generation, flow and VAr control, DER, and DR resources will be available to system operators by the near real-time data provided by integrated communications.

The grid will be more secure from outside threats when integrated communications have been implemented. All will benefit when the availability of near real-time data over a secure communications infrastructure provides detection and mitigation of both cyber and physical threats to the grid (See "Appendix A3: Resists Attack"). As an additional bonus, the integrated communications system will facilitate security monitoring of even non-grid infrastructures because the electric grid physically reaches virtually all other sensitive societal systems.

Another way in which Integrated communications will make the grid more secure is by providing the key data needed by emergency response organizations in a timely manner, which will reduce restoration times following major grid events.

As a further benefit, the environmental impact of producing power will be significantly reduced by the modern grid's integrated communications technologies. Providing the needed data will enable DER to be dispatched as a system resource, leading to an increased investment in DER (single units as well as larger DER "farms"), particularly those units that are environmentally friendly, such as wind, solar, and geothermal. The wide use of renewable DER and DR depends on the ability of the grid to address their intermittency and effectively integrate them with grid operations.

Significant economic benefits will follow implementation of integrated communications and the other key technologies of the modern grid.

The following list shows some of the economic benefits that will be enabled by integrated communications:

* The overall reliability of distribution and transmission systems will improve, leading to decreased costs and increased revenues.

* The grid will operate more economically for all stakeholders by making available the collection and transfer of market information, prices, and conditions to participants.

* The high-speed data needed for identifying and correcting power quality issues will be provided, leading to a reduction in quality-related costs currently incurred by consumers and grid operators. At the same time, equipment condition data needed by asset management processes will be provided, leading to a reduction in failure-related maintenance and outage costs.

* The need for new and costly hard assets will lessen as integrated communications technologies provide an alternative way to increase grid reliability rather than adding new and costly hard assets.

* Consumers will profit as well when the integrated communications infrastructure enables them to make financially smart energy choices. Providing price signals to consumers will motivate them to participate in the electricity market based on real supply and demand influences. Also, the integrated communications will link end users with communications options for non-utility applications, such as home security.

* Major long-term investments needed to increase system capacity will become more cost effective when asset-utilization data is integrated into the distribution and transmission planning models.

* The data and information made available to the modern grid using integrated communications technologies will also greatly benefit other enterprise-wide processes and technologies, such as asset management, work management, outage management, and also GIS-based systems.

BARRIERS TO DEPLOYMENT

Lack of an industry vision for integrated communications and a lack of understanding for the benefits of this technology are the greatest barriers to their deployment.

Research and development efforts are yielding communications media technologies that will support the needs of the modern grid. However, successful deployment of these technologies has not yet been achieved. Some of the gaps that must be overcome are described below:

- **There are no universal communications standards** that promote interoperability and enable the various communication technologies to work as an integrated suite.
- **So far, stakeholders have not developed and endorsed** a clearly defined communication architecture that will meet the requirements of the modern grid, or the transition plan needed to achieve such architecture. The transition plan needs to illustrate how to reach the desired future state without significant loss due to stranded investments.
- **Regional and national demonstrations of communications technologies are needed** to create interest, excitement, and the societal, political, and economic stimuli that will accelerate their deployment. It is likely, however, that different solutions will be required to address differing regional landscapes.
- **Regulatory and policy-setting bodies have not yet provided the regulations that will ensure that investments in new technologies will not lead to losses.** Deployment of modern grid technologies is costly, and without such incentives, utilities and energy providers are reluctant to invest in the needed technology areas even though these efficiency improvements will benefit the consumer and will provide great societal benefits, such as a cleaner, safer environment. Creative regulatory solutions are needed to assure that utilities and energy providers are protected financially (i.e., remain "revenue neutral.")
- **We do not yet have effective consumer education to create interest and motivation among the consumer groups.** Consumers can realize substantial benefits when the modern grid vision is achieved. Currently these benefits are not clear to the consumer. In order for consumers to value investment in communications systems, they must have a stronger link to grid operators and energy providers.
- **Vendors who supply sensors, IEDs, DER, and other end-use devices are hesitating to invest in these products until universal standards are adopted.** To compensate for the lack of universal standards, some vendors are creating their own proprietary solutions and protocols to enable them to bring specific products to market. This approach has the potential to create stranded investments and rework in the future as universal standards are ultimately adopted. There is also the danger that these vendor-specific protocols will

become industry standards by default rather than standards set through conscious and intelligent evaluation of what is most advantageous for all stakeholders.

POSSIBLE SOLUTIONS

The answer to overcoming these barriers to deployment lies in gaining buy-in from all stakeholders of the modern grid. Regulation and legislation, such as the Energy Policy Act of 2005, may serve as a catalyst for technology deployment, but more is needed.

Only by motivating all stakeholders to invest in the modern grid vision will widespread deployment of integrated communications be hastened. Here are three steps toward that goal:

- **Energy consumers must be informed about the cost of energy and the benefits of an integrated communication system.** An understanding of real-time pricing will motivate them to demand an integrated communications system that will support their ability to manage energy consumption.

  The technologies needed to motivate the consumer to invest include a cost-effective communications system that enables consumer-portal functionality and possibly broadband Internet service.

- **Energy companies need to clearly see the improved reliability, reduced cost, and increased revenues that integrated communications and the modern grid will bring.** This understanding will motivate them to work more quickly toward the universal standards needed to allay the natural fear of having large investments stranded due to changes in technology over time.

  Cost-effective and universally accepted standard communication technologies need broad acceptance. These will motivate energy companies to invest in applications that can satisfy their interests. Regional and national demonstrations of these technologies would bring the needed exposure to energy company executives.

- **Vendors will be motivated to invest in new products when they see a market for them.** As consumers and energy companies catch the vision, they will demand from vendors the next generation of products needed to support the modern grid.

In addition to increasing stakeholder demands for a communications infrastructure, a specific schedule of requirements needs to be established through regulation or legislation to accelerate completion of the universal standards and the deployment and marketing of associated communications technologies.

SUMMARY

Achievement of the modern grid vision is fully dependent on integrated communications technologies.

Implementation of these technologies, the first step toward achieving a truly modern power grid, will lead to major gains in reliability, security, economy, safety, efficiency, and improved environmental performance.

In general terms, an effective integrated communications system will provide the information and data necessary to optimize the reliability, asset management, maintenance, and operations required by a modern power grid.

The acceptance of universal standards will encourage the continued development and effective deployment of the needed communication infrastructure and other technologies. In addition, it is likely that demand will drive prices down.

Without a modern communications infrastructure, however, the modern grid cannot become a reality. Our nation's power grid must be updated through implementation of five key technology areas: Integrated Communications, Sensing and Measurement, Advanced Components, Advanced Control Methods, and Improved Interfaces and Decision Support. Of these five, integrated communications is of first importance since this technology enables the other four.

The electric utility industry has lagged behind other industries in taking advantage of the enormous strides in communication technology that have been made in the past decades. While the technologies needed to establish a modern grid are within reach, the industry has yet to focus on this opportunity.

Until these barriers are overcome, our power grid remains vulnerable to costly large-area blackouts such as was experienced in the Great Lakes region in 2003. Action is needed on the part of all stakeholders for integrated communications to be fully deployed and the multiple societal benefits of a modern grid to be realized. Integrated communications will open the way for the other key technology areas to be accepted and implemented, leading to the full modernization of our power grid.

For more information

This document is part of a collection of documents prepared by the Modern Grid Initiative (MGI) team. For a high-level overview of the modern grid, see "A Systems View of the Modern Grid." For additional background on the motivating factors for the modern grid, see "The Modern Grid Initiative."

MGI has also prepared five papers that support and supplement these overviews by detailing more specifics on each of the key technology areas of the modern grid. This paper has described the first key technology area, "Integrated Communications."

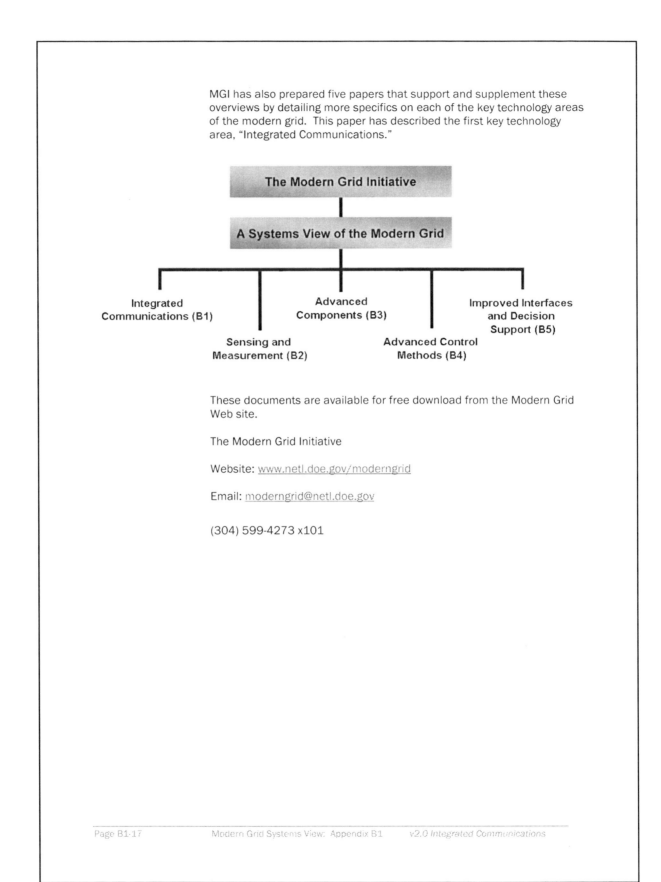

These documents are available for free download from the Modern Grid Web site.

The Modern Grid Initiative

Website: www.netl.doe.gov/moderngrid

Email: moderngrid@netl.doe.gov

(304) 599-4273 x101

BIBLIOGRAPHY

Electric Power Research Institute. 2004. Integrated energy and communications architecture: Volume IV: Technical analysis, appendix D, technologies, services, and best practices. Palo Alto, CA: EPRI.

Gunther, E. 2004. A strawman reference design for demand response information exchange. Report prepared for the California Energy Commission (draft).

Institute of Electronic and Electrical Engineers. 2003. IEEE-SA TR1550 Communication Requirements, Version 5.

Schmidt, R. and T. Lebakken. 2005. Broadband power line communications: Where is it? What to consider? Utility University course presented prior to Distributech Conference and Exhibition, San Francisco, CA.

Schwarz, K. 2004. IEC 61850 and UCA™ 2.0: A discussion of the history of origins.

Sumic, Z. and J. Spiers. 2004. The grid is becoming smarter: How about you? Stanford, CT: META Group.

Thorpe, J. 2004. Session V: Countermeasures. Synopsis presented at CRIS International Workshop on Power System Blackouts – Causes, Analyses, and Countermeasures, Lund, Sweden.

Yeager, K. E. and C. W. Gellings. 2004. A bold vision for T&D. Paper presented at the Carnegie Mellon University Conference on Electricity Transmission in Deregulated Markets, Pittsburgh, PA.

313

ACRONYMS LIST

AMI	Advanced Metering Infrastructure
AMR	Automatic Meter Reading
BPL	Broadband over Power Line
DA	Distribution Automation
DER	Distributed Energy Resources
DR	Demand Response
EMS	Energy Management System
GIS	Geographic Information System
IEC	International Electrotechnical Commission
IED	Intelligent Electronic Device
SCADA	Supervisory Control and Data Acquisition
VAr	Volt-amperes reactive

Appendix B2:
A Systems View of the Modern Grid

SENSING AND MEASUREMENT

Conducted by the National Energy Technology Laboratory
for the U.S. Department of Energy
Office of Electricity Delivery and Energy Reliability
March 2007

Office of Electricity
Delivery and Energy
Reliability

v2.0

TABLE OF CONTENTS

EXECUTIVE SUMMARY

The urgent need for major changes to our power delivery system requires that key technologies be developed and implemented in our nation's power grid. A wide range of technologies are needed and can generally be grouped into five key technology areas.

- Integrated Communications
- **Sensing and Measurement**
- Advanced Components
- Advanced Control Methods
- Improved Interfaces and Decision Support

Figure 1: The Modern Grid Systems View provides an "ecosystem" perspective of modern grid development.

Sensing and Measurement, the key technology area discussed in this paper, is an essential component of a fully modern power grid. Advanced sensing and measurement technologies will acquire and transform data into information and enhance multiple aspects of power system management.

These technologies will evaluate equipment health and the integrity of the grid. They will support frequent meter readings, eliminate billing estimations, and prevent energy theft. They will also help relieve congestion and reduce emissions by enabling consumer choice and demand response and by supporting new control strategies.

In the future, new digital communication technologies, combined with advanced digital meters and sensors, will support more complex measurements and more frequent meter reading. They also will facilitate direct interaction between the service provider and the consumer. Broadband over Power Line (BPL) and digital wireless communications are examples of technologies that can accomplish this interaction.

The core impacts of the sensing and measurement transformation further strengthen the case for their implementation. These technologies will fully empower the electric power market, allowing customer choice and input and resulting in savings in capital and operating costs, benefits to the environment through improved efficiency, and benefits to the economy and the public from enhanced safety, reliability, and power quality.

In this paper, we will look at the following important features:

- The current state of sensing and measurement
- The future state of sensing and measurement
- Benefits of implementation
- Barriers to deployment

Although it can be read on its own, this paper supports and supplements "A Systems View of the Modern Grid," an overview prepared by the Modern Grid Initiative team.

CURRENT STATE

The transformation of grid sensing and measurement technologies is happening today. In this section, we review how far industry and other investors have moved toward the realization of truly advanced capabilities.

In recent decades, the power industry has begun a move toward modernization by employing digital electronics in metering, typically with large customers whose usage and interval measurement requirements justify the added expense. In addition, a number of utilities have deployed automatic meter reading via a variety of communication channels. However, meter-reading intervals have not been shortened.

At the customer level, the transformation of the electric meter is still in its infancy. There have been a number of pilot projects and a few commercial installations using advanced meters to enable demand response (DR). However, the utility industry as a whole has not yet begun to deploy this strategy on a large scale. In addition, technology suppliers have been reluctant to invest in research and development before the market's scope and size is fully understood. Unless they can be retrofitted later, technologies invested in today may ultimately be "thrown away" if different technologies become the standard.

At the utility grid level, advancements are occurring more extensively on transmission than on distribution. Wide-area monitoring systems and dynamic line rating are helping achieve real transmission progress, while technologies such as electromagnetic signature measurement/analysis have only begun to be used to identify distribution equipment problems.

Today's challenge is to accurately envision the modern grid and successfully design the advanced sensing and measurement infrastructure that can enable it.

The technologies, processes, and research described here will contribute to the realization of a modern grid. Deployment times will vary, given differing states of development and their relative impacts on grid modernization.

Advances focused on both the customer and on the utility grid, including the related area of protection systems, are covered next in greater detail.

319

CUSTOMER-FOCUSED ADVANCES

Utilities can employ various pricing tools to shape customer usage patterns, resulting in benefits for electric energy users and providers alike. For example, some utilities have already implemented unique DR programs that combine conventional time-of-use pricing, real-time super-peak pricing, advanced metering, and appliance control. With such programs, the energy company controls the price signal while its customers decide how and when they will modify their usage patterns.

Table 1 and the figures that follow provide more detail on a number of customer-focused applications that are either currently available or under development.

Table 1: Customer-Focused Applications

Customer-Focused Applications	
Application	**Description**
Consumer gateway	• Bi-directional communications between service organizations and equipment on customer premises.
	• Advanced meter reading.
	• Time-of-use and real-time pricing (RTP).
	• Load control.
	• Metering information and energy analysis via website.
	• Outage detection and notification.
	• Metering aggregation for multiple sites or facilities.
	• Integration of customer-owned generation.
	• Remote power-quality monitoring and services.
	• Remote equipment performance diagnostics.
	• Theft control.
	• Building energy management systems.
	• Automatic load controls integrated with RTP.
	• Monitoring of electrical consumption of total load and, in some cases, various load components.
	• Functions embodied in meters, cable modems, set-top boxes, thermostats, etc.
	• The Electric Power Research Institute (EPRI) has performed substantial conceptual work in this area.

Customer-Focused Applications	
Application	**Description**
Residential consumer network See Figure 2 for a representative home network.	• Subset of the consumer gateway concept. • Reads the meter, connects controllable loads, and communicates with service providers. • End-users and suppliers monitor and control the use and cost of various resources (e.g., electricity, gas, water, temperature, air quality, secure access, and remote diagnostics). • Consumers monitor energy use and determine control strategies in response to price signals. • Products are available to a limited extent today. • Pacific Northwest National Laboratory, together with the Bonneville Power Administration, has piloted a comprehensive residential consumer network.
Advanced meter See Figure 3 for a representative advanced meter system	• Employs digital technology to measure and record electrical parameters (e.g., watts, volts, and kilowatt hours). • Communication ports link to central control and distributed loads. • Provides consumption data to both consumer and supplier. • In some cases, switches loads on and off. • Products are available from commercial vendors today.

Table 1: Customer-focused applications

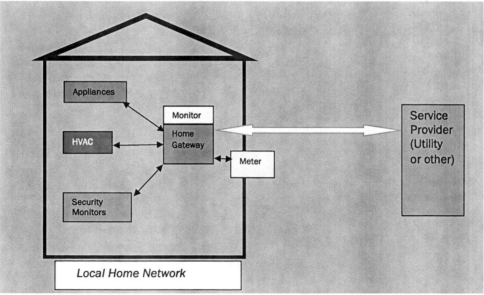

Figure 2: Example of a residential consumer network. Home gateway reads the meter, is tied to various end-use devices, and communicates with a service provider.

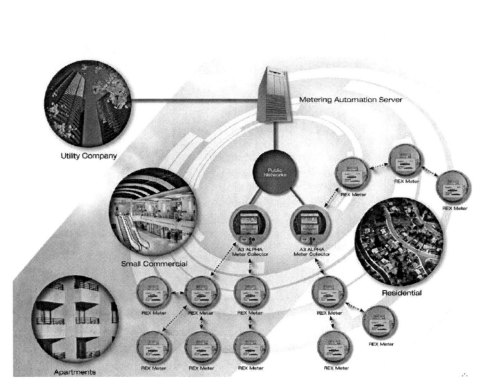

Figure 3: An example of meter-reading architecture. The electric meter measures a variety of electrical parameters and may also control metered load. Image courtesy of Elster Metering.

What is the Consumer Portal?

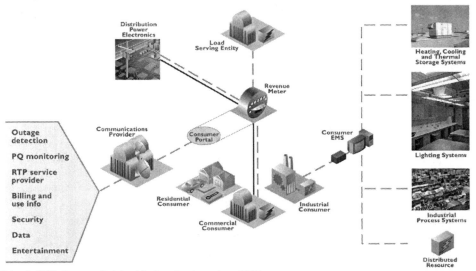

Figure 4: EPRI's Consumer Portal architecture. Image courtesy of EPRI.

Utility-Focused Advances

Utility Monitoring Systems

For most utilities, real-time monitoring systems provide up-to-date information on major substation equipment and some transmission line conditions. However, this is not true for most distribution facilities.

EPRI and the Tennessee Valley Authority have identified several requirements for advanced condition sensors, such as devices that determine the instantaneous condition of switches, cables, and other grid components.

They have found that first of all, costs must be low for the sensors, including their installation and maintenance. Second, inspections must be easily implemented, with special attention to hard-to-access locations, such as energized conductors on structures and inside cabinets. Third, the sensors must be small in size and secure from damage. Finally, they must not create problems related to electromagnetic compatibility (EMC).

Table 2 describes some advanced utility monitoring systems that are either commercially available or are currently under development.

Table 2: Advanced Utility Monitoring System

Advanced Utility Monitoring System	
System	Description
Wide-area monitoring system (WAMS) See Figure 5	• GPS-based phasor monitoring unit (PMU). • Measures the instantaneous magnitude of voltage or current at a selected grid location. • Provides a global and dynamic view of the power system, automatically checks if predefined operating limits are violated, and alerts operators. • Provides a global view of disturbances, as shown in Figure 5. • Compares generator operation points to allowable limits to keep generators in a safe state. • Tracks inter-area, low-frequency power oscillations and presents results to system operators; also used to tune damping controllers. • Combines phasor data with conventional SCADA (supervisory control and data acquisition) data for enhanced state estimation. • The Consortium for Electric Reliability Technology Solutions is one of several groups striving to develop WAMS technology for North America. • Of the many sensing and measurement technologies currently under development, WAMS may have the greatest potential for enabling grid reliability improvement.
Dynamic line rating technology See Figure 6 for an example	• Measures the ampacity of a line in real time. • The Power Systems Engineering Research Center has sponsored development of a computer program that calculates line sag and current-carrying capacity in real time using inputs from both direct and indirect measurements. • One university, with support from the National Science Foundation and a local utility, plans to develop a wireless network for dynamic thermal rating of a line at a target cost of $200 per sensor; this would allow the determination of dynamic thermal line rating for all spans, eliminating the need to identify a critical span.
Conductor/ compression connector sensor Image courtesy of EPRI.	• Measures conductor temperature to allow accurate dynamic rating of overhead lines and line sag, thus determining line rating. • Measures temperature difference between conductors and conductor splices. • Interrogation via helicopter, ground, BPL, or wireless. • Battery-free. • Unique serial number tied to asset GPS location.
Insulation contamination leakage current sensor Image courtesy of EPRI.	• Continually monitors leakage current and extracts key parameters. • Critical to determining when an insulator flashover is imminent due to contamination. • Clip-on, wireless, battery-free. • Unique identification number. • On-board storage of key parameters. • On-board storage of solar power. • Ability for interrogator to reset data.

Advanced Utility Monitoring System	
System	**Description**
Backscatter radio technology Image courtesy of EPRI.	• Provides improved data and warning of transmission and distribution component failure. • Communicates data back to a substation or other data collection point. • Small, low cost, reliable. • Battery-free, with minimal electronics and long service life. • Radiation-free for reduced EMC concerns. • Uses inexpensive, off-the-shelf components. • Always "awake," enabling fast inspection speeds.
Electronic instrument transformer	• Replaces precise electromagnetic devices (such as current transformers and potential transformers) that convert high voltages and currents to manageable, measurable levels. • Fiber-optic-based current and potential sensors, available from several venders, accurately measure voltage and current to revenue standards over the entire range of the device.
Other monitoring systems	• Fiber-optic, temperature-monitoring system: Provides direct, real-time measurement of hot spots in small and medium transformers, thus addressing utility concerns about the safety and reliable operation of high-voltage equipment. • Circuit breaker real-time monitoring system: Measures the number of operations since the last time maintenance was performed, as well as operation times, oil or gas insulation levels, breaker mechanism signatures, etc. • Cable monitor: Determines changes in buried cable health by trending partial discharges or through periodic impulse testing of lines. • Battery monitor: Minimizes battery failure by assessing cell health, specific gravity liquid level, cell voltage, and charge/discharge characteristics. • Sophisticated monitoring tool: Combines several different temperature and current measurements to dynamically determine temperature hot spots; measure dissolved gases in oil; evaluate the high-frequency patterns and signatures associated with faulty components and report the health of transformers and load tap changers in real time. • Many of these systems are commercially available.

Table 2: Advanced Utility Monitoring System

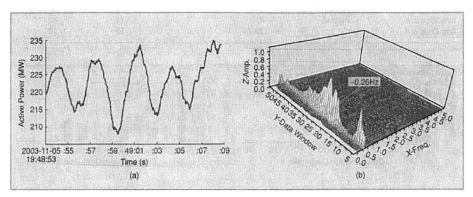

Figure 5: This sample image of a WAMS record-and-replay function display provides a global view of disturbances. Graphic under IEEE copyright. It appeared in the article, "WAMS Applications in Chinese Power Systems" in the *IEEE Power & Energy Magazine*, V4:1 (Jan/Feb 2006). Image courtesy of *IEEE Power and Energy Magazine*.

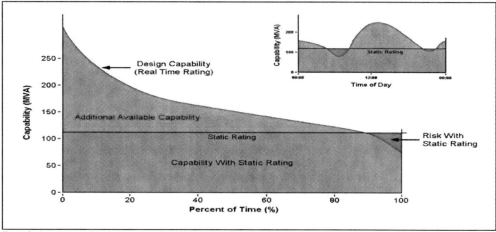

Figure 6: A real-time rating characteristic. Historically, line ratings have been based on assumed static conditions, but in-service conditions can differ substantially from those assumptions. This figure illustrates the dynamically increased power rating achievable with actual conditions provided by tower-mounted weather stations, line tension monitors, and/or visual cameras. Image courtesy of EPRI.

Advanced Protection Systems

In the past, more than 70 percent of major system disturbances have involved protective relaying systems, not necessarily as the initiating event but as a contributor to the cascading nature of the event. Today, products to reduce this problem are increasingly available. Compared to the electromechanical and analog relays of the past, new digital relays include many value-adding functions, such as fault location, high-impedance distribution fault detection, more sophisticated transformer and bus fault detection, self checking diagnostics, adaptive relaying and greater use of networked digital communications. As these new digital devices continue to be deployed throughout the grid, reliability will be significantly enhanced.

Table 3 describes two advanced protection technologies.

Advanced Protection System	
System	**Description**
Fault-testing recloser	• Applies a very fast, low-energy pulse to the line to determine if a fault is still present. • Minimizes damage caused by reclosing into faulted lines. • Significantly reduces damaging fault-current and voltage sags on the faulted line and adjacent feeders. • Substation transformers experience fewer through-faults, thus extending service life. • Cables, overhead conductors, splices, and terminations experience less thermal and mechanical stress from through-fault currents.
Special protection system	• Real-time monitoring of key generation assets or transmission lines and their associated power flows. • Upon a change of status (like loss of generation and/or loss of transmission), a pre-programmed set of actions takes place (e.g., wide-area load shed, generator redispatch, separation of interties). • Allows power transfers across the grid that would not comply with single or multiple contingencies under normal criteria. • Allows operators to load transmission lines closer to thermal limits or beyond normal voltage or system stability limits.

Table 3: Advanced protection systems

RELATED RESEARCH AND DEVELOPMENT

Research and development will support the integration of the sensing and measurement capabilities discussed previously as they come to market. In addition to utilities, EPRI and various equipment vendors are actively engaged in important R&D efforts.

Table 4: Research and Development Related to Sensing and Measurement.

Research and Development Related to Sensing and Measurement	
Name	**Description**
The Consortium for Electric Reliability Technology Solutions (CERTS)	• Is working to accelerate meaningful opportunities for customers to participate voluntarily in competitive electricity markets. • Studies focus on determining the effect of demand response on market efficiency and on demonstrating advanced demand-response technologies and strategies that will improve the reliability of the grid. • Work on WAMS contributes to this key technology area.

Research and Development Related to Sensing and Measurement	
Name	**Description**
The Power Systems Engineering Research Center (PSERC): Transmission and Distribution Technologies stem	• Reliability-based vegetation management through intelligent system monitoring. • Digital protection systems using optical instrument transformers and digital relays interconnected by an IEC 61850-9.2 digital process bus. • Optimal placement of PMUs for state estimation.
The California Energy Commission	• Advanced metering design, costs, and benefits. • Tariff design. • The evaluation of dynamic rates and programs for small customers. • The evaluation of DR programs for large customers.

Table 4: Research and Development Related to Sensing and Measurement

REQUIREMENTS AND REGULATIONS

Customer metering has always fallen within the purview of state regulatory bodies. So, too, have the tariffs that determine how and what a customer will pay. Hence, a metering transformation cannot occur without the support and encouragement of these regulators.

The Energy Policy Act of 2005 (EPAct) is very clear in this regard. The following sums up the spirit of this new law:

> It is the policy of the United States that time-based pricing and other forms of demand response, whereby electricity customers are provided with price signals and the ability to benefit by responding to them, shall be encouraged, and the deployment of such technology and devices that enable electricity customers to participate in such pricing and demand response systems shall be facilitated, and unnecessary barriers to demand response participation in energy, capacity, and ancillary service markets shall be eliminated.

The law is also very proactive in requiring electric suppliers to employ advanced metering and communications technology. It will certainly motivate regulators to more seriously address the need for these technologies.

FUTURE STATE

We have examined the current state of the sensing and measurement technologies. Now, we will look at how this key technology area will develop in the future.

At the customer level, the modern grid will have no electromechanical customer meters or meter readers. Instead, modern solid-state meters will communicate with both the customer and the service provider.

Microprocessors in these advanced meters will offer a wide range of functions. At a minimum, they will record usage associated with different times of day and costs of production. Most will also be able to register a critical peak-pricing signal sent by the service provider, charging at that critical rate while it is in effect. At the same time, the meter will notify the customer that the critical rate has been implemented.

A still more sophisticated version will adhere to a desired-usage profile preprogrammed by the customer. In response to fluctuating electricity prices, the unit will automatically control the customer's loads in accordance with that schedule.

The most sophisticated versions will even provide non-utility services, such as fire and burglar alarms.

This new metering approach will be built on the digital communications capabilities of the Internet, will employ standard Internet protocols, and will use reliable, ubiquitous communications media, such as wireless, BPL, or even fiber to the home. The customer interface will be user friendly, with increasing levels of sophistication as product features are added. Security will be designed to prevent tampering or disruption.

At the utility level, advanced sensing and measurement tools will supply expanded data to power-system operators and planners. This will include information about the following:

- Power factor
- Power quality throughout the grid
- Phasor relationships (WAMS)
- Equipment health and capacity
- Meter tampering
- Vegetation intrusion
- Fault location
- Transformer and line loading

- Circuit voltage profiles
- Temperature of critical elements
- Outage identification
- Power consumption profiles and forecasting
- Curtailable load levels

New host software systems will collect, store, analyze, and process the abundance of data that flows from these modern tools. The processed data will then be passed to the existing and new utility information systems that carry out the many core functions of the business (e.g., billing, planning, operations, maintenance, customer service, forecasting, statistical studies, etc.).

As the requirements for a modern grid crystallize, additional parameters will need to be calculated in the meters, and other measurement and sensing points will be desired. The architecture of a modern grid must allow retrofitting of advancements without the need for massive infrastructure change-out.

Future digital relays that employ computer agents will further enhance reliability. A computer agent is a self-contained software module that has properties of autonomy and interaction. Wide-area monitoring, protection and control schemes will integrate digital relays, advanced communications and computer agents. In such an integrated distributed protection system, the relays will be capable of autonomously interacting with each other. This flexibility and autonomy adds reliability because even with failures in parts of the system, the remaining agent-based relays continue to protect the grid.

The primary assumption underlying the realization of sensing and measurement is that the benefits of developing and implementing these technologies will exceed the cost. Hence, an important variable is the value assigned to those benefits, some of which reach beyond the utility function to impact society as a whole.

There is little doubt that modern digital technology can produce low-cost, highly effective solutions, with all such technological developments depending on two major factors:

- Scale of deployments
- The continued reduction in the price of digital integrated circuits

The projected scale of various deployments is enormous. A global metering transformation would employ hundreds of millions of intelligent, communicating meters. But, as Moore's Law consistently shows, the price of chips will continue to drop even as their processing power grows.

Also, as history has shown us, the associated requirement of ubiquitous, reliable, inexpensive communications will become increasingly available as the revolution in digital communications continues to play out (see "Appendix B1: Integrated Communications").

BENEFITS OF IMPLEMENTATION

The benefits of realizing the sensing and measurement key technologies are many. Some of the most important are listed below.

METER TRANSFORMATION

The "transformed meter," in the form of the consumer portal and other gateway technologies, will provide a wealth of information to customers and utilities alike.

Resulting benefits to consumers:

- The ability to make informed usage decisions
- A direct, real-time connection to the electricity market
- The motivation to participate in that market
- Reduced energy costs

Benefits to utilities:

- Greater load control
- Reduced operational costs
- Congestion relief
- Reduced energy theft

DATA COLLECTION

Advanced sensors and new measurement techniques will collect valuable information about electrical conditions throughout the grid. Advanced tools will then analyze system conditions, perform real-time contingency analyses, and initiate a necessary course of action, as needed.

Benefits of improved data collection:

- More effective asset utilization and maintenance
- Constant awareness of in-service equipment health and capacity
- Identification and prevention of potential failures and the rapid assessment of emergent problems
- Safer operation with alerts to operators when devices are about to fail

CONTROL INSTRUMENTATION

Advanced monitoring, control, and protection systems, as well as DR tools, are integral to a reliable, self-healing grid. Below are some benefits that will be realized:

- Reduction of cascading outages
- Prevention of rapidly developing instability situations

- Control of slowly developing instability situations
- Deterrence of organized attacks on the grid
- Full utilization of existing assets
- Reduced congestion
- More targeted and efficient maintenance programs
- Fewer equipment failures and thus reduced costs from catastrophic equipment failures
- Minimization of environmental impact
- Maximum use of the most efficient generators and reduced emissions associated with power generation
- Reduced power-delivery energy losses

BARRIERS TO DEPLOYMENT

Considerable research, development, and deployment are needed to fully realize the sensing and measurement key technologies. Unfortunately, an unintended consequence of restructuring in the electric power industry has been reduced research and development.

Engineering-oriented power industry managers, having a long term view, created a U.S. grid that was world-class for most of the twentieth century. But today's business-oriented managers, operating in a restructured utility environment, have adopted a shorter term perspective. Incremental improvements are still sought, but the break-through technologies (such as advanced monitoring systems) that are frequently more costly have lost their appeal.

The cost to create and deploy many of the sensing and measurements technologies discussed in this paper is high, and private industry has been reluctant to invest in costly, long-term developments. The federal and state funds needed to augment private investment have been very limited.

A lack of understanding of the fundamental value of a modern grid, and of the societal costs associated with an antiquated one, has created the misperception that today's grid is "good enough."

Meanwhile, the technical experience base is graying, so there are fewer and fewer advocates for the modern grid technology.

POSSIBLE SOLUTIONS

Fortunately, the core measurement (communication and information) technologies are within practical reach, and economic projections are favorable. In addition, the regulatory impetus provided by the 2005 EPAct can be expected to drive this area so that, a decade from now, extensive advanced sensing and measurement technologies could be commonplace.

Regulators can be the change-agent to lead the way in grid modernization. They must act now to correct the unintended drawback that deregulation has caused. R&D must be significantly encouraged, supported, and increased in the utility sector.

Successfully realizing sensing and measurement technologies requires progress in the following areas:
- The broad development and deployment of supporting communication technologies, such as wireless and BPL

- User-friendly customer interfaces and agents
- Tariffs that are effective from both the consumer and the utility perspective
- New, low-cost sensing and measurement techniques and central information technology systems to process, analyze, and take action on the large volume of collected data
- Additional proof-of-concept demonstrations to explore and illustrate the benefits to consumers and energy companies

A number of advanced sensing and measurement technologies that could have a major impact on grid modernization are available today. However, the current high production, installation, and maintenance costs, along with limited market exposure, have hindered their adoption. These technologies include the following:

- Advanced communicating revenue meters
- Consumer portals and agents
- Major equipment health monitors
- Electronic instrument transformers
- PMUs
- Line sag monitors

Broader deployment will take place as technical performance is proven, costs drop, and as societal value is recognized. It is the role of government to ensure that the proper value is placed on these extended societal benefits.

Several other components that could have a major impact remain in research and development. With the proper incentives, they could ready for deployment in two to four years. These are some examples:

- WAMS analysis systems
- Advanced low-cost communicating transmission and distribution sensors
- Electromagnetic interference detectors
- Sag monitors that utilize a global position system or other advanced, low-cost technology

The full integration of multiple key technologies must occur before the complete modern grid vision can be realized, and this could take ten to fifteen years or more. As these technologies are being incorporated into the power grid, many other synergies among the five key technology areas will surface. For example, the implementation of integrated communications, advanced control methods, sensing and measurements, and advanced components will allow a new series of special protective systems that can be customized to a region's unique characteristics.

The U.S. economy will suffer in many ways if we cannot develop and employ the technologies needed for a world-class power grid. Without the development and deployment of key technologies like sensing and measurements, our power grid will remain at high risk for widespread blackouts, such as the one that occurred in 2003 affecting 40 million people in the United States and 10 million in Canada.

SUMMARY

Advanced sensing and measurement technologies are essential components of a modern power grid.

When deployed, they will perform crucial functions that will transform both the customer and utility levels of the electric power industry. For instance, they will enable consumer choice and demand response, eliminate billing estimations, transform data into information, help relieve grid congestion, and support advanced protective relaying.

The consumer interface will be user-friendly, with increasing levels of sophistication as product features are added. New digital metering systems will be built on the communications capabilities of the Internet and will employ standard Internet protocols. They will use reliable and ubiquitous communications media, such as wireless, BPL, or even fiber to the home.

At the utility level, new sensing and measurement tools will supply expanded data to power system operators. This data will include voltage, power factor, phasor relationships (WAMS), power quality, equipment health and capacity, as well as timely information about meter tampering, vegetation intrusion, fault location, and outage occurrences.

Development in this key technology area is projected to bring savings in capital and operating costs throughout the grid. Another benefit is a reduction in environmental impact, realized due to the improved efficiency and asset utilization brought by sensing and measurement. Additionally, the economy and the public will benefit from enhanced safety, reliability, and power quality. These, as well as other advantages all give weight to the need to develop this key technology area.

Our ability to implement other key elements of the modern grid will be significantly limited without the successful realization of sensing and measurement.

Today's challenge is to accurately envision the grid of the future and successfully design the advanced sensing and metering infrastructure that will enable it.

For More Information

This document is part of a collection of documents prepared by the Modern Grid Initiative (MGI) team. For a high-level overview of the modern grid, see "A Systems View of the Modern Grid." For additional background on the motivating factors for the modern grid, see "The Modern Grid Initiative."

MGI has also prepared five papers that support and supplement these overviews by detailing more specifics on each of the modern grid key technology areas. This paper has described "Sensing and Measurement."

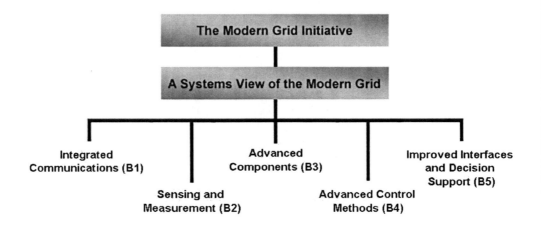

These documents are available for free download from the Modern Grid Web site.

The Modern Grid Initiative

Website: www.netl.doe.gov/moderngrid

Email: moderngrid@netl.doe.gov

(304) 599-4273 x101

BIBLIOGRAPHY

Borenstein, S., M. Jaske, and A. Rosenfeld, October 2002, Dynamic Pricing, Advanced Metering, and Demand Response in Electricity Markets. Hewlett Foundation Energy Series.

California Energy Commission. http://www.energy.ca.gov/.

Consortium for Electric Reliability Technology Solutions. http://certs.lbl.gov/.

Elster Metering. www.elstermetering.com.

Energy Policy Act of 2005.

Graziano, Joe. October, 2004, Private Communication with TVA.

Power Systems Engineering Research Center. http://www.pserc.wisc.edu/.

Schweitzer Engineering Laboratory. www.SELinc.com.

Thorp, J. S. Advanced Relaying White Paper. jsthorp@vt.edu.

Xiaorong Xie et al, January 2006, WAMS Applications in Chinese Power Systems. IEEE Power and Energy Magazine.

ACRONYMS

BPL	Broadband over Power Line
CERTS	The Consortium for Electric Reliability Technology Solutions
DR	demand response
EMC	electromagnetic compatibility
EPAct	Energy Policy Act of 2005
EPRI	Electric Power Research Institute
GPS	Global Positioning System
MGI	Modern Grid Initiative
PMU	phasor measurement unit
PSERC	Power Systems Engineering Research Center
R&D	Research and Development
RTP	real-time pricing
SCADA	Supervisory Control and Data Acquisition
WAMS	wide-area monitoring system

Appendix B3:
A Systems View of the Modern Grid

ADVANCED COMPONENTS

Conducted by the National Energy Technology Laboratory
for the U.S. Department of Energy
Office of Electricity Delivery and Energy Reliability
March 2007

Office of Electricity
Delivery and Energy
Reliability

V2.0

TABLE OF CONTENTS

EXECUTIVE SUMMARY

The United States urgently needs a fully modern power grid if we are to meet our country's growing requirements for reliability, security, efficiency, cost of service, and environmental responsibility.

To achieve a modern grid, a wide range of technologies must be developed and implemented. These technologies can generally be grouped into five key technology areas as shown in Figure 1 below.

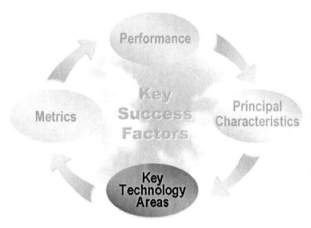

- Integrated Communications
- Sensing and Measurement
- **Advanced Components**
- Advanced Control Methods
- Improved Interfaces and Decision Support

Figure 1: A Systems View of the Modern Grid provides an "ecosystem" perspective that considers all aspects and all stakeholders.

Advanced components play an active role in determining the electrical behavior of the grid. They can be applied in either stand-alone applications or connected together to create complex systems such as microgrids. These components are based on fundamental research and development (R&D) gains in power electronics, superconductivity, materials, chemistry, and microelectronics.

Unfortunately, the needed grid-related R&D in the United States has dropped to unacceptably low levels, particularly since the drive began to restructure the industry. Should this trend continue, the U.S. economy will suffer severely from the absence of a suite of advanced components that would elevate the existing North American grid to world-class status.

Another barrier to the development and implementation of advanced components is the high cost involved in developing them. This, combined with the lack of clearly articulated argument for them, has had a chilling effect on the investment community.

Stakeholders must come to understand the worth of implementing key technologies such as advanced components. These technologies are critically important to the supply of electric power, allowing greater economy, safety, cleanliness, and reliability than is currently possible. To achieve a truly modern grid, we must have buy-in from all stakeholders.

This paper will cover the following important topics:
- Current state of advanced components
- Future state of advanced components
- Benefits of implementation
- Barriers to deployment

Although it can be read on its own, this paper supports and supplements "A Systems View of the Modern Grid," an overview prepared by the Modern Grid Initiative team.

CURRENT STATE

We will now discuss the current state of various advanced components, as well as the core technologies upon which they depend.

We must keep in mind, however, that while all of these technologies and components are needed for a modern grid, the timetable of expected availability varies.

Power Electronics in Transmission and Distribution Systems

Flexible alternating current transmission system devices (FACTS devices include UPFC, DVAR, SVC, etc. See Table 1) are good examples of advanced components that are based on power electronic technologies. FACTS have already demonstrated their worth in a number of transmission and distribution (T&D) applications, including the following (*1*).

* Voltage control at various load conditions
* Power quality enhancement
* Reactive power balance
* Stability problems with energy transfer over long distances

High voltage direct current (HVDC), a mature technology, also relies on power electronics to resolve many issues involving the power grid, such as these:

* Coupling of asynchronous systems
* Stability problems with energy transfer over long distances
* Increase of short-circuit currents in meshed systems

All of the advanced components in the following tables are either available today or are under development.

Table 1 shows some of the power electronic devices that are vital to the modern grid vision:

Table 1: Examples of power electronic devices

Power Electronic Devices	
Advanced component	**Description**
Unified power flow controller (UPFC)	• Can fully manage most requirements of reactive power compensation and flow control. • A 345 kV UPFC that incorporates an interline power flow controller addressing congestion issues has been commissioned at NYPA's Marcy substation. • Are available today but, because of their high cost and limited experience base, they have been deployed only sparingly. Broader deployment will take place as technical performance is proven, as costs drop, and as their societal value is recognized.
DVAR or DSTATCOM	• Mobile, relocatable, insulated gate bipolar transistor (IGBT) device that can be sited at system or industrial interfaces. • Provides voltage support, reduces industrial flicker, provides improved power quality, mitigates wind generator impact on transmission lines, and a variety of other applications. • Has had fairly wide acceptance and deployment.
Medium voltage static voltage regulator (MV SVR)	• Boosts whole-facility load voltage during source voltage sags caused by faults in the utility distribution grid or in the transmission system. • Load voltage boost performed within a quarter to half cycle, enabling even the most sensitive facility equipment to ride through sag events without operational disruptions.
Static VAr compensator (SVC)	• Perhaps the most important FACTS device, SVCs has been used for a number of years. • Improves transmission line performance by resolving dynamic voltage problems. • Provides high performance steady-state and transient voltage control far superior to classical shunt compensation. • Dampens power swings, improves transient stability, and reduces system losses by providing optimized reactive power control.
Solid state transfer switch	• Provides undisturbed power using two independent feeders. • Mitigates power quality events. • Is available today, but because of high cost and limited experience base, has been deployed only sparingly. Broader deployment will take place as technical performance is proven, as costs drop, and as its societal value is recognized.

Power Electronic Devices	
Advanced component	**Description**
Dynamic brake	• Rapidly extracts energy from a system by inserting a shunt resistance into the network. • Adding thyristor controls to the brake permits addition of control functions, such as on-line damping of unstable oscillations. • BPA has installed such a dynamic brake on their system. • Is available today, but because of high cost and limited experience base, has been deployed only sparingly. Broader deployment will take place as technical performance is proven, as costs drop, and as its societal value is recognized.
AC/DC inverter	• Mature technology that can be further improved to meet future needs. • Provides the grid interface for a variety of distributed generation sources. • Future improvements in high-power semiconductors may make it economically viable to convert large areas of the grid to DC operation.

Table 1: Examples of power electronic devices

Superconductivity

Several DOE-sponsored cable demonstration programs are now underway at AEP/Southwire/Columbus, Ohio; National Grid/Sumitomo-SuperPower/ Albany, NY; and LIPA/Nexans-AMSC/Long Island, NY

The bulk of U.S. research and development related to superconductivity is currently supported by the Department of Energy (DOE) (2). Several projects demonstrating pre-commercial utility applications of high temperature superconducting (HTS) technology have emerged, and new projects are being developed.

National Laboratories are engaged in research aimed at investigating underlying principles of superconductivity and to address fundamental technological issues. A close working relationship between the national labs and academia ultimately benefits both organizations through the use of university expertise and facilities that, in turn, strengthen and expand the national laboratories' capabilities.

Part of the program is aimed at completing research needed for U.S. industry to scale up new superconducting-wire manufacturing processes. Innovative approaches discovered at national laboratories are being developed into commercially viable processes by public companies. Only short lengths of second-generation wire have been produced thus far, but the performance is far better than any existing wire and the cost-savings potential is significant. The goal is to enable U.S. industry to manufacture long-length wire, suitable for widespread use in industrial and commercial settings.

Examples of advanced superconducting materials and components currently being researched and developed may be found in Table 2.

Superconductivity	
Technology	**Description**
First generation (1G) wire	• Can be manufactured today. • Used in short line segments as exits from congested substations or in urban areas and as fault current limiters. • Applied as a very low impedance path to help control power flows on congested parallel lines. • Reduces pollution from electric generating facilities. • Raises electric system reliability. • Improves power delivery systems in urban areas without new rights-of-way.
HTS cable	• Transmits large quantities of power at reduced voltages and high currents. • Lower voltages reduce HVDC terminal costs by 25% to 50%. • May be competitive with UG cables employing large quantities of high-priced copper. • Can reduce urban transmission congestion or allow for more intensive urban development. • Can be manufactured today in small quantities. • Still in R&D; could have a huge impact and could, with the proper motivation, move from the laboratory to broad application in the short term.
Second generation (2G) superconducting wire	• Longer term development needed (5-10 years), with huge potential grid impact. • Can be manufactured today in small quantities. • Cost and performance of HTS devices will significantly improve. • Provides lower-cost control of flicker, voltage, and transient stability. • Prices could be 3 to 10 times lower than 1G wire and have 10 times lower losses. • Will penetrate the replacement market for large industrial motors, power plant auxiliary motor drives, and power plant generators. • Long-distance, low-impedance underground transmission of power is the ultimate goal. • 2G wire fault current limiters (FCLs) can be developed that have ten times lower losses, limit currents by a factor of three to ten, and have small footprints. • Other issues will need to be anticipated and resolved, like the changing dynamic characteristics of customer and plant auxiliary loads, and increased fault currents.

Table 2: Examples of superconducting devices

Advancements in materials are needed to accelerate development and implementation of superconductivity for the grid. As these advances occur, superconductivity will increasingly be used for short-line segment exits from congested substations, superconducting magnetic energy storage (SMES), superconducting synchronous condensers, fault current limiters (FCLs), high-efficiency motors and generators, and, ultimately, long-distance lossless transmission lines.

Generation and Storage Distributed Energy Resources

Distributed energy resources (DER) are small-scale power generation and storage technologies, typically in the range of 3 to 10,000 kW, located close to where electricity is used (e.g., a home or business) to provide an alternative to, or an enhancement of, the traditional electric power system. DER will play a large role in the modern grid. But an interface using power electronics, along with new local control and protection schemes, is needed to mate DER sources to the grid. Further development of fundamental DER energy-conversion technology is needed to lower cost and improve performance. Eventually, key technology improvements will allow remote management of multiple, diverse DER devices operating as an integrated system.

The development of advanced components, such as the low-cost power electronic interfaces needed for a variety of DER sources, combined with the associated communications, protection, and control systems, is a work in progress. DER will also profit by advances in the underlying core technologies such as the chemistry involved with distributed storage devices.

Distributed Generation Devices

Enabled by improvements in chemistry, materials, and power electronics, these new advanced components are all in various stages of development today. With sufficient industry support, they will become available in large quantities at attractive prices in the future.

A portfolio of distributed generation (DG) technologies (3) is summarized in Table 3.

Distributed Generation	
Advanced component	**Description**
Microturbine	• An emerging class of small-scale distributed power generation in the 30-400 kW size range. • Consists of a compressor, combustor, turbine, and generator. • Most microturbine units are designed for continuous-duty operation, fueled using natural gas. • A number of companies are currently field-testing demonstration units, and several commercial units are available for purchase. • Combined heat and power applications offer very high overall efficiency.
Fuel Cell	• Very low levels of NO_x and CO emissions. • Many types of fuel cells currently under development, including phosphoric acid, proton exchange membrane, molten carbonate, solid oxide, alkaline, and direct methanol. • One company currently manufacturers a 200 kW phosphoric acid fuel cell for use in commercial and industrial applications. • A number of companies are close to commercializing proton exchange membrane fuel cells, with marketplace introductions expected soon. • Cost-effective, efficient fuel reformers that can convert various fuels to hydrogen are necessary to allow increased flexibility and commercial feasibility.
Photovoltaic (PV): "Solar Panel"	• Solar panels are made up of discrete cells connected together that convert light radiation into electricity. • Produce no emissions, are reliable, and require minimal maintenance to operate. • Those deployed by NASA for space applications have efficiencies of 25% in actual use. • Currently available from a number of manufacturers for both residential and commercial applications; manufacturers continue to reduce installed costs and increase efficiency. • Applications for remote power are quite common.
Wind Turbine	• In the United States alone, 8 million mechanical wind generators have been installed. • Considered the most economically viable choice within the renewable energy portfolio. • Environmentally sound and convenient alternative. • Are currently available from many manufacturers, and improvements in installed cost and efficiency continue.

Table 3: Examples of distributed generation

Older, more mature categories of distributed generation include the reciprocating diesel engine and the reciprocating natural gas engine.

These are the most common machines for power generation, mechanical drive, or marine propulsion. While improvements in operation, environmental responsibility, and power production are still being made, these cannot truly be called advanced components.

Combustion gas turbines, which range from one MW to several hundred MW, are another older distributed generation technology that is still undergoing improvements. Here are some features of the combustion gas turbine:

- Based on jet propulsion engine technology designed specifically for stationary power generation or compression applications in the oil and gas industries.

- Relatively low installation costs, low emissions, and infrequent maintenance requirements.

- Low electric efficiency has limited turbines to primarily peaking units and combined heat and power applications.

- Co-generation distributed-generation installations are particularly advantageous when a continuous supply of steam or hot water is desired.

Table 4 compares various distributed generation technologies (3).

Distributed Generation: A Comparison					
Technology	Recip Engine: Diesel	Recip Engine: NG	Microturbine	Combustion Gas Turbine	Fuel Cell
Size	30kW - 6+MW	30kW - 6+MW	30-400kW	0.5 - 30+MW	100-3000kW
Installed Cost ($/kW)	600-1,000	700-1,200	1,200-1,700	400-900	4,000-5,000
Elec. Efficiency (LHV)	30-43%	30-42%	14-30%	21-40%	36-50%
Footprint (sqft/kW)	.22-.31	.28-.37	.15-.35	.02-.61	.9

Table 4: Comparison of distributed generation

Distributed Storage Devices

Energy storage systems employ such chemical formulations as the vanadium redox flow battery (VRB) or the sodium sulfur (NaS) battery to provide long-life systems that improve load factor (8 hours of storage or more) while enhancing power quality, and at the same time offer voltage and transient stability mitigation and frequency regulation. Advanced, low cost, high-energy density flywheel energy-storage systems with 15 to 60 minutes of storage can provide customer power during grid outages.

Table 5 describes some distributed storage devices.

Distributed Storage	
Advanced component	**Description**
NaS battery	• Consists of liquid (molten) sulfur at the positive electrode and liquid (molten) sodium at the negative electrode as active materials separated by a solid beta alumina ceramic electrolyte. • Is efficient (about 89%). • Is economic to use in combined power quality and peak shaving applications. • Has been demonstrated at over 30 sites in Japan totaling more than 20 MW with stored energy suitable for 8 hours daily peak shaving. • Combined power quality and peak shaving applications in the U.S. market are under evaluation. • Pilot installation at AEP.
Vanadium Redox Battery (VRB)	• A flow battery with external tanks containing vanadium aqueous solutions. • Can be used for peak shaving, frequency regulation, voltage and transient stability support, and customer ride-through. • Number of hours of storage can be increased by simply increasing the size of the external tanks. • Low power density of the electrolyte and the space requirements are drawbacks. • VRB flow batteries with 8 hours of storage are available in small sizes.
Ultracapacitors	• Stores energy like a battery. • Can quickly discharge the energy in seconds like a capacitor.
Superconducting Magnetic Energy Storage (SMES)	• Most commonly devoted to improving power quality. • Power is available almost instantaneously and very high power output can be provided for a brief time. • Loses the least amount of electricity in the energy storage process compared to other methods of storing energy. • Distributed SMES units have been deployed to enhance stability of a transmission loop.

Table 5: Distributed storage

Complex Systems

Various distributed energy resources can be integrated to create power supply systems with unique characteristics, as shown in Table 6.

Complex Systems Employing Advanced DER Components	
System	**Description**
Microgrid	• Small local power systems that can stand alone or be integrated with a larger conventional distribution feeder. • Includes energy storage and DG to establish a small independent control area. • Can employ various types of DG. • One design employs SmartSource™, which provides plug-and-play functionality without relying on communications. • One design employs SmartSwitch™, which provides a single interface to the power system allowing smooth transition between parallel and islanded operation. • Phased field trials at American Electric Power Company.
Premium Power park	• Employs uninterruptible power supplies, such as battery banks, ultracapacitors, or flywheel energy storage; can also employ high-speed transfer switches, DVRs and other power electronic devices, and DG. • Premium Power parks can attract high-tech industry to a region by providing the ultra-clean power needed for sensitive industrial processes.

Table 6: Complex Systems

In addition to DER, the modern grid must continue to support a wide variety of large central generation units, including fossil, hydro, wind, geothermal, and nuclear plants.

Composite Conductors

New materials are opening up new ways for advanced components to improve the performance of the grid.

For instance, composite conductors such as new high-temperature, composite transmission-cable designs will enable increased utilization of right of way (ROW), allowing a doubling of amperage limits with little change to the line support or towers.

The description of several composite conductors is found in Table 7.

353

Composite conductors	
Advanced Component	**Description**
Aluminum Conductor Composite Core Cable (ACCC™ Cable)	• Is superior to existing T&D cable in a number of key performance areas (4). • Offers double the current-carrying capacity when compared to most standard conductors. • Can dramatically increase system reliability by virtually eliminating problematic high-temperature cable sag. • ACCC™ cable will be as easy to install as conventional utility cable. • In operation at HV.
Aluminum Conductor Composite Reinforced Cable (ACCR)	• Can increase transmission thermal capacity 150% to 300%. • In operation at over a dozen utilities beginning in 2001.
Annealed aluminum, steel supported, trapezoid cross section conductor wire (ACSS/TW)	• Can operate at 200 C. • Carries 100% more current. • Reduces line losses at normal loads. • Can be handled like normal aluminum cable steel reinforced (ACSR) conductor wire. • In operation at HV.

Table 7: Composite conductors

Grid-Friendly Appliances

Advances in microelectronics are making possible the production of grid-friendly appliances (GFAs) that will help stabilize the grid in times of system stress. These localized controllers, installed in a wide range of home appliances, will make it possible to automatically switch home electric appliances off and on in order to modulate load during system disturbances. They may also be designed to respond to electricity pricing signals.

The addition of voltage- and/or frequency-sensing chips to a wide range of home appliances could offer substantial power system benefits. Properly conceived GFA control algorithms can impact the power grid in profound ways. Frequency-sensitive GFAs are ready now, but manufacturers need to be convinced of the value of adding them to appliances such as washing machines, dryers, refrigerators, and heating, ventilating, and air conditioning units. Government incentives, similar to the Energy Star program, would help advance this product offering,

Improved modeling of loads and power system performance is also needed to verify the potential of this new advanced component. A proliferation of grid-friendly appliances could create the ability to use distributed load itself as a major system-level control element. This

would fundamentally alter electric system behavior and place more of
the control actuation into the loads themselves.

FUTURE STATE

The modern grid will employ a range of advanced components that will greatly enhance the performance of transmission and distribution systems.

Power quality will be improved through new technology and by seeking an optimal balance between grid and load characteristics. Transmission capacity and reliability will be enhanced through the application and retrofitting of a variety of advanced components, many based on advanced power electronics and new types of conductors. Distribution systems will incorporate many new storage devices and sources and will employ new topologies, including microgrids.

Economical FACTS devices will make use of new low-cost power semiconductors having far greater energy-handling capacity than today's semiconductors. Distributed generation will be widely deployed and multiple units will be linked by communications to create dispatchable virtual machines. Superconductivity will be applied to fault current limiters, storage, low loss rotating machines, and lossless cables. Advanced metering and communications will enable a suite of demand response (DR) applications, including the integration of GFAs and plug-in hybrid electric vehicles (PHEVs).

New energy storage technologies will be deployed as DER and as large central plants. The mix of generation will include large central power plants having a range of characteristics (e.g., heat rates, emissions, inertia, ramp rates, etc), in addition to distributed energy resources (many of the green variety) having a different set of performance characteristics. The combination of generation types will operate in a coordinated manner so as to optimize cost, efficiency and reliability and minimize environmental impact.

THE ROLE OF POWER ELECTRONICS

Further developments in semiconductor technology will allow new advanced components, based on power electronics, to be reliably and economically applied to a variety of T&D solutions. Greater energy-handling capabilities of individual power semiconductors will lead to more economical applications.

Material developments in SiC or GaN can lead to advanced higher current and higher voltage power electronic devices than are available today. These could operate directly at line voltage and require fewer electronic switches than is possible in today's grid.

Power electronic technology will also be applied to advanced power quality devices, switching devices, transformation devices (e.g. transformers with little or no magnetic material) and frequency-conversion devices (e.g., for microturbines, fuel cells, wind turbines, or solar panels interfaced to the grid).

Farther into the future, power electronic components could employ diamonds (chemical-vapor deposition polycrystalline diamond tip or edge anodes) operating as ultra-high current, voltage, and frequency switches. FACTS, HVDC, high-speed transfer switches, and DER could then be available at much lower cost.

A bold new concept is the application of distributed FACTS, as exemplified by the distributed series impedance device. These devices might be integrated with insulator strings and shoes at transmission towers (5). Coupled with the transmission system to obtain operating power, they would insert inductance or capacitance in series with the transmission system to increase or decrease series impedance. This would allow for more effective control of the transmission network, reduce fault currents, and balance line voltages. These concepts and potential benefits are illustrated in Figure 2 below:

Figure 2: Concept for a distributed flow control device: Distributed series impedance (DSI) or Smart Wires device for instant power flow control. Image courtesy of IPIC at Georgia Institute of Technology.

In addition, power electronic devices could begin to replace iron and copper transformers, initially in the distribution system and eventually in the transmission system. These power electronic systems, first based on SiC devices and eventually based on diamond devices, would not only be able to control the voltage, but also be able to inject reactive power into the T&D system based on automated distributed controls, as shown in Figure 3 below:

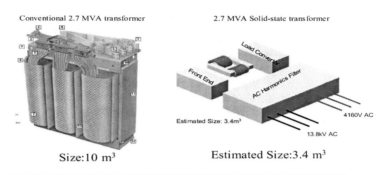

Conventional 2.7 MVA transformer

2.7 MVA Solid-state transformer

Size:10 m³

Estimated Size:3.4 m³

3x size reduction compared to conventional transformer.

Figure 3: Future concept of solid-state power electronic transformer: Solid-state transformer (SiC based), from concepts by North Carolina State University's Semiconductor Power Electronics Center (SPEC). Image courtesy of NC State University.

THE ROLE OF SUPERCONDUCTIVITY

Commercial HTS products are beginning to reach the market.
SuperVAR™ synchronous condenser is a good example. SuperVAR™ dynamic synchronous condensers help alleviate voltage problems in many applications including the following:

- Reactive compensation for T&D systems
- Steady state voltage regulation of long radial delivery systems
- Dynamic power factor correction at large industrial sites
- Flicker mitigation for sensitive power quality
- Grid stability

Figure 4 below shows the first installation of this new technology.

Figure 4: Voltage and transient stability control technology, high temperature superconducting (HTS) synchronous condenser. American Superconductor at TVA. Image is courtesy of TVA and AMSC.

It is important to note that superconducting technology will, in general, increase the severity of fault currents. This intensifies the need for higher-rated circuit breakers and HTS fault current limiters. However, second Generation (2G) wire FCLs could be developed that have ten times lower losses, limit currents by a factor of three to ten, and have small footprints.

With the commercial deployment of HTS synchronous condensers, the realization of new HTS industrial electric motors and HTS generators becomes more likely. This will impact the dynamic characteristics of loads, especially when 2G wire becomes available in 1 km lengths.

The increased efficiency from superconducting machines and cable has the potential to produce huge environmental benefits.

THE ROLE OF PLUG-IN HYBRID ELECTRIC VEHICLES

Plug-in Hybrid Electric Vehicles (PHEVs) could turn out to be the most important new electrical load in a century. The potential benefits to the grid and to the nation in general are impressive. The National Renewable Energy Laboratory concluded the following (6):

- Hybrid electric vehicles, with the capability of being recharged from the grid may provide a significant decrease in oil consumption. These "plug-in" hybrids will affect utility operations, adding additional electricity demand. Because many individual vehicles may be charged in the extended overnight period, and because the cost of wireless communications has decreased, there is a unique opportunity for utilities to directly control the charging of these vehicles at the precise times when normal electricity demand is at a minimum.

- Based on existing electricity demand and driving patterns, a 50 percent penetration of PHEVs would increase the per capita electricity demand by around 5 to 10 percent, depending on the region evaluated. While increasing total electrical energy consumption (but without requiring additional generation capacity), the optimal dispatch of the additional PHEV demand would increase loading of baseload power plants built to meet the normal demand. This also would substantially decrease the daily "cycling" of power plants, both of which would translate into lower operational costs.

- While it appears that PHEVs are much better suited for short-term ancillary services such as regulation and spinning reserve, a large fleet of PHEVs could possibly replace a moderate fraction (perhaps up to 25 percent) of conventional low-capacity factor (rarely used) generation for periods of extreme demand or system emergencies. Overall, the ability to schedule both charging and very limited discharging of PHEVs could significantly increase power system utilization.

BENEFITS OF IMPLEMENTATION

Installation of advanced components will lead to a significantly enhanced grid that provides power to meet the increasingly diverse needs of the twenty-first century.

At the transmission level, FACTS or HTS synchronous condensers will provide instantaneous support of voltage to reduce the sags that are the biggest customer power-quality problem. Fault current limiters will reduce the voltage depressions created by transmission system faults, while synchronous switching will limit transient over-voltages.

At the distribution level, high-speed transfer switches will instantly remove disturbed sources and replace them with clean, backup power supplies. FACTS (e.g., DVAR), DER, and microgrids will provide voltage support and load isolation to aid grid reliability and minimize power-quality events.

The modern grid's reliability will greatly increase due to its self-healing characteristic. Self-healing will be enabled by several complementary key technology areas:

- Very rapid and sophisticated sensing and measurements will be enabled by technologies such as the instantaneous phasor measurements of a wide-area monitoring system (WAMS).

- Advanced components (such as FACTS, HTS synchronous condensers, and distributed power-flow control devices) will give the grid the ability to respond quickly to an emergent problem by using strategies like changing flow patterns and voltage conditions.

- Decision support systems will enable a modern grid to "know" when there is a need to shed load on the distribution system. A modern grid could also immediately call for increased real and/or reactive power output from DER to support transmission needs.

- Additional reliability will result when low-cost, power-electronic interfaces for a variety of DER sources (along with the associated communications, protection, and control systems) are developed to provide built-in local control and protection.

- In addition to DER, the modern grid will continue to support a wide variety of large central generation units (fossil, hydro, wind farm, geothermal, nuclear, etc.). Advanced components will help ensure the stability and the efficiently integrated utilization of these many diverse generation sources.

In addition to more reliability, the grid will be more secure when advanced components contribute to its self-healing characteristic. The security of the grid is integrally related to its ability to heal itself. A grid that is highly diversified, with multiple sources, is all the more secure from a concerted physical or cyber attack. The modern grid will be designed with hardened integrated communications systems that are less vulnerable to such attacks than the grid that exists today. Additionally, the measurement, protection, and control systems associated with advanced components will all communicate through highly encrypted digital channels that are extremely difficult to overcome.

Grid efficiency is another characteristic that will greatly improve as advanced components maximize asset utilization and reduce electrical losses. As an example, superconducting lines and machines will produce major efficiency gains throughout the electric power system, including even the customers' loads. Flow and voltage control, as well as DER, will reduce electrical losses on the grid, DER by reducing the need to transport power over long distances. And DER that employs combined heat and power will operate much more efficiently than conventional central generation.

Advanced components will enable the grid to become more environmentally friendly. To the degree that advanced components make the grid more efficient, they make it cleaner. For instance, many DG technologies, such as solar and wind farms, fuel cells, and superconducting machines, are less polluting than conventional energy-producing methods. In addition, improved power-transfer capability means fewer lines are needed, which also lessens environmental impact. And the environmental damage associated with power outages is reduced when the grid becomes more reliable.

Another benefit of advanced components is the wide variety of ways they will foster grid economy. The following are a few specific examples:

- Advanced components improve the linkage between buyers and sellers of electric energy, and thus create a more robust market and greater access to lower-priced power.
- The proper application of FACTS devices will allow the deferral of costly major line additions
- FCLs will reduce the need to replace entire systems that are unable to handle increasing fault levels.
- DER storage will lessen expense by making the addition of peaking generation unnecessary.

Advanced components will be employed to reduce transmission congestion costs, saving billions of dollars each year. The above are but a few examples.

BARRIERS TO DEPLOYMENT

An unintended consequence of restructuring in the electric power industry has been reduced research and development related to advanced components.

Engineering-oriented power industry managers, having a long term view, created a U.S. grid that was world-class for most of the twentieth century. But today's business-oriented managers, operating in a restructured utility environment, have adopted a shorter term perspective. Incremental improvements are still sought, but the break-through technologies, such as superconducting transmission lines, that are frequently more costly have lost their appeal.

The R&D cost to create many advanced components (e.g., superconducting transmission lines, advanced power electronics, etc.) is high and private industry has been reluctant to invest in costly, long-term developments. Federal and state funds needed to augment private investment have been very limited.

Still another barrier is the lack of integration testing that demonstrates the benefits technologies that are incorporated into functional systems.

A lack of understanding of the fundamental value of a modern grid, and of the societal costs associated with an antiquated one, has created the misperception that today's grid is "good enough."

Meanwhile, the technical experience base is graying, so there are fewer and fewer advocates for the modern grid technology.

The U.S. economy will suffer in many ways if we cannot develop and employ the technologies needed for a world-class power grid.

Without the development and deployment of key technologies like advanced components, our power grid will remain at high risk for widespread blackouts, such as the one that occurred in 2003 affecting 40 million people in the United States and 10 million in Canada.

Possible Solutions

A first step toward achieving the modern grid is prioritization of advanced components' development. An understanding of the benefits of implementing these components will stimulate investment and governmental support. A consensus among all stakeholders is needed regarding the value of advanced components. So, too, is their enthusiastic and vocal support.

Regulators can be the change-agent to lead the way in grid modernization. They must act now to correct the unintended drawback that deregulation has caused. R&D must be significantly encouraged, supported, and increased in the utility sector.

SUMMARY

Achieving the modern grid is absolutely necessary to provide our country with reliable, secure, economic, and efficient power that is safe and environmentally responsible.

To do this, the modern grid requires a wide range of advanced components based on new developments in power electronics, superconductivity, chemistry, materials, and microelectronics.

For example, one important need is the development of economical FACTS devices that will employ low-cost power semiconductors having far greater energy-handling capacity than today's semiconductors. Also, it is necessary that DER be widely deployed, with multiple units being linked by communications to create dispatchable virtual machines.

Superconductivity needs to be economically applied to fault current limiters, storage, low loss rotating machines and lossless cables. And new storage technologies must be deployed for both DER and large central plants.

New kinds of electrical loads can enhance grid performance by responding to momentary problems (GFAs) and by improving load factors (PHEVs).

The mix of power generation must include large central power plants and DER. Environmental emissions will be reduced when many of the DER technologies, such as wind, fuel cells, and solar, are incorporated into the power grid.

To achieve a modern grid, a combination of advanced components must operate in a coordinated manner so as to optimize efficiency and reliability and lessen environmental emissions.

All of these advanced components are necessary to build the modern grid our country must have to support the energy needs of our modern society.

For More Information
This document is part of a collection of documents prepared by the Modern Grid Initiative (MGI) team. For a high-level overview of the modern grid, see "A Systems View of the Modern Grid." For additional background on the motivating factors for the modern grid, see "The Modern Grid Initiative."

MGI has also prepared five papers that support and supplement these overviews by detailing more specifics on each of the key technology areas of the modern grid. This paper has described the third key technology area, "Advanced Components."

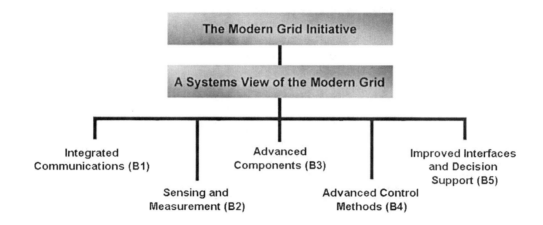

These documents are available for free download from the Modern Grid web site.

The Modern Grid Initiative

Website: www.netl.doe.gov/moderngrid

Email: moderngrid@netl.doe.gov

(304) 599-4273 x101

REFERENCES

1. Dale T. Bradshaw, March 7, 2006. Results of the High Level Survey of Transmission R&D in North America. Presentation to the CA Public Interest Energy Research (PIER) Transmission Research Program (TRP) Policy Advisor Committee (PAC)

2. Oak Ridge National Laboratory, Superconductivity Program for Electric Systems Research and Development, http://www.ornl.gov

3. Distributed Generation, Distributed Resources, Distributed Energy resources. http://www.distributed-generation.com

4. Composite Technology Completes First Overhead Installation of Its New ACCC™ Power Cable, http://www.compositetechcorp.com

5. Dr. Deepak Divan at Georgia Tech's Intelligent Power Infrastructure Consortium (IPIC) and Dr. Alex Huang at North Carolina State University's Semiconductor Power Electronics Center (NCSU SPEC): Personal communication and unpublished presentation.

6. P. Denholm and W. Short. An evaluation of utility system impacts and benefits of optimally dispatched plug-in hybrid electric vehicles. NREL/TP-620=40293. Revised October 2006.

Acronyms

1G	First generation
2G	Second generation
AC	Alternating current
ACCC	Aluminum conductor composite core
ACCR	Aluminum conductor composite reinforced
ACSR	Aluminum cable steel reinforced
AEP	American Electric Power Co.
ACSS/TW	Annealed aluminum steel supported trapezoidal Wire
BPA	Bonneville Power Administration
DC	Direct current
DER	Distributed energy resources
DG	Distributed generation
DR	Demand response
DSI	Distributed series impedance
DSTATCOM	Dynamic static compensator
DVAR	Dynamic VAr
ETO	Emitter turn off
FACTS	Flexible alternating current transmission system
FCL	Fault current limiter
GaN	gallium-nitrogen
GFA	Grid- friendly appliances
HTS	High temperature superconducting
HVDC	High voltage direct current
IGBT	Insulated gate bipolar transistor
IGCT	Insulated gate commutated transistor
MW	Megawatt
NaS	Sodium sulphur
NYPA	New York Power Authority
PHEV	Plug-in hybrid electric vehicles
PV	Photovoltaic
R&D	Research and development

ROW	Right of way
SiC	Silicon-carbon
SMES	Superconducting magnetic energy storage
STATCOM	Static compensator
SVC	Static VAr compensators
T&D	Transmission and distribution
UG	Underground
UPFC	Unified power flow controller
VAr	Volt-amperes reactive
VRB	Vanadium redox battery
WAMS	Wide area monitoring system
WBG	Wide band gap

Appendix B4:
A Systems View of the Modern Grid

ADVANCED CONTROL METHODS

Conducted by the National Energy Technology Laboratory
for the U.S. Department of Energy
Office of Electricity Delivery and Energy Reliability
March 2007

Office of Electricity
Delivery and Energy
Reliability

v2.0

TABLE OF CONTENTS

EXECUTIVE SUMMARY

It is becoming increasingly difficult today to meet our nation's 21st century power demands with an electric grid built on yesterday's technologies.

A fully modernized grid is essential to provide service that is reliable, secure, cost-effective, efficient, safe, and environmentally responsible. To achieve the modern grid, a wide range of technologies must be developed and implemented. These technologies can be grouped into five key technology areas as shown in **Figure 1** below.

- Integrated Communications
- Sensing and Measurement
- Advanced Components
- **Advanced Control Methods**
- Improved Interfaces and Decision Support

Figure 1: The Modern Grid Systems View provides an "ecosystem" perspective that considers all aspects and all stakeholders.

The advanced control methods (ACM) featured in this paper comprise one of the five key technology areas that must be developed if we are to have a truly safe, reliable, and environmentally friendly modern grid.

ACM technologies are the devices and algorithms that will analyze, diagnose, and predict conditions in the modern grid and determine and take appropriate corrective actions to eliminate, mitigate, and prevent outages and power quality disturbances. These methods will provide control at the transmission, distribution, and consumer levels and will manage both real and reactive power across state boundaries.

To a large degree, ACM technologies rely on and contribute to each of the other four key technology areas. For instance, ACM will monitor essential components (Sensing and Measurements), provide timely and appropriate response (Integrated Communications; Advanced Components), and enable rapid diagnosis (Improved Interfaces and

371

Decision Support) of any event. Additionally, ACM will also support market pricing and enhance asset management.

The analysis and diagnostic functions of future ACM will incorporate predetermined expert logic and templates that give "permission" to the grid's software to take corrective action autonomously when these actions fall within allowable permission sets.

As a result, actions that must execute in seconds or less will not be delayed by the time required for human analysis, decision-making, and action. Significant improvement in grid reliability will result due to this self-healing feature of the modern grid.

ACM will require an integrated, high-speed communication infrastructure and corresponding communication standards to process the vast amount of data needed for these kinds of system analyses. ACM will be utilized to support distributed intelligent agents, analytical tools, and operational software applications.

This paper covers the following four important topics:

- Current state of ACM
- Future state of ACM
- Benefits of implementation
- Barriers to deployment

Although it can be read on its own, this paper supports and supplements "A Systems View of the Modern Grid," an overview prepared by the Modern Grid Initiative team.

CURRENT STATE

The communication infrastructure supporting today's control systems consists of a wide spectrum of technologies patched together. The required information is transmitted from the sensor to the control systems, processed by the control systems, and then transmitted to the controlling devices.

This current communication infrastructure is too limited to support the high-speed requirements and broad coverage needed by ACM, and it does not provide the networked, open architecture format necessary for the continued enhancement and growth of the modern grid. Additionally, today's grid lacks many of the smart sensors and control devices including consumer portal devices that need to be deployed to measure the required data and provide the control mechanisms to manage the electric system.

Some progress is being made. For instance, distribution automation (DA) technologies are presently being integrated with supervisory control and data acquisition (SCADA) systems to provide rapid reconfiguration of specific sections of the distribution system. This will minimize the impact of system faults and power quality disturbances on customers. DA provides the ability to monitor and operate devices that are installed throughout the distribution system, thereby optimizing station loadings and reactive supply, monitoring equipment health, identifying outages, and providing more rapid system restoration. However, this integration needs to happen more quickly and on a much wider scale.

Some of today's ACM technologies are locally based, such as at a substation, where the necessary data can be collected in near real time without the need for a system-wide communication infrastructure. But these control algorithms act autonomously at a local substation level and hence do not benefit from a system-wide perspective. Often, these algorithms are integrated with centralized systems to enable others not located at the substation to have access to the data. Substation automation technologies provide this functionality and are in their early phases of implementation at most utilities. Numerous vendors provide modern substation automation technologies today using architectures similar to that shown in Figure 2.

Architecture with process bus

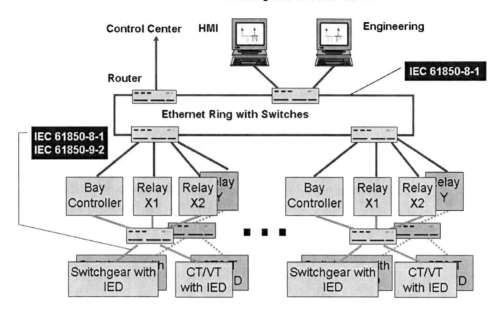

Figure 2: Example schematic of substation data architecture. Image courtesy of the International Electrotechnical Commission.

ACM technologies depend heavily on data sensing and some form of data transmission (See "Appendix B2: Sensing and Measurement" and "Appendix B1: Integrated Communications"). Today's sensors that measure system parameters (e.g.., watts and watt-hours, VArs and VAr-hours, volts, amperes, power factor, phase angles, harmonics, etc.) are only beginning to evolve from the traditional electric/electromechanical design to a solid-state, electronic-based technology of higher accuracy, more intelligence, and with the capability to interface with digital communication systems. The widespread deployment of intelligent electronic devices (IEDs) at the system, equipment, and consumer levels must occur to support ACM in the future.

Significant advances have been made in software-based control algorithms in nearly every industry and much has been done in the area of ACM. Some of the ACM technologies needed for the modern grid are currently available or are in research and development. These technologies are slowly being integrated into three important areas: distributed intelligent agents, analytical tools, and operational applications. Some of the technologies in these areas are described in the three tables which follow.

DISTRIBUTED INTELLIGENT AGENTS

Distributed Intelligent Agents are adaptive, self-aware, self-healing, and semi-autonomous control systems that respond rapidly at the local level to unburden centralized control systems and human operators. Several of these agents are often combined to form a multi-agent system with peer-to-peer communication. These multi-agent systems are capable of reaching goals difficult to achieve by an individual system. Some of these technologies are described in Table 1 below.

Distributed Intelligent Agents	
Agent	**Description**
Digital protective relay	• Senses electric system parameters, analyzes data, and initiates control actions autonomously to protect system assets • Communication-enhanced coordination ensures only last device feeding a faulted section clears the fault • Protection coordination can be automatically updated as circuits are reconfigured • Provides post-disturbance data for analysis of event • New design not yet universally deployed across the grid
Intelligent tap changer	• Senses both high- and low-side voltages to perform advanced control • Minimizes draw of reactive power from transmission system
Dynamic circuit rating tool	• Determines the safe and accurate dynamic rating of lines • Interfaces with advanced sensors that monitor weather parameters, line sag, and conductor temperature to obtain the required inputs • Normally provides additional line capacity except during times when weather conditions and line loadings are not favorable
Energy management system	• Monitors electric system parameters and marketing information; considers consumer preset settings and acts on the behalf of the consumer to manage energy costs, comfort, and health • Supports demand-response (DR) programs based on real-time pricing
Grid-friendly appliance controller	• Senses grid conditions by monitoring the frequency or voltage of the system and provides automatic DR in times of system distress • Can be installed in household appliances such as refrigerators, washers, dryers, stoves, etc., to turn them off or on as required to allow the grid to stabilize
Dynamic distributed power control devices	• Increases or decreases line impedance • Improves utilization of under-utilized lines • Can manage flexible alternating current transmission system (FACTS) devices installed at substations to provide instantaneous and autonomous control of line flow and voltage • Low-cost, mass-produced, distributed power-flow devices can be installed on each phase of a line to provide 10% or more instantaneous control of power flow

Table 1: Distributed intelligent agents

ANALYTICAL TOOLS

The heart of the ACM analytical tools are the software algorithms and the high-speed computers needed to process and analyze the information. This feature is a key part of the overall ACM control loop. Some of these tools are described in Table 2 below.

Table 2: Analytical Tools

Analytical Tools	
Tool	**Description**
System performance monitoring, simulation, and prediction	• Monitors frequency, voltage, congestion, and market power to detect abnormal operating patterns • Predicts how system will respond if critical equipment is forced out of service • Validates quality of real-time data and off-line system models • Optimizes plans for system stabilization and restoration based on real-time or simulated disturbances
Phasor measurement analysis	• Monitors the instantaneous value of voltage or current • Determines whether a transient swing in the power system is stable or unstable • Detects imminent grid emergencies • Supports more rapid state estimation • Improves dynamic modeling and analysis • Online model with better visualization still needs to be developed so that control room operators can effectively interpret phasor measurement unit (PMU) phasor data.
Weather prediction and integration	• Improved accuracy of weather forecasts leading to improved load forecasts • Better at detecting the possibility of extreme events at long range • Various methodologies available, including artificial intelligence, neural networks, fuzzy logic, etc.
Ultra-fast load flow analysis	• Provides visualization tools showing regions of secure operations limited by voltage constraints, voltage instability, thermal limits, and flow gate constraints • Optimal mitigation measures can be applied online to expand the boundary of the operating region, reduce transmission congestion, optimize outage management, and support improved system planning analyses • Ultra-fast load flows that solve a 40,000-bus system in less than a second are available

Analytical Tools	
Tool	**Description**
Market system simulation	• Analyzes engineering and market aspects of the grid – links physical performance and control with economics • Provides open-source environment where independently developed software components can be shared by other people and organizations • Spans energy systems currently analyzed in isolation (e.g., transmission grid, distribution systems, and customer systems) • Under development at PNNL
Distribution fault location	• Will use data from digital relays or other monitoring systems along with circuit databases to determine the location of a fault on the distribution circuit • A new traveling wave system is being developed that is expected to be cost-effective for distribution systems. • Technologies have not yet been adopted due to complexity and high cost.
High-speed computing	• Essential because of the vast amount of data and the complexity of the analyses performed by ACM • Takes advantage of multiple networked computers to create a virtual computer architecture capable of distributing process execution across a parallel infrastructure • Work is being done to create a universal medium for information exchange. • New technologies are under development

Table 2: Analytical tools

OPERATIONAL APPLICATIONS

The modern grid will rely on local intelligence, automation, and decentralized control for selected applications, particularly those with primarily local impact. Centralized ACM will be utilized in other applications that provide a broader and more integrated perspective, such as the prediction of overall system capability and health. Some of these applications are described in Table 3 below.

Table 3: Operational Applications

Operational Applications	
Application	**Description**
SCADA	• Supports energy management systems (EMS) and transmission operations but has limited deployment at the distribution level • Today's SCADA systems are too slow and do not acquire data at the speed needed for ACM technologies. • More recently, Regional Transmission Organizations and Independent System Operators have expanded SCADA capabilities.
Substation automation	• Provides local control, remote control, and monitoring at the substation level • IEDs utilized for protection and control are normally integrated with a station computer, providing human-machine interface for local control, monitoring, and system configuration. • Makes the substation information available for retrieval by substation planners, protection engineers, maintenance personnel, and others as needed • The IEDs and the local network are linked to various other users to lay the foundation for higher-level remote functions such as advanced power system management and equipment condition monitoring while it is in service.
Transmission operations, energy management systems, and market operations	• Transmission SCADA systems provide system data to advanced state estimators, which solve large networks to determine system conditions every five minutes • Determines needed changes in system generation and, based on economic factors, provides EMS signals to participating units • Optimizes the economic dispatch of energy and at the same time mitigates the congestion on transmission lines • Location-dependent, real-time prices are calculated based on the re-dispatch of generation every five minutes to provide a price signal to generators, transmission owners and operators, and consumers. • Control methods currently in use by some of the Regional Transmission Organizations have advanced substantially over the past few years. • Need to incorporate advanced flow control, distributed energy resources (DER), and demand response (DR) options
Distribution automation	• IEDs have been integrated with distribution SCADA systems on a limited basis to provide rapid reconfiguration to minimize system impacts from faults and other power-quality disturbances • Currently, cost of DA technology has limited its widespread deployment. • New low-cost, high-speed, and reliable digital communications systems will eliminate many of the economic hurdles faced by DA deployment.

Page B4-9   Modern Grid Systems View: Appendix B4   *v2.0 Advanced Control Methods*

Operational Applications	
Application	**Description**
Demand response	• Used by system operators as a tool for mitigating congestion and peak-loading issues • Consumers give permission to system operators to interrupt loads under specific conditions • Consumers interact with system operators using the consumer's energy management system • Load can be interrupted autonomously using technologies embedded in grid-friendly appliances (GFAs) when specific conditions are detected.
Condition-based maintenance (CBM)	• Monitors and trends key asset characteristics, analyzes the information, and predicts when maintenance or replacement should be performed to prevent failure • Enables more effective and efficient maintenance practices, reducing occurrences of unexpected component failures as well as consumer and system outages • Becoming an accepted practice for managing health and maintenance of system assets
Outage management	• Integrates customer outage information with the up-to-date status of the distribution network • Helps operators rapidly determine causes of distribution outages • Enables more rapid restoration, including remote reconfiguration • Gives accurate information to customers regarding the status of power interruptions
Asset optimization	• Integrates plant operations, fuel management, and maintenance processes • Collects, verifies, and analyzes operational data using facility-specific parameters, and informs operators in real-time when a system is malfunctioning or running below expectations • Identifies conditions that could lead to a problem, determines the root cause, and prioritizes recommended solutions • Provides actual and what-if load data for devices, feeders, and substation transformers at the system level • Reconciles hourly SCADA data to provide an accurate view of asset loading system-wide and hourly-load profiles for each device • Assists operators in understanding which assets are over- or under-utilized and performs a risk analysis for each asset • Various technologies currently exist

Table 3: Operational applications

FUTURE STATE

The advanced control methods of the future require an advanced and integrated communication system to operate effectively (see "Appendix B1: Integrated Communications").

Many control functions are performed today to some degree and in limited locations. In the future, however, ACM will become significantly more sophisticated, will consider regional and national perspectives in addition to local ones, and will be fully deployed throughout the national grid. Where appropriate ACM will be distributed and where necessary, it will be centralized.

FUNCTIONS ACM WILL PERFORM

Collect data and monitor grid components – In the future, low-cost, smart instrument transformers, IEDs, and analytical tools will measure system and consumer parameters for every significant data point needed by ACM. New, low-cost devices will provide the condition of grid components and will be deployed and integrated with ACM to provide an overall assessment of the system's condition. These data will be presented to ACM for analysis on a near real-time basis by an integrated communication system. In addition, phasor measurement units (PMU), integrated with global positioning system (GPS) time signals will be deployed nationwide to provide a perspective of grid status and an early warning of developing instabilities.

Analyze data – The availability of near real-time data for all needed data points, and more powerful processors to analyze this data, will make possible rapid expansion and advancement in the capability of software-based analytical tools. Here are some specific examples:

* State estimators and contingency analyses will be performed in seconds rather than minutes, giving ACM and human operators additional time to react to emerging problems. This will also support the use of real-time transmission system optimization tools.
* Expert systems will convert the data to information that can be used for decision making. This information can then be input into probabilistic risk analyses.
* Load forecasting will take advantage of the system-wide distribution of near real-time data as well as improved weather forecasting technologies to produce highly accurate load forecasts at the system, component, and consumer levels.
* Probabilistic risk analyses will be performed routinely to determine the level of risk when taking equipment out of service for repair, during periods of high system stress, and following unexpected

outages. Indicators that present real time operating risk will be in place at regional and local operations centers to assist operators with the decision-making process.

- Grid modeling and simulations will enable operators to perform accurate "what-if" scenarios from a deterministic as well as probabilistic perspective.

Diagnose and solve – The availability of near real-time data processed by powerful high-speed computers will enable expert diagnostics to identify solutions for existing, emerging, and potential problems at the system, subsystem, and component levels. The probability of success for each solution will also be identified and the results made available to the human operator. This function of ACM will be carried out at local, regional, and system-wide levels based on the perspective needed or desired.

Take autonomous action when appropriate – Protective relaying schemes have acted autonomously in response to system faults for many years and will continue to do so in the future. The modern grid, however, will make significant advances by incorporating real-time communication systems with advanced analytical technologies. These advances will make possible autonomous action for problem detection and response. They also will mitigate the spread of existing problems, prevent emergent problems, and modify system configurations, conditions, and flows to prevent predicted problems.

Autonomous action will continue to be performed at the local level but will also be expanded to the regional and national level as control methods become more integrated with local control systems and centralized in the overall structure. Protective relay settings will be adapted to meet actual system conditions in real time.

Provide information and options for human operators – In addition to providing actuating signals to the control devices, ACM will provide information to the human operator. This information will be useful in two different ways.

First, the vast amount of data collected by the control system for its own use is of great value to the human operator. This data will be filtered and presented to sophisticated visualization programs to create an effective man-machine interface. (See "Appendix B5: Improved Interfaces and Decision Support"). These visualization programs will reduce the large amount of data to a format that allows the human operator to understand system conditions at a glance.

Second, the data will provide decision assistance. When the control algorithms determine a corrective action needs to be taken by a human operator (i.e., an action not appropriate for autonomous control), it will provide options to the human operator, giving probabilities for success for each option. In addition, when the controls take autonomous actions, those actions and their results will be reported to the operator.

Integrate with other enterprise-wide processes and technologies – Much of the data collected by ACM and the results obtained through the analyses they perform are of significant value to numerous other enterprise-wide processes and technologies. Equipped with this new data, these other processes and technologies can be significantly enhanced. Feedback from these secondary results will enable the advanced control methodologies to gain additional intelligence that will further refine the self-healing nature of the modern grid. The following are some examples where ACM can enhance existing processes and technologies.

- **Load forecasting and system planning** – Having extensive near real-time load data will eliminate the need to estimate past load and will provide accurate coincident load data from which more accurate forecasting will result. More accurate load forecasting will optimize the decision-making process concerning when and where new capacity additions are needed.

- **Maintenance** – Near real-time component condition and loading information will make possible a significant reduction in the number of equipment failures and the cost of reactive maintenance. The results of the maintenance process (including condition-based maintenance) will be fed back to the ACM technologies to improve their probabilistic risk analysis capabilities.

- **Market operations with RTOs** – ACM at the control area level will improve the interface with advanced control algorithms at the RTO level, resulting in the improvement of economic dispatch, the mitigation of transmission congestion, and the enhancement of system reliability.

- **Work management** – Near real-time consumer and system component data will enable work management and scheduling processes to determine the most effective timing for performing scheduled work. For example, "what if" scenarios will be performed to determine the risk in taking equipment out of service for performing work.

- **Outage management** – ACM will assist operators and storm-response personnel by sectionalizing, isolating, and providing recovery status on a near real-time basis. The outage management system will take advantage of the status information of all consumers and system components (integrated and analyzed by ACM) to precisely locate the outage and its cause. This information will also allow more accurate prediction of return-to-service times.

- **Simulation and training** – The increased level of sophistication that ACM technologies bring to the modern grid requires a corresponding increase in sophistication in the training for the human operator. The online controls and data will be interfaced to the training simulator to provide realistic system conditions and responses for various training scenarios.

- **Geographic information systems (GIS) for Spatial Analysis** – Near real-time data will be imported into GIS technologies to enable spatial analyses of various types to be performed. Locations of movable assets such as trucks, equipment, and personnel will be provided to the ACM to give the operators a better understanding of where these assets are located and to incorporate the personnel safety component into the self-healing feature of the modern grid.

- **Automatic meter reading** – Manual meter reading will be eliminated as meter reading and billing will be performed using accurate near real-time data collected by modern grid technologies.

BENEFITS OF IMPLEMENTATION

The wide acceptance and implementation of the modern grid's advance control methods will benefit all involved – the power industry, businesses, and industry as well as consumers and society in general.

Here are some of the many advantages to be realized:

- **The overall reliability** of the distribution and transmission systems will be generally improved, leading to decreased costs and increased revenues.
- **The self-healing vision** for the Modern Grid will be achieved. Appropriate actions will be taken to prevent or minimize adverse consequences. The scope of cascading events will be limited to prevent wide-area outages.
- **Sophisticated analytical capabilities** will prevent, detect, and mitigate the consequences of security attacks.
- **Integration with consumers and their loads** will provide energy price signals to encourage them to participate in the electricity market based on real supply-and-demand influences. The markets will then be more efficient and the result will be the lowest possible price for electricity.
- **Restoration times** following major grid events will be reduced by the provision of key and timely information and strategies needed by emergency response organizations.
- **Transmission congestion** will be minimized, contributing to further reductions in energy prices and more robust energy markets.
- **Supply-side and demand-side conditions will be monitored** to identify both emerging and actual power quality issues. Appropriate corrective actions will be taken to address power quality challenges before they become significant or lead to loss of reliability.
- **Utilization of DER and DR** to displace spinning reserve and increase system efficiency will reduce environmental impacts.
- **Integration of asset utilization data** into transmission and delivery (T&D) planning models will aid the planning of major long-term investments needed to increase system capacity.
- **Providing the material-condition data** for assets to condition-based maintenance (CBM) programs will improve the overall health and reliability of assets, reduce their out-of-service times, reduce the cost of maintenance, and improve the repair vs. replace decision-making process.
- **Integration of ACM** with work management and outage management systems will improve the efficiency in performing system and trouble work and will reduce the outage time and cost to consumers.

BARRIERS TO DEPLOYMENT

Significant barriers exist that impede the development and implementation of advanced control methods, but deployment of ACM is necessary to ensure safe, reliable, clean, economic, and environmentally responsible power in the future.

The move forward will remain limited until system data are available from a much wider area in near real time and a high-speed communication system is in place so that these ACM technologies can act. In addition, faster and more powerful computers are required so that ACM can respond immediately to rapidly forming power system events.

Another barrier is the lack of broad consensus for the modern grid vision among stakeholders. A greater understanding of the advantages of the modern grid – especially its self-healing function and its huge environmental benefits – is lacking. Conflicting objectives among stakeholders impede the full implementation of these control methods and their integration with other important processes and technologies. For example, the possibility of reduced revenues to suppliers of electricity impedes the full utilization and dispatch of consumer DER, much of which is far cleaner than central fossil-based generation. New regulatory models may be a solution to this conflict.

State regulatory bodies and current regulations do not fully support the vision for the modern grid. Increased cooperation between state and federal regulators is also necessary. New regulations that stir and motivate the vision for a modern grid must be created and existing regulations that impede progress must be modified. Regulated utilities need incentives for investing in ACM that provide societal benefits.

The perspective for ACM lacks breadth. Existing control methods are primarily focused at the local level. A greater deployment of local controls is needed and must be encouraged in the future; however, a wider, more centralized perspective is also needed. Effective integration of distributed controls to support a regional and even national monitoring and control perspective is lacking. An integrated, system-wide (region-wide or greater) control perspective needs to be formulated.

The cost of sensors is too high. The widespread deployment of IEDs is currently limited because of cost. In addition, a method to retrofit existing components to make them IED-ready is needed to keep implementation costs down. Otherwise, the placement of IEDs into the current electric system could take decades since components are not replaced today until they fail. Economies of scale and design innovation are needed to drive costs down.

385

The data today are incomplete and not available fast enough. As long as only limited data are available, many needed features of the modern grid, like the self-healing characteristic, are not possible.

The infrastructure for integrated communication is missing. Deployment of the needed communication systems, including supercomputers, is needed to support the processing and analysis of the large data volumes that will be supplied by advanced technologies of the modern grid.

Summary

Advanced control methods are technically achievable.

The needed software and hardware systems can be developed relatively easily following the development of a comprehensive set of control-system specifications.

But first the lack of a clear vision, the problem of insufficient data, the absence of a comprehensive communications infrastructure, and the inadequacy of IED deployment must be addressed for ACM to be universally accepted and implemented.

In addition, conflicting objectives among stakeholders must be addressed and answers found to benefit all involved.

Development of clear specifications for ACM and other key technologies is an important step in advancing the modern grid. Specifications for data acquisition, communication standards, and retrofits of existing components to convert them to IED functionality, all in support of these ACM technologies, will provide the basis for rapid progress.

But the most important step is the development of a national vision for the modern grid, endorsed by the great majority of stakeholders.

For More Information

This document is part of a collection of documents prepared by The Modern Grid Initiative (MGI) team. For a high-level overview of the modern grid, see "A Systems View of the Modern Grid." For additional background on the motivating factors for the modern grid, see "The Modern Grid Initiative."

MGI has also prepared five papers that support and supplement these overviews by detailing more specifics on each of the key technologies of the modern grid. This paper has described the fourth key technology area, "Advanced Control Methods."

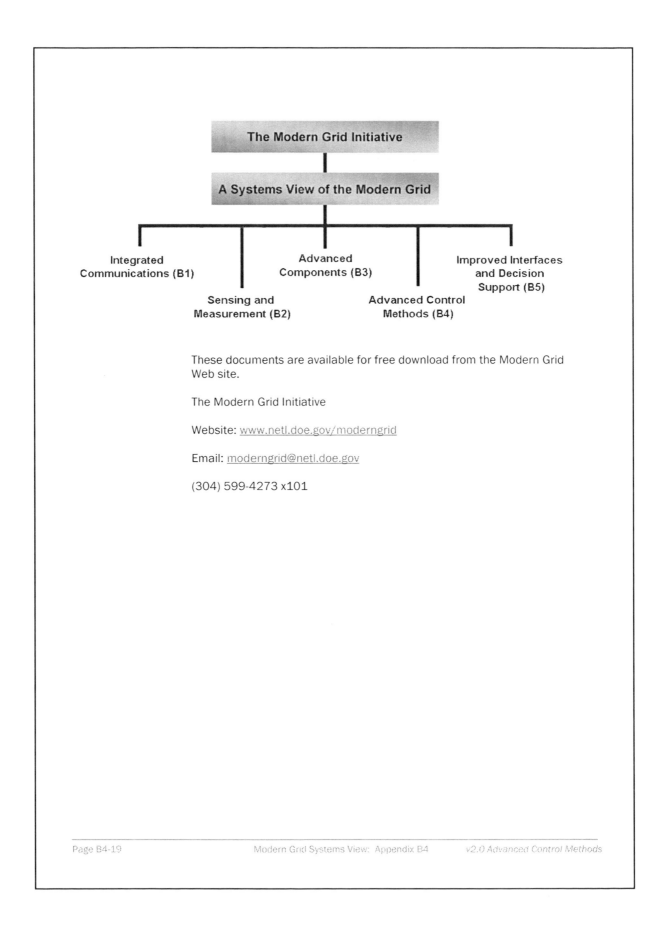

The Modern Grid Initiative

A Systems View of the Modern Grid

Integrated Communications (B1)

Sensing and Measurement (B2)

Advanced Components (B3)

Advanced Control Methods (B4)

Improved Interfaces and Decision Support (B5)

These documents are available for free download from the Modern Grid Web site.

The Modern Grid Initiative

Website: www.netl.doe.gov/moderngrid

Email: moderngrid@netl.doe.gov

(304) 599-4273 x101

BIBLIOGRAPHY

Ackermann, T., P. Lund, P. B. Eriksen, S. Cherian. 2005. Development of a new operation approach based on cell architecture for distributed generation in the Danish electric power system. Paper presented at the FPS 2005 International Conference on Future Power Systems, Amsterdam, the Netherlands.

Advanced Grid Applications Consortium. http://www.gridapp.org

Coley, B. 2001. Opportunities for 21st century meteorology: New markets for weather and climate information. Brief presented at the First AMS Presidential Policy Forum, American Meteorological Society (AMS) Annual Meeting, Albuquerque, NM. http://www.ametsoc.org/atmospolicy/presforums/albq2001/coley.pdf.

Dukart, J. R. 2003. The future of distribution. Transmission and Distribution World (January), http://tdworld.com/mag/power_future_distribution/

Electric Power Research Institute. 2004. Integrated energy and communications architecture: Volume IV: Technical analysis, Appendix D, technologies, services, and best practices. Palo Alto, CA: EPRI.

Electric Power Research Institute. 2005. IntelliGrid: SM transmission fast simulation and modeling (T-FSM) – business case analysis. Palo Alto, CA: EPRI. Product no. 1012152. http://www.epri.com/IntelliGrid/ http://intelligrid.info/

Grid-Friendly Appliance Controller. http://gridwise.pnl.gov/technologies/transactive_controls.stm

Power System Engineering Research Center. http://www.pserc.org

U.S. Department of Energy Office of Electric Transmission and Distribution. 2004. National electric delivery technologies roadmap: Transforming the grid to revolutionize electric power in North America.

Yeager, K. E. and C. W. Gellings. 2004. A bold vision for T&D. Paper presented at the Carnegie Mellon University Conference on Electricity Transmission in Deregulated Markets, Pittsburgh, PA.

ACRONYMS

ACM	Advanced Control Methods
CBM	Condition Based Maintenance
DA	Distribution Automation
DER	Distributed Energy Resources
DR	Demand Response
EMS	Energy Management System
FACTS	Flexible Alternating Current Transmission System
GFA	Grid-Friendly Appliance
GIS	Geographic Information System
GPS	Global Positioning System
IED	Intelligent Electronic Device
MGI	Modern Grid Initiative
PMU	Phasor Measurement Units
PNNL	Pacific Northwest National Laboratory
SCADA	Supervisory Control and Data Acquisition
T&D	Transmission and Distribution
VAr	Volt-amperes reactive

Appendix B5:
A Systems View of the Modern Grid

IMPROVED INTERFACES AND
DECISION SUPPORT

Conducted by the National Energy Technology Laboratory
for the U.S. Department of Energy
Office of Electricity Delivery and Energy Reliability
March 2007

Office of Electricity
Delivery and Energy
Reliability

v2.0

391

TABLE OF CONTENTS

EXECUTIVE SUMMARY

We urgently need a fully modern power grid if we are to meet our country's growing requirement for power that is reliable, secure, efficient, economic, safe, and environmentally responsible.

To achieve a modern grid, a wide range of technologies must be developed and implemented. These technologies can generally be grouped into five key technology areas as shown in Figure 1.

- Integrated Communications
- Sensing and Measurement
- Advanced Components
- Advanced Control Methods
- **Improved Interfaces and Decision Support**

Figure 1: A Systems View of the Modern Grid provides a direct approach to a total system solution.

Improved interfaces and decision support (IIDS), the focus of this paper, are essential technologies that must be implemented if grid operators and managers are to have the tools and training they will need to operate a modern grid.

IIDS technologies will convert complex power-system data into information that can be understood by human operators at a glance. Animation, color contouring, virtual reality, and other data-display techniques will prevent "data overload" and help operators identify, analyze, and act on emerging problems.

In many situations, the time available for operators to make decisions has now shortened from hours to minutes, sometimes even seconds. Thus, the modern grid will require the wide, seamless, real-time use of applications, tools, and training that equip grid operators and managers to make decisions very quickly.

Here are some areas where IIDS technologies will make a significant difference in the modern grid:

- **Visualization** – IIDS will take vast amounts of data (gathered by other advanced key technologies) and reduce it into the format, timeframe, and technical categories most crucial to system operators. Visualization techniques will present this information in a quickly-grasped visual format to support operator actions and decisions.
- **Decision Support** – IIDS technologies will identify existing, emerging, and predicted problems and provide what-if analyses for decision support. For situations requiring system operator action, multiple options and the probabilities of success and risk for each will be presented.
- **System Operator Training** – Dynamic simulators utilizing IIDS tools and industry-wide certification programs will significantly improve the skill sets and performance of today's system operators.
- **Customer Decision Making** – Demand Response (DR) systems will provide information to customers in easily understood formats that allow them to make decisions about how and when to purchase, store, or produce electric power.
- **Operational Enhancements** – As IIDS technologies are integrated with existing asset management processes, managers and users will be able to improve the efficiency and effectiveness of grid operation, maintenance, and planning.

The IIDS technologies in use today fall short of accomplishing these tasks. Improvements are needed at the human – machine interface to assist operators in comprehending the growing volume of data collected, its availability on a near real-time basis, and the complexity and speed of the advanced control methods that analyze and process this wealth of information (see "Appendix B4: Advanced Control Methods").

In this paper, we will look at the following important topics:

- Current state of IIDS technologies
- Future state of IIDS
- Benefits of implementation
- Barriers to deployment

Although it can be read on its own, this paper supports and supplements "A Systems View of the Modern Grid," an overview prepared by the Modern Grid Initiative team.

CURRENT STATE

On August 14, 2003, North America experienced the largest power blackout in its history.

Affecting an estimated 10 million people in Canada and 40 million people in the United States, this outage shut down more than 100 power plants and caused financial losses estimated at 6 billion dollars.

The internal control room procedures, protocols, and technologies of the day did not adequately prepare system operators to identify and react in time to prevent the August 14 emergency. In fact, throughout the afternoon of August 14, there were many clues that one of the control areas had lost its critical monitoring functionality and that its transmission system's reliability was becoming progressively more compromised. Clearly, the technologies and training in place at that time did not provide the visualization or decision support needed to manage that scenario.

The August 14, 2003, blackout task force, in analyzing causes of the blackout, emphasized the need for Improved Interfaces and Decision Support (IIDS) technologies. This necessity for better visualization capabilities and decision support tools over a wide geographic area has been a recurrent theme in blackout investigations. However, not enough progress has been made to date: Our power grid remains at risk for incidents of this nature until key technologies of the modern grid are in place.

In the first place, data are not available in the quantity and quality needed for operation of a modern grid. Advancement in grid management and operations is handicapped by these limitations in available data and tools for converting it into timely information.

IIDS technologies need data to function and vast amounts of it to be effective. However, data of the desired type and quality are not yet widely available from as broad an area as needed. In many cases today, the data acquired by multiple isolated systems stay isolated. To increase the effectiveness of this information, it must be shared with and utilized by other subsystems.

One more area needed by the modern grid is visualization for system operator use. Recently, research has been conducted to determine how various techniques, such as color contouring, animation, and virtual reality environments as well as other data-aggregation techniques, can prevent data overload and help operators identify problems, analyze them, and quickly take corrective action. Some of these techniques are already being used for operations, and others have been used for several years by such power system simulators as POWERWORLD™, which uses a power-system simulation engine to present the information calculated by its complex power-system analysis tools.

Another visualization tool has been released by the North American Electric Reliability Council (NERC): the Area Control Error (ACE)-Frequency Real Time Monitoring System. Visualization provided with this tool improves the reliability coordinator's monitoring of ACE and frequency at the interconnection level over a wide area. Additionally, it provides early and automatic notification and alarming for abnormal frequencies, improves the analysis of interconnection ACE-frequency relationships, and reduces from months to minutes the time it takes to search for the root causes of abnormal frequency.

Significant progress is being made by independent system operators (ISOs), who use IIDS technologies to manage the great volume of reliability- and market-information they receive; however, this is only occurring at the transmission level. The use of IIDS applications at the ISOs could be a model for the entire transmission system including those systems not currently operated by ISO's and for addressing the needs of the distribution system.

Overall, the current state of development of IIDS leaves much to be desired. This is of special concern because, as the other key technology areas are developed and implemented, critical IIDS technologies must be in place to enable the human operators to fully utilize the wealth of information and tools that will be available to them.

FUTURE STATE

As we move toward the modern grid, significant additional advances in IIDS technologies will become available to prevent the kind of data overload that would hinder system operators from understanding true grid conditions.

Supporting these technologies will be supercomputers capable of accepting the extensive flow of near real-time data, processing these data through a variety of analytic programs, and presenting recommended actions to the operator in a timely fashion.

Software systems will take more autonomous control actions both at the centralized and decentralized levels when advanced control methods operate synergistically with IIDS technologies. By allowing advanced controls to take on more of the analyses and control responsibilities, the human operators will be free to focus on grid conditions at a higher level. System operators will analyze system-wide parameters and review the automatic actions of new advanced controls, taking action themselves only when needed to override these controls or to handle issues outside the controls' permission sets.

Modern grid technologies will provide the data needed to assess current system status and conditions and predict the probability of future problems. It is the availability and presentation of this quantity of data by IIDS technologies that will allow operators to significantly improve grid performance.

Future operators will have more control options. They will have flexible alternating current transmission system (FACTS) devices on key transmission lines, high temperature superconducting synchronous condensers, adaptive relay settings, distributed energy resources (DER), and demand response (DR) dispatch. They will use IIDS technologies to guide them on how best to utilize these new assets.

Additionally, the diagnosis of power quality (PQ) issues requires that relevant data and sophisticated analyses be available to determine the cause and location of PQ disturbances (whether on the transmission system, distribution system, or on the consumer's side of the meter).

The ability of the modern grid to collect and analyze the necessary real-time data and, through advanced visualization techniques, present the cause of the condition to the operator will greatly reduce the time it takes to resolve PQ issues.

IIDS technologies will also assist in improving the economics of electricity markets. As consumers become engaged in market operations, they too will need access to more information. IIDS technologies will supply the visualization and analysis tools that will interface with consumer-based agents and portals.

Generation, transmission, and distribution operators, as well as consumers down to the residential level will be able to better understand the state of the grid and current conditions of the energy market, allowing them to more effectively participate in it. This widespread participation will improve the economics of the modern grid by creating more efficient markets.

Integration with other enterprise-wide technologies will lead to substantial improvements in information sharing among users. For example, IIDS technologies will utilize a standard information architecture that enables such areas as outage management, weather forecasting, transmission congestion, work management, condition-based maintenance, DR, dispatch of decentralized energy sources, and geographic information systems (GIS) to effectively and efficiently share information.

As IIDS technologies are integrated with existing asset management processes and technologies, system operators and users will be enabled to improve the efficiency and effectiveness of grid operation, maintenance, and planning. Substantial improvements will be seen in the following areas:

- **Spatial analyses** will give a better understanding of the location of human and material assets, leading to improved emergency response.
- **Identification of stressed equipment** will allow action to be taken to reduce stress and prevent equipment outage loss.
- **Assessments of asset conditions** will optimize the utilization of assets and reduce out-of-service times.
- **Understanding the actual loading conditions of equipment** will lead to more accurate predictions of when and where capacity additions will be needed.
- **The application of artificial intelligence** will facilitate the transfer of knowledge from an aging workforce to a new generation of power system workers.

As managers, operators, and users become equipped with advanced IIDS technologies, still more improvements in efficiency will be realized.

We will now look in more detail at three areas where the technology of human interface with power system tools will be further developed and implemented in the modern grid of the future.

VISUALIZATION, DECISION SUPPORT, AND OPERATOR TRAINING

Visualization

The vast amount of information collected and utilized by the modern grid will be one of its greatest assets. These data alone, however, are of little value to the operator in their as-received condition. To be effective, appropriate rule-based algorithms must reduce and simplify the data into the format, time frame, and technical categories most important to the power system operator. This is the role of advanced visualization tools.

System operators will judge the reasonableness of the results generated by the technologies, and they must rapidly intervene when they detect a technology malfunction. IIDS technologies will rely heavily on visualization to provide the essential link between human and machine. These human interface technologies engage the operators and allow them to ensure that modern grid parameters remain within prescribed limits.

Although advances in data visualization have been made, they are yet to be put to power system use on a widespread basis. In fact, because visualization tools remain largely unavailable, it is more and more difficult for operators and others to gain an intuitive understanding of the actual real-time operations and control of the grid. For instance, the August 14, 2003, blackout gave the system operators only 19 seconds to analyze the data and take corrective action. Yet conditions leading up to that event could have been better understood through the use of advanced analysis and visualization tools combined with enhanced training. These tools could have provided the support needed for operators to take sufficient and timely corrective action

Tomorrow's grid will have the needed advanced controls and IIDS technologies to prevent such events. With the modern grid of the future, system information will be presented to the operator using proven human-factors engineering techniques that will incorporate the latest two- and three-dimensional visualization tools and performance dashboards in conjunction with advanced control room designs.

Recent advances in computer hardware and software technology have made it possible to move beyond the simple tabular displays and one-line diagrams in common use today. The ability to draw relatively complex displays at frame rates close to, or even at, full-motion video speeds opens new possibilities for dynamic online displays. For example, animation can produce visually appealing displays of real and reactive flows and line loadings to point out overloads.

The following are examples of several technologies that will aid in the presentation of useful information to the operator:

- **Artificial Intelligence- (AI) driven data reduction** – AI will be utilized in conjunction with advanced control methods to minimize data volume without losing the information needed by the operator and to create the format most effective for human comprehension.

- **Color contouring and animation** – The use of color contouring and animation will be further researched and applied as part of IIDS technologies to increase not only operator speed of recognition but also speed of diagnosis.

- **Rapid refresh** – Advancements in communication and processor speeds will enable rapid refresh and display of real-time conditions so operators can quickly understand rapidly moving trends.

- **Voice recognition** – Advances in this area will increase the speed and effectiveness of the human-machine interface.

- **Virtual reality environments** – Virtual reality techniques will be applied to grid control centers to closely integrate the "thinking" of the operator and the IIDS technologies.

Of these, it is the contour displays that will illustrate spatial data such as bus voltage magnitudes and angles, line flows, or even derived values such as line power transfer distribution. Contours provide a natural encoding mechanism for displaying large amounts of spatial data. The advantage of the contour is that it allows the user to rapidly process large amounts of information and quickly spot developing trends. Thus, through the combined use of contours and animation, the operator will be able to quickly assess how the system state has been changing over a specified period.

Some examples of these visualization techniques are shown in Figures 2, 3, and 4, which follow:

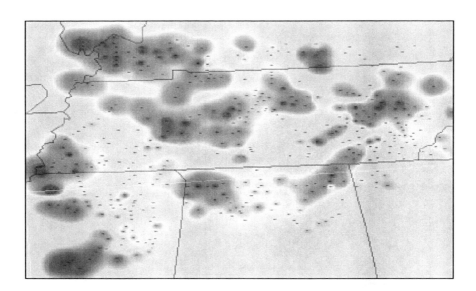

Figure 2: Bus voltage contours. Image courtesy of Tom Overbye University of Illinois at Urbana-Champaign.

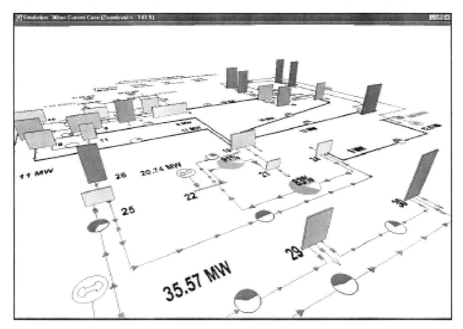

Figure 3: 3-D Visualization. Image courtesy of Tom Overbye University of Illinois at Urbana-Champaign.

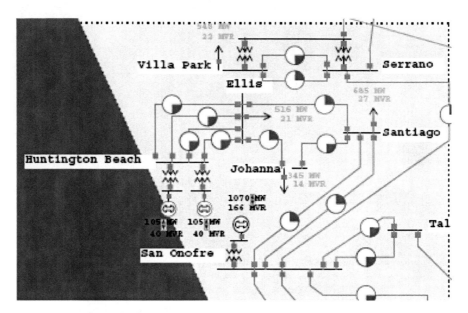

Figure 4: POWERWORLD™ display. Image courtesy of Tom Overbye University of Illinois at Urbana-Champaign.

Before data can be put into visual form, however, data optimization – the reducing, combining, and categorizing of data to eliminate unnecessary clutter – must be performed so that the visualization processes can present the data to an operator using the most effective visual interface.

Online optimization software is currently being demonstrated that transforms complex calculations into an easily visualized graphic format. In the future, software of this kind is expected to be widely accepted and implemented. Figure 5, below, depicts the results of such an online transmission-system optimization program.

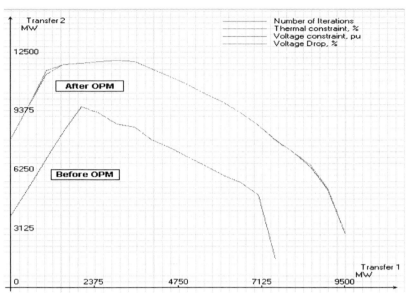

Figure 5: Visualization of online transmission system optimization. Image courtesy of V&R Energy Systems Research, Inc.

Without optimization, the secure region of operation is limited by voltage constraints, but after optimization, the secure region of operation increases until lines become limited by voltage stability.

Using visualization technologies, the system operator may optimize the transmission system by applying control measures such as these:

- MW dispatch
- MVAR dispatch
- Capacitor and reactor switching
- Operation of FACTS devices
- Transformer tap changes
- Line switching
- Adjustment of phase shifter settings
- Load curtailment
- Defined operating procedures
- Switching not-affected lines

In the near future, many new applications will be available that will improve visualization and thus increase the human operator's understanding and speed of comprehension. Here are some examples:

- **Advanced pattern recognition** – Used for intruder detection, forgery detection, biometrics, next-generation computer interfaces, and automatic paraphrasing, translation and language understanding.

- **Advanced speech recognition** – Speech recognition systems are composed of three major functions: Words are captured and translated into a digital signal; speech-recognition algorithm then compares those signals to words and phrases from a pre-set dictionary; and finally, the software offers the most likely match for the spoken phrase. The goal of the speech-to-speech translation (S2S) research is to enable real-time, interpersonal communication via natural spoken language for people who do not share a common language.

- **Haptic interfaces** – Will enable users to generate inputs through hand movements and provide users with tactile and force feedback consistent with what the user is seeing. These interfaces permit users to sense and manipulate 3D virtual objects with respect to features such as shape, weight, surface textures, and temperature. Haptic devices such as haptic gloves, joysticks, and tactile arrays have advanced rapidly and can generate a wide range of force and tactile feedback.

- **Holographic video** – Enables users to observe high-resolution spatial images that can be viewed from any angle as the user moves around the display.

- **Geographic information systems (GIS)** – will create and manage spatial data and associated attributes. The power of spatial information and location has been underutilized as a vital resource for improving economic productivity, decision-making, and delivery of services. Open interface specifications for GIS will enable users to freely exchange and apply spatial information, applications, and services across networks, platforms, and products. Advancements are expected in the integration of GIS technologies with virtual reality simulation and modeling.

- **Dashboard presentations** – Simplified displays of specific parameters, similar to the dashboard in today's modern vehicles, will assist users to rapidly detect the status of key variables.

In addition, human-factor engineering principles and techniques currently used in air traffic control and other industries, as well as those available from military applications, will be incorporated into the advanced control centers of the modern grid. IIDS technologies will also utilize the sophisticated analytical capabilities of the modern grid's advanced control methods to generate visualizations.

Visualization methodologies will differ somewhat, depending on the time scale of the data displayed. However, when possible, they will be standardized to simplify operator training. This will enable the system operator to rapidly identify and analyze issues and take corrective action within seconds.

Decision Support

The autonomous controls performed by advanced control methods will do much to stabilize the grid, but system operators will still be

responsible for monitoring these controls and making decisions that are beyond the scope of ACM.

Advanced control methods will only perform autonomous control actions when those actions are within their prescribed "permission sets" (i.e., a set of circumstances or conditions where technologies are given permission to act without first "checking" with operators). When other control actions are required of the operator, these will be presented to the operator in a way that maximizes the probability of success. IIDS technologies will recognize and address the relatively limited capability of the operator to make rapid decisions as compared to the speed of advanced control methods.

One of the principal causes of the extensive August 14, 2003, blackout was a lack of situational awareness, which was in turn the result of inadequate reliability tools and backup capabilities. Failures by control computers and alarm systems contributed directly to this lack of situational awareness. Incomplete tool sets and the failure to supply state estimators with correct system data also contributed.

In the future, however, there will be many tools to alert operators to existing, emerging, and predicted problems: AI, operator decision tools (such as alerting, what-if, and course-of-action tools), semi-autonomous agent software, visualization and systems tools, performance dashboards, advanced control-room design, and real-time dynamic simulators for training are some of these tools. In addition, for each issue and opportunity requiring operator action, multiple options for resolution and the probabilities of success and risk for each will be presented to assist in the decision-making process.

Some wide-area tools to aid situational awareness, such as real-time phasor measurement systems, have been tested in some regions but are not yet in general use. Improvements in this area will require significant new investments involving existing or emerging technologies.

Over the past few years, research has been done to incorporate a risk-based approach into the decision-making process. This approach recognizes that credible contingencies have different probabilities of occurrence. By understanding the consequences of each contingency and its probability of occurrence, decision support systems can quantify the relative risk and severity. These relative risks can be integrated into a composite risk factor and presented to the operator to assist with decision-making. This is more effective than the current approach to making security assessments and operating decisions. The current approach is to review past and current states of the system and to take actions based on the single most severe credible contingency.

Risk as a computable quantity will be used to integrate security with economics in formal decision-making algorithms. By applying probabilistic risk assessment (PRA) techniques into decision support technologies, grid operations can be made more secure and more economic.

In addition to risk-based decision support technologies, other forms of decision support are being developed. For example, intelligent user interfaces, utilizing techniques from the field of autonomous agents, provide a new complementary-style of human-computer interaction in which the computer becomes an intelligent, active, and personalized collaborator.

Some other examples of decision-support technologies are as follows:

- Practical real-time applications for wide-area system monitoring, using phasor measurements and other synchronized measuring devices, including post-disturbance applications.
- Enhanced techniques for the modeling and simulation of contingencies, blackouts, and other grid-related disturbances
- Improved visibility of the grid's status beyond an operator's own area of control – will aid the operator in making adjustments in operations and thereby mitigate potential negative impacts elsewhere on the grid.
- Extensive use of what-if technologies, including PRA techniques, to enable system operators to forecast the future state of the grid, identify potential issues, and take preemptive action to prevent future negative consequences.
- As part of the IntelliGrid program, EPRI has undertaken the Transmission Fast Simulation and Modeling (TFSM) project to develop a technical vision that models and simulates system behavior based on real-time data. TFSM will anticipate changing conditions and provide timely response during system disturbances, including prevention, containment, and support for recovery. A key concept incorporated in TFSM is the recognition of the different time frames associated with the various aspects of the modern grid that need to be controlled. TFSM is expected to also address the decision-support and human-interface needs of the modern grid.

System Operator Training

Since system operators will supply both the intelligence and intuition necessary to ensure the modern grid's successful operation, training programs and simulators in the future will be adjusted to incorporate the new features of the modern grid.

Visualization and decision support technologies, once developed, will need to be incorporated into the training and certification programs that support the operation of the modern grid.

IIDS technologies, together with system operator training, will be continuously evaluated and improved through control-room design reviews, human-factor reviews, and human-performance analysis, as well as through corrective actions based on critiques of the operators' responses to system events.

Operators will need a deep understanding of power system theory and will need to be trained in relevant power system engineering disciplines. Operator training will advance substantially beyond that provided today. Processes, procedures, and technologies used by NASA, the U.S. military, and nuclear power plants will be applied where appropriate to improve training programs.

The August 14 blackout task force investigation team found that some reliability coordinators and control area operators had not received adequate training in recognizing and responding to system emergencies. Most notable was the lack of realistic simulations and drills to train and verify the capabilities of operating personnel. Such simulations are essential to prepare operators and other staff to respond adequately to emergencies. This training deficiency contributed to the lack of situational awareness on August 14.

In the future, the development and use of real-time dynamic simulators and industry-wide certification programs will significantly improve the skill sets and human performance of system operators. Simulator training will be conducted using both normal and emergency scenarios and will be complemented by power-system engineering course work to ensure operators have a thorough understanding of the theories behind dynamic system performance.

Power-system simulators, specific to each operating area and region, will also be used. Training will include not only routine operations but also emergency operations and restoration. To the extent practical, the simulators will mirror the actual control room and will be tested to ensure their scenarios are realistic when compared to actual events.

Certification programs will be in place to verify that the simulators perform correctly and to certify that operators have completed specific training requirements prior to assuming duty.

Several new applications are available, or will be in the near future, to improve the training of system operators. Some examples include the following:

- PNNL's Integrated Energy Operations Center (IEOC) is a new user-based facility dedicated to energy and hydro-power research, operations training, and back-up resources for energy utilities and industry groups.
- Advanced simulators will give operators a real-time or historic view of the power system and its various parameters quickly, accurately, and in a format that increases situational awareness. These new technologies are extremely visual and efficiently perform power-flow analysis on systems containing up to 100,000 busses.

Significant progress must be made in the area of training and the operator's use of advanced decision support technologies. As the complexity of the transmission and distribution (T&D) systems and their connected generation sources grows, as the operating margins shrink, and as loads requiring higher power quality and improved reliability increase, additional stresses will be placed on the human operator that only first-rate, systematic training will be able to relieve.

BENEFITS OF IMPLEMENTATION

Overall, with improved interfaces and decision support technologies in place, the grid will experience more reliable operation and fewer incidences of outage from natural events and human error.

With IIDS, complex and extensive system information will be rendered into formats quickly understood by trained system operators so that they can accomplish the following:

- Understand the overall status of the grid at a glance and thus lend support to the self-healing aspect of the grid
- Maintain grid security and integrity by quickly detecting and mitigating threats against it
- Monitor and control a large number of new, decentralized energy sources (such as DER, DR, and advanced storage)
- Deal quickly with emerging PQ issues
- Identify stressed equipment so that relief can be provided or equipment replaced before a breakdown can cause a costly outage
- Identify the location of system assets, human resources, portable equipment, and physical landmarks such as roads, bridges, and city streets, thus enabling system operators to significantly improve worker and public safety and to create a safer environment for completing restoration work
- Better understand the environmental impact of grid resources and thus balance that impact with economics in the dispatch of centralized generation and DER.
- Improve overall operation and maintenance of the entire power delivery system

In addition, IIDS technologies will assist other stakeholders in the following ways:

- Provide an effective interface to allow consumers to actively participate in the energy market and grid operations, thereby incorporating load as an active factor in grid operations
- Assist in the communication of grid information to stakeholders, thereby providing the needed level of transparency of operations and market information
- Support the integration of key technology areas with other enterprise-wide processes and technologies to improve their overall understanding of grid conditions

BARRIERS TO DEPLOYMENT

One obvious barrier is the extent of work that remains to be done to gain consensus among all stakeholders on the national vision for the modern grid.

The vision needs to be communicated, examined, refined and accepted by all stakeholders, including federal, state, and local regulators. It will provide the fundamental framework around which the necessary supportive regulatory and other environments will emerge.

Another barrier to overcome is the lack of sufficient power system data. IIDS technologies cannot be effective regionally and nationally at either the transmission or distribution levels until sufficient quantities of data are made available by advancements in the technology areas listed below:

- Integrated Communications (Appendix B1)
- Sensing and Measurement (Appendix B2)
- Advanced Control Methods (Appendix B4)

More research is needed to improve grid reliability, with particular attention to improving the capabilities and tools for system monitoring, management, and operations.

Although some progress has been made, there still remains a wide gap between today's IIDS technologies and what is required to meet the vision for the modern grid. For example, many of the recommendations from the August 14, 2003, blackout task force have yet to be implemented, including improvements needed in IIDS technologies that will give system operators the capability to detect and prevent cascading outages.

Here are some additional barriers that need to be addressed:

- **Development is needed in applications** that integrate advanced visualization technologies with geospatial tools to improve speed of comprehension and real-time decision-making.
- **Advances in computing power** are needed to support the processing of complex, near real-time applications and in presenting it to the operators.
- **Development of low cost sensors and an integrated communications infrastructure** are needed to acquire the type of data needed by the modern grid.

- **Consumer-based agents and portals** need to be equipped with decision support algorithms and visualization technologies that empower consumers to participate in the energy markets.

In addition to the needed technologies, effective and successful system operator training programs – which depend on a corporate commitment to training – need to be in place. Adequate funding is needed to maintain the power system models and provide for instructors. Also, operating procedures, documentation, and training must be reviewed and updated each time a new technology is introduced that will impact control room operations.

SUMMARY

Improved Interfaces and Decision Support is one of five key technology areas that need to be developed and implemented to reach the goal of achieving a modern power grid that is reliable, secure, economic, efficient, environmentally friendly, and safe.

IIDS technologies are essential if system operators are to have the tools and training necessary to most effectively operate and maintain the modern grid. IIDS will equip managers, operators, and even consumers with the needed applications and tools.

IIDS technologies are also needed to present the vast and wide-ranging volume of data collected by the modern grid in a quickly comprehensible format. Often this information will have to be presented in ways that can be very quickly assimilated and understood, such as when operators have only minutes or even seconds to make decisions.

Today, however, sufficient information is not available to grid operators and managers to enable them to provide consistently reliable, economic, clean, and safe power throughout the grid.

Development is needed in several areas before IIDS technologies can be fully implemented in the grid:

- **Visualization** – to enable system operators to grasp system changes quickly and thus respond very quickly to emerging changes and problems
- **Decision support** – to present system operators and other stakeholders with the best decision options for actions that fall outside of the system's autonomous controls
- **Training and certification of system operators** – to adequately prepare operators to utilize all tools available to assist in maintaining grid integrity

For more information

This document is part of a collection of documents prepared by the Modern Grid Initiative (MGI) team. For a high-level overview of the modern grid, see "A Systems View of the Modern Grid." For additional background on the motivating factors for the modern grid, see "The Modern Grid Initiative."

MGI has also prepared five papers that support and supplement these overviews by detailing more specifics on each of the key technology areas of the modern grid. This paper has described the fifth key technology area, "Improved Interfaces and Decision Support."

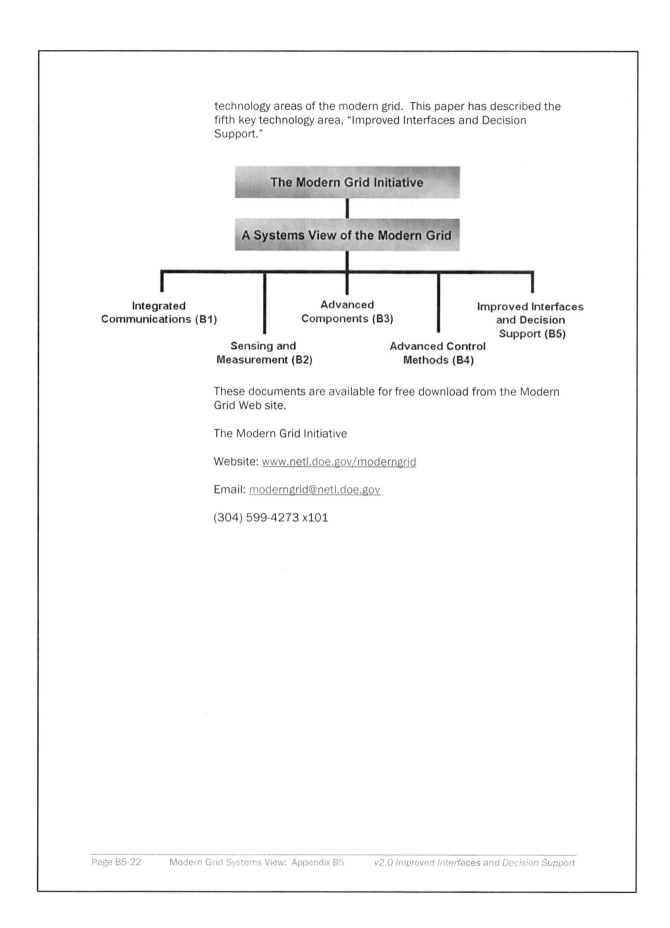

These documents are available for free download from the Modern Grid Web site.

The Modern Grid Initiative

Website: www.netl.doe.gov/moderngrid

Email: moderngrid@netl.doe.gov

(304) 599-4273 x101

BIBLIOGRAPHY

Advanced Grid Applications Consortium. http://www.gridapp.org/.

Bradshaw, Dale. 2006. Presentation on NRECA Cooperative Research Network R&D in Support of a Modern Grid, MGI SE Summit, Nashville, TN, August 10, 2006.

Consortium for Electric Reliability Technology Solutions (CERTS). http://certs.lbl.gov/.

Electric Power Research Institute. 2005. IntelligridSM transmission fast simulation and modeling (T-FSM) – business case analysis. Palo Alto, CA: EPRI. Product no. 1012152.

Lee, S. T. Industry-wide power delivery reliability initiative bears fruit. Morgan Hill, CA: Hoffman Publications.

Martinez, C. 2004. NERC tools – wide area real time monitoring. Presented to the Consortium for Electric Reliability Technology Solutions, Washington, D.C. Pasadena, CA: Electric Power Group.

McCalley, J. D. 2005. Security assessment: Decision support tools for power system operators. Ames, IA: Iowa State University.

Overbye, T. J., P. W. Sauer, G. Gross, M. J. Laufenberg, and J. D. Weber. 1997. A simulation tool for analysis of alternative paradigms for the new electricity business. Proceeding of the 30th Annual Hawaii International Conference on System Sciences (HICSS) Volume 5: Advanced Technology Track: 634.

Overbye, T. J. and J. D. Weber. 1999. Visualization of power system data.

Overbye, T. J., J. D. Weber, and M. Laufenberg. 1999. Visualization of flows and transfer capability in electric networks. Proceedings of the 13th Power Systems Computation Conference: 420–426. Zürich, Switzerland: Executive Board of the Thirteenth Power Systems Computation Conference.

Power Systems Engineering Research Center (PSERC). http://www.pserc.wisc.edu/.

Talukdar, S., J. Apt, M. Ilic, L. B. Lave, and M. G. Morgan. 2003. Cascading failures: Survival versus prevention. Electricity Journal 16 (November): 25–31.

U.S.–Canada Power System Outage Task Force. 2004. Final report on the August 14, 2003, blackout in the United States and Canada: Causes and recommendations.

U.S. Department of Energy Office of Electric Transmission and Distribution. National Electric Delivery Technologies Roadmap (Workshop), Washington DC, July 8–9, 2003, January 2004.

Wickens, C.D., M. S. Ambinder, and A. L. Alexander. 2004. The role of highlighting in visual search through maps. Proceedings of the Human Factors and Ergonomics Society 48th Annual Meeting, Santa Monica, CA.

Wiegman, D. A., A. M. Rich, T. J. Overbye, and Y. Sun. Human factors aspects of power system voltage visualizations. Proceedings of the 35th Hawaii International Conference on System Sciences (HICSS'02)-Volume 2: 58. Piscataway, NJ: Institute of Electrical and Electronics Engineers.

Yeager, K. E. and C. W. Gellings. 2004. A bold vision for T&D. Paper presented at Carnegie Mellon University's Conference on Electricity Transmission in Deregulated Markets: Challenges, Opportunities, and Necessary R&D, Pittsburgh, PA.

ACRONYM LIST

ACE	Area Control Error
AI	Artificial Intelligence
DER	Distributed Energy Resources
DR	Demand Response
EPRI	Electric Power Research Institute
GIS	Geographic information system
IEOC	Integrated Energy Operations Center
IIDS	Improved interfaces and decision support
ISO	Independent System Operator
MW	Megawatt
NERC	North American Electric Reliability Council
PNNL	Pacific Northwest National Laboratory
PQ	Power Quality
PRA	Probabilistic risk assessment
S2S	Speech-to-speech
T&D	Transmission and Distribution
TFSM	Transmission Fast Simulation and Modeling

Appendix A1:
A Systems View of the Modern Grid

SELF-HEALS

Conducted by the National Energy Technology Laboratory
for the U.S. Department of Energy
Office of Electricity Delivery and Energy Reliability
March 2007

Office of Electricity
Delivery and Energy
Reliability

417

TABLE OF CONTENTS

EXECUTIVE SUMMARY

The systems view of the modern grid features seven principal characteristics. One of those characteristics is 'Self heals'. What that means and how we might attain that characteristic is the subject of this paper.

• Self-heals

  • Motivates and includes the consumer

   • Resists attack

   • Provides power quality for 21st-century needs

   • Accommodates all generation and storage options

  • Enables markets

 • Optimizes assets and operates efficiently

Figure 1: The Modern Grid Systems View provides an "ecosystem" perspective that considers all aspects and all stakeholders.

In the context of the modern grid, "self-healing" refers to an engineering design that enables the problematic elements of a system to be isolated and, ideally, restored to normal operation with little or no human intervention. These self-healing actions will result in minimal or no interruption of service to consumers. It is, in essence, the modern grid's immune system.

The modern, self-healing grid will perform continuous, online self-assessments to predict potential problems, detect existing or emerging problems, and initiate immediate corrective responses. The self-healing concept is a natural extension of power system protective relaying, which forms the core of this technology.

A self-healing grid will frequently utilize a networked design linking multiple energy sources. Advanced sensors on networked equipment will identify a malfunction and communicate to nearby devices when a fault or other problem occurs. Sensors will also detect patterns that are precursors to faults, providing the ability to mitigate conditions before the event actually occurs.

The self-healing objective is to limit event impact to the smallest area possible. This approach can also mitigate power quality issues; sensors can identify problematic conditions and corrective steps can be taken, such as instantly transferring a customer to a "clean" power quality or source.

A simplified example of the self-healing concept, illustrated in Figure 2 below, shows two power lines having many "intelligent switches" (noted as "R") located along the circuit. This diagram illustrates the intelligent switching feature of self-healing, which can maintain power to a maximum number of customers by instantaneously transferring them to an alternate energy source.

Alternate energy sources may include circuit ties to other feeders or to distributed energy resources (DER) such as energy storage devices and small electrical generators (powered by both renewable and non-renewable fuels). Demand response (DR) can also be a tool in matching load to generation in the self-healing process.

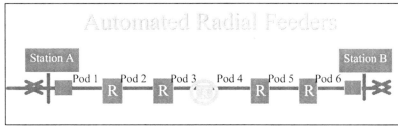

Figure 2: Automated radial feeders schematic. Image courtesy of DV2010.

The modern electrical grid will know a great deal about problems affecting its operation. One of the keys to self-healing is the utilization of a wide assortment of information gathered from modern grid devices to enable rapid analysis and initiation of automatic corrective actions,

Fault locations, circuit configuration changes, voltage and power quality problems and other grid abnormalities can be quickly discovered and corrected. High-risk areas, as well as individual pieces of equipment, can be analyzed for immediate action. Also advanced models can provide new visualization tools revealing congestion issues, overlays of failure probabilities, and resulting threat levels.

Another element of self-healing is the avoidance of high-risk situations. When impending weather extremes, solar magnetic disturbances, and real-time contingency analyses are incorporated into a probabilistic model, grid operators will be better able to understand the risks of each decision they may make, as well as ways to minimize those risks. In such applications, the expected volume of real-time data is high. And it will be necessary to integrate those data up to the control area, regional transmission organization level, NERC Region level, or the entire national grid, including its interconnections with Canada and Mexico.

Figure 3 below illustrates one way to convey broad information (relative energy prices, in this case) at a glance. Similar presentation techniques supporting self-healing are possible, showing relative risk, overloads, voltage violations, or other applicable metrics.

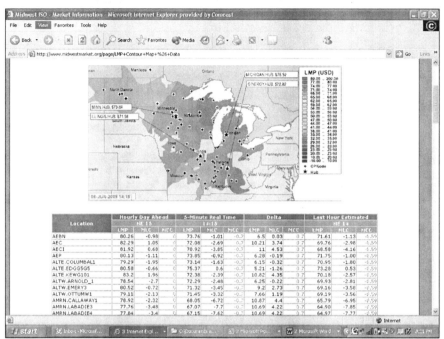

Figure 3: Advanced visualization gives operators detailed data about the workings of the modern grid. Image courtesy of Midwest Independent Transmission System Operator, Inc. For informational purposes only.

The modern grid will enhance self-healing functions in the transmission grid and will extend those functions to the distribution level. For example, Wide Area Monitoring Systems (WAMS), utilizing phasor measurement units, combined with advanced state estimation algorithms run on high-speed computers, could control Flexible Alternating Current Transmission System (FACTS) devices to prevent or mitigate a developing system collapse. And dynamic islanding using DER devices and intelligent switching is but one example of self-healing at the distribution level—there are many others.

While advanced sensing, analysis, protection and control are important elements of a self-healing grid, so too is a robust T&D infrastructure. High-capacity circuit ties joining major Regional Transmission Organizations (RTOs) allow for inter-region power flows in an emergency. But if this power transfer capability is not adequate, then upgrades to higher capacity or the construction of new tie lines is required. This infrastructure improvement would also result in more robust energy markets, allowing less expensive generating unit power to flow to areas of high-cost congestion.

This paper explores how the modern grid will act to reduce the number and duration of outages, minimize restoration times and reconfigure the grid to produce optimum reliability and quality of service. **All these features are rolled up to a common name that we call the 'Self-healing' characteristic of the Modern Grid.**

In this paper we address these following important topics:
- The current and future states of the grid
- The requirements of a "Self-healing' grid
- The barriers to implementing a "Self-healing" grid
- The benefits in achieving a "Self-healing" grid
- Recommendations for moving forward in this endeavor

Although this whitepaper can be read on its own, it supports and supplements "A Systems View of the Modern Grid," an overview prepared by the Modern Grid Initiative (MGI) team.

CURRENT AND FUTURE STATES

CURRENT STATE

Transmission

Today's transmission grid was designed with many self-healing features. Auto-reclosing and auto-sectionalizing are common techniques employed to maintain loads under adverse conditions. The network mesh design of the transmission system is in itself self-healing due its built-in redundancy and such protective relaying features as high-speed reclosing and single-phase tripping.

System planners have historically modeled the transmission system to verify that, under a normal system configuration, assumed loads could be met even during expected peak conditions. In addition, planners ensured that these same loads could be met even with the failure of single, and in some cases, multiple lines or components.

Sophisticated protective relaying schemes are in place to monitor system conditions and take corrective action should specific parameters exceed their limits. Transmission lines and equipment are relayed out (i.e., tripped) when conditions require, and most loads normally are not impacted by a single fault because the system can tolerate a single contingency. Substation automation and new intelligent electronic devices have taken transmission protection to the next level.

The design of the current transmission system incorporated the notion of self-healing many years ago and utilized the technologies, processes, and techniques available at the time. Significant advances in digital technologies, correctly applied, will dramatically improve this self-healing capability.

Distribution

At the distribution level, new distribution automation (DA) technologies are being deployed to increase reliability and efficiency. DA applications improve the efficiency of system operation, reconfigure the system after disturbances, improve reliability and power quality, and identify and resolve system problems. Many DA applications can also be extended to coordinate with customer services, such as demand-response, and distributed energy resources (DER). In addition, distribution systems that include feeder-to-feeder backup allow enhanced DA functionality. These new approaches are directionally consistent with the vision of the self-healing feature of the modern grid. DA is integral to the concept of a self-healing grid.

The current distribution system, without distributed resources and without an intelligent networked configuration, has been handicapped from a self-healing perspective. Today most DA and substation automation (SA) systems are applied at a local level, using local information for decision-making. The basic design of the integrated transmission grid—many geographically diverse generation sources feeding a high-voltage networked transmission system—is conducive to self-healing. On the other hand, the fundamental design of today's distribution systems cannot, in most cases, incorporate the depth of self-healing found on today's transmission systems.

FUTURE STATE

The self-healing feature of the modern grid, at both the transmission and distribution levels, will advance from its current state by integrating advanced capabilities in the following areas:

Look Ahead Features

- Analytical computer programs, using accurate and near real-time state estimation results, will identify challenges to the system, both actual and predicted, and take immediate automatic action to prevent or mitigate the event. Where appropriate, and when time allows, these algorithms will also provide options for the system operator to manually address the challenge.

- Probabilistic risk analysis, also in near real time, will identify risks to the system under projected normal operating conditions, single failures, double failures, and out-of-service maintenance periods.

- Load forecasting will be greatly improved to support more accurate look-ahead simulations. These simulations will be performed over various time horizons—minutes, hours and days in support of operations; monthly, quarterly, and annually to support O&M planning activities; and longer range to support investment decisions.

Monitoring Features

- Real-time data acquisition, employing advances in communication technology and new, lower-cost smart sensors, will provide a significantly larger volume and new categories of data, such as wide-area phasor measurement information. This dramatic increase in the volume of real-time data, combined with advanced visualization techniques (see Appendix B 5, Improved Interfaces and Decision Support), will enable system operators to have an accurate understanding of the power delivery system's health.

- By analyzing equipment condition data - including high frequency emission signatures—condition monitoring technologies will provide additional perspectives on the consequences of potential equipment failures.

- State estimators will take advantage of advanced data acquisition technologies and powerful computers that enable them to solve problems in seconds or less.

- Advanced visualization techniques will consolidate data and present the appropriate information to operators in easily understood formats.
- Command and control centers at the regional level for transmission operations and at more local levels for distribution operations will serve as hubs for the new self-healing features.

Protection and Control Features

- Advanced relaying will be employed to communicate with central systems and adapt to real-time conditions.
- High-speed switching, throttling, modulating, and fault-limiting devices will dynamically reconfigure the grid, including faster isolation and sectionalization as well as rapid control of real and reactive power flows in response to system challenges.
- Intelligent control devices, such as grid friendly appliances, will modulate load requirements in response to dynamic grid changes

Distributed Technology Features

- Distributed generation and energy storage technologies will be widely deployed, particularly at the distribution level, and dispatched as system resources in response to self-healing needs. DER will also be used to support local circuit needs.
- Transformation of the distribution system from a radial design to an intelligent network design, through the addition of circuit-to-circuit ties, the integration of DER and DR and the application of advanced communication technology will create a self-healing infrastructure.
- DR programs will be widely expanded and utilized as system resources to assist in the management of system overloads, voltage issues, and stability issues. DR will also be used to support local circuit needs.
- DA will be further expanded and integrated with widespread DER/DR and, in conjunction with new operating and visualization tools, will enable successful dynamic islanding.
- Critical system components will be "hardened" where appropriate, including redundant designs and in-place spares.

These advances will together create a sophisticated self-healing capability in the modern grid that will dramatically improve its overall reliability, efficiency, safety and will also increase its tolerance to a security attack.

REQUIREMENTS

KEY SUCCESS FACTORS

The self-healing principal characteristic is essential to achieving each of the modern grid's key success factors. The ability to detect, analyze and respond to undesirable conditions and events supports these key success factors in the following ways.

Reliable

The predictive nature of the modern grid, coupled with its ability to implement corrective actions in real time, will provide a major improvement in reliability at the transmission, distribution, and consumer level. Advancements in the following areas will enable this:

- Real-time data acquisition of needed parameters.
- High-speed analytical tools that can determine system state, identify system challenges both deterministically and probabilistically, and determine options for preventing or mitigating negative consequences.
- High-speed switching and "throttling" devices that can correct system parameters prior to the occurrence of negative consequences.
- Advanced relaying that adjusts to real-time conditions.
- Redundancy and hardening of critical components.

The self-healing feature of the modern grid will go beyond the prevention and mitigation of outages and will include monitoring of system equipment and consumer portals to identify both emerging and actual power quality issues.

- If a low or unbalanced voltage condition occurs on a distribution network, that condition will be monitored and an appropriate corrective action will be taken.
- If harmonics or other sustained or intermittent power quality issues are detected, these conditions will likewise be corrected.

Secure

The same features of the self-healing grid that enable it to improve reliability also enable it to better tolerate security attack and natural disaster.

- Probabilistic analytical tools will identify weaknesses in the modern grid that can be integrated into the overall security plan.
- Self-healing's intelligent networking and DER features make the grid far more difficult to attack.
- The real-time data acquisition capability of the modern grid will immediately detect challenges to its security.
- The real-time response of high-speed control devices will provide rapid response to security attacks.

- Following security challenges, real-time data acquisition and control will greatly enhance the damage assessment process, and significantly reduce restoration times.

Economic

The self-healing feature of the modern grid will optimize the economics for all stakeholders:

- System reliability and power quality will improve, leading to a substantial reduction in losses incurred by business and individual consumers when power is lost.
- Generators, transmission owners and operators, and distribution companies will benefit from a reduction in lost revenues that now occur when the grid experiences high congestion or unplanned outages. Greatly improved restoration times will also provide these stakeholders with economic benefits.
- Consumers will benefit from more efficient energy markets.
- More efficient operation will reduce electrical losses and maintenance costs.

Efficient and Environmentally Friendly

Much of the same data acquired to support the self-healing feature of the modern grid will also provide value to the stakeholders' asset management programs. In addition, the self-healing characteristic supports a range of environmental benefits.

- Real time data will be used to more effectively load assets in real time and manage their condition.
- Equipment failure prediction/prevention will reduce the environmental impact associated with such events as transformer fires and oil spills.
- The self-healing grid will accommodate all forms of generation, including many green technologies that produce zero emissions. Both health and environmental stresses are diminished as emissions are reduced.

Safe

The self-healing feature of the modern grid includes the capability and intelligence that promotes the safety of workers, consumers, and other stakeholders. In addition, by reducing outages and area blackouts, associated safety issues are mitigated as the exposure of hazards to workers and the public is reduced.

OBSERVED GAPS

The gap between the current and future states of the self-healing modern grid can be summarized as follows:

- Self-healing in the current transmission system is more advanced than in the distribution system, but opportunities exist to significantly improve both. While individual vendor applications exist for certain self-healing features, no previous initiative has integrated a full complement of transmission and distribution technologies to create a fully self-healing power delivery system.

- The cost to develop and implement the needed changes is high. Addressing this cost will require the alignment of all stakeholders, including the federal government, because many benefits of a self-healing grid are societal in nature. Utilities alone cannot justify the investment to attain the societal benefits.

- Advances are needed in many technical areas (as represented by the modern grid's five Key Technology Areas), including the following:
  - Development and deployment of intelligent electronic devices, including advanced sensors.
  - Development and deployment of DER and DR, as well as their integration and utilization by reliability coordinators.
  - Deployment of DA managed by local distribution reliability centers.
  - Installation of circuit-to-circuit ties to move the distribution system toward a networked topology.
  - Deployment of a ubiquitous communication infrastructure to support the self-healing feature.
  - Development and deployment of new visualization techniques to help operators understand system risk levels.
  - Development and deployment of new control algorithms and new control devices to execute self-healing actions.

DESIGN CONCEPT

The self-healing grid will employ multiple technologies to identify threats to the grid and immediately respond to maintain or restore service.

Probabilistic risk assessment technologies will identify equipment, and systems that are most likely to fail. For example, inadequate vegetation control creates a higher probability of transmission and distribution line failures. Power plants with poor material conditions are at high risk for forced shutdown or de-rating. Such assessments will not just be based on historical records, but on real-time measurements and probabilistic analysis as well.

Given probabilistic assessments for equipment, weather, and load as inputs, real-time contingency analysis engines can determine the overall prognosis of the grid's health. Equipment with high risk of failure can be identified for immediate investigation and even deeper analysis. New operator visualization displays will create a clear understanding of grid capability and levels of risk.

High speed, reliable communications and computing capability is an essential ingredient of the modern grid. The self-healing grid will employ extensive voltage and flow control, along with fault current limiting capabilities. Appropriate local and remote devices, running real-time analyses of electrical events, will issue control signals that address emerging problems. Frequently, the short time interval of such events will require all this to happen without human intervention, requiring improved communication and computing.

DESIGN FEATURES AND FUNCTIONS

The features and functions of the self-healing grid, as described in the following, will be present at all levels of the power system from generating source to load, including regional transmission organizations and distribution utilities.

Probabilistic Risk Assessment

State Estimation and Real-Time Contingency Analysis results will be available within seconds. These results will drive Probabilistic Risk Assessment algorithms, which incorporate real-time condition monitoring, combined with short-term weather and load forecasts, to produce easily interpreted descriptions of impending risks at the interconnect level, Regional Transmission Operator (RTO) level and control area level. Such visualization tools provide an unprecedented understanding of the consequences of multiple failures and by their very nature offer clues to resolution options.

Power Stabilization Techniques

New power stabilization software and hardware will be developed to look for the early signs of, and then prevent, a spreading blackout. While alarms will be initiated for human intervention, automation may take mitigating actions, as determined by control algorithms. Split-second decisions, such as opening tie lines, changing flow patterns or shedding load must be taken before an instability becomes a blackout.

Additional inter-RTO Tie Line capacity will maximize needed power transfer capability during emergency conditions. While many tie lines exist today, some require upgrade to higher capacities and more need to be built.

Distribution System Self-healing Processes

Distribution circuits will have many isolating elements that communicate with each other. By sensing circuit parameters and applying internal logic, these circuits will determine when and where to isolate a fault and restore service to others. This can be done through the closing of optimally selected switches, the injection of distributed generation and energy storage devices, and the management of load levels using DR tools. These actions will take place in a timeframe not possible by human operators. Further, these same isolating elements will monitor and control voltage and power quality.

USER INTERFACE

The self-healing grid is comprised of many sub-systems. Those features and functions dealing with state estimation, probabilistic risk assessment, and major equipment reliability will be accessible to the North American Electric Reliability Corporation (NERC), RTO, and Control Area operators. Distribution Centers will have access to distribution information and relevant transmission information. They will also aggregate energy storage, distributed generation and curtailable load information and pass it along to the relevant RTO and Control Areas for bulk power supply applications.

FUNCTIONAL ARCHITECTURE STANDARDIZATION

Figure 4 shows how substation data can be collected at the substation and used to deliver new self-healing applications involving cascading blackout protection, fast state estimation, real-time problem identification and power quality analysis.

Architecture with process bus

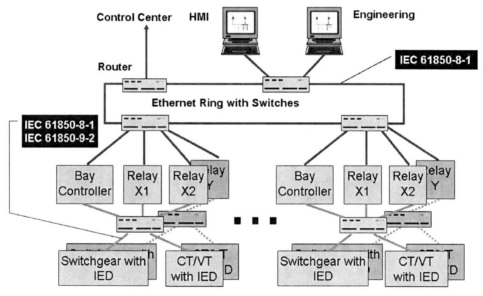

Figure 4: Example schematic of substation data architecture. Image courtesy of International Electrotechnical Commission.

The collection of modern grid data will be placed in a Common Information Model (CIM) format to clearly correlate parameters with the equipment they represent. Standardization on the use of CIM architecture will enable wide transportability as well as fast and easy access to results. The application of CIM Architecture will also extend into the distribution space.

Standardization of communication protocols is another essential enabler of grid modernization. The specific communication technology can be different depending on the application requirements, such as cost and location, but each communications channel must satisfy required security, transit time and quality of transmission.

Reclosers and sectionalizers presently exist on the Distribution System to isolate faults. These switching devices can often be retrofitted with standardized communications, data processing, and actuating devices to make them compatible with the modern grid's requirements.

PERFORMANCE REQUIREMENTS

A number of performance measures are required to validate the effectiveness of the self-healing grid. State estimation resolution time and the subsequent real time contingency analysis time are two important metrics. A target of less than 10 seconds is desirable. Additionally, the time to achieve a probabilistic risk assessment with its attendant contingency analysis could follow within a minute.

Existing performance measures recommended for this characteristic include:

- The Customer Average Interruption Duration Index (CAIDI) measures how long it takes a utility to restore service after an interruption and is scored by adding up the durations of each service interruption in a year and dividing the total by the total number of customer service interruptions, thereby deriving the average outage duration for that year.
- The Momentary Average Interruption Frequency Index (MAIFI) is the total number of momentary customer interruptions (less than 5 minutes in duration) divided by the total number of customers, expressed as momentary interruptions per customer per year. MAIFI characterizes the average number of momentary electric service interruptions for each customer during the time period.

Within the modern grid vision, distribution systems become an asset of the control area and the regional transmission organization. The distribution system's load forecast is an important parameter used by control area operators and RTO's. By using distributed generation, energy storage and demand response to manage its deviation from its load forecast, the distribution system would enhance the performance of the transmission system. A metric could be developed to measure how effective the distribution system operators are in meeting their load forecasts.

The identification of distribution reserve is another possible metric and perhaps a new ancillary service. A large reserve would signal a more robust distribution system. The distribution reserve would include both energy (Kwh) and capacity (KW) ratings. The accumulation of many small energy and capacity resources could result in a significant combined energy resource for bulk power system operators.

BARRIERS

Major change usually faces substantial barriers. The modern grid is no exception.

This section discusses the barriers to achieving a self-healing grid.

- **Financial Resources** – The business case for a self-healing grid is good, particularly if it includes societal benefits. But regulators will require extensive proof before authorizing major investments based heavily on societal benefits.

- **Government Support** -- The industry may not have the financial capacity to fund new technologies without the aid of government programs to provide incentives to invest. The utility industry is capital-intensive, with $800 billion in assets, but it has undergone hard times in the marketplace and some utilities have impaired financial ratings.

- **Compatible Equipment** – Some older equipment must be replaced as it cannot be retrofitted to be compatible with the requirements of the self-healing characteristic. This may present a problem for utilities and regulators since keeping equipment beyond its depreciated life minimizes the capital cost to consumers. Early retirement of equipment may become an issue.

- **Speed of Technology Development** – The solar shingle, the basement fuel cell, and the chimney wind generator were predicted 50 years ago as an integral part of the home of the future. This modest historical progress will need to accelerate. Specific areas that will need to be more fully developed and deployed include:
  - Integrated high-speed communications systems.
  - Intelligent electronic device's (both front end sensors and back end control devices)
  - Distribution automation schemes to provide distribution level self-healing capabilities, to accommodate all forms of DER and to act as an asset to the transmission system.
  - Cost-effective environmentally acceptable Distributed Energy Resources, including energy storage devices capable of existing among residential populations.
  - DR systems using real-time pricing.

- **Policy and Regulation** – Utility commissions frequently take a parochial view of new construction projects. A critical circuit tie crossing state boundaries has historically met significant resistance. The state financing the project may not always be the one benefiting most from it. Unless an attractive return on self-healing and other modern grid investments is encouraged, utilities will remain reluctant to invest in new technologies.

- **Cooperation** – The challenge for 3,000 diverse utilities will be the cooperation needed to install critical circuit ties and freely exchange information to implement modern grid concepts.

BENEFITS

The benefits of implementing a self-healing grid are many and diverse, providing benefits to consumers, utilities, employers and government.

The following list is representative of the types of gains that can be expected.

- **Improved Reliability** - Resolving the gaps noted previously will enable a substantial improvement in grid reliability. The cost of power disturbances to the U.S. economy is significant (on the order of $100 billion). The savings from a massive blackout is estimated on the order of $10 billion per event as described in the *'Final Report on the Aug.14, 2003 Blackout in the United States and Canada'*. Since blackout events are increasing in frequency, it is not unreasonable to assume another one will occur within a few years.

- **Improved Security** - A self-healing grid is almost, by definition, the most secure grid. A grid that self heals is a less attractive target since its resiliency reduces the impact an attack can inflict. Also, the consequences of an attack are reduced because energy sources are distributed and self-healing technologies can restore service during and after an attack.

- **Safety** - Increased public safety will be a benefit of the modern grid. Grid re-configurations will quickly de-energize downed wires. Restoring power faster to more people will reduce the impact to customers who rely on the grid for medical necessities as well as maintaining HVAC to elder care facilities. Also, fewer outages reduce the opportunities for criminal acts and civil disturbances.

- **New Revenue** - The installation of DER and DR will create peak shaving and the accumulation of reserves. Both are commercial products in the energy market that can produce revenue streams for their owners.

- **Quality** - The self-healing grid will detect and correct power quality issues. Power quality issues represent another large cost to society, estimated to be in the tens of billions of dollars. In addition, the quality of decisions will improve and autonomous control will occur more quickly.

- **Environmental** - The self-healing grid will accommodate multiple green resources, both distributed and centralized, resulting in substantial reductions in emissions. In addition, the environmental impact associated with outages and major equipment failures will be dramatically reduced. And a more efficient grid means lower electrical losses (hence lower emissions).

RECOMMENDATIONS

Thoughtful, deliberate, concise actions of change are required to enable the self-healing grid to become a reality. This section outlines the recommendations that will help to achieve the self-healing vision.

Many of the individual components, hardware and software, already exist for self-healing features to become a reality. But the integration of all the elements to form a unified single purposed entity still remains to be done.

A clear vision for the modern grid and a transition plan to accomplish it is needed to successfully implement the self-healing feature.

The self-healing characteristic is enabled by each of the five Key Technology Areas. Hence progress must be encouraged in all five. Most essential is the Integrated Communications key technology area, which provides the foundation for all self-healing features.

Demonstration projects of untested and previously never-before integrated technologies are necessary to provide a platform for broader deployment. Technologies that have never been integrated with other technologies in a system context need to be integrated and tested to provide the realistic, business-case quality data needed to cause broader deployment of the technologies.

Many benefits of a self-healing grid accrue to the society in general. The public is the beneficiary whether the benefits are environmental, national security, safety, economic or other. Legislators and regulators must recognize these public goods so that the utility industry has the incentive to move forward.

SUMMARY

The health of an electric system, like that of the human body, is determined in large part by the strength of its immune system—by its ability to heal itself. And in that context, the North American grid's immune system is not especially strong.

Today, there are ways to strengthen this system, to improve its ability to detect and fight off stress. Modern technology can make it much more resistant to the challenges of a 21st century society. Today's advances in computers, communications, materials and chemistry have yet to be applied in a meaningful way to this task. That is what can and must be done.

There can be no doubt that a prosperous society is built upon a healthy electric power infrastructure. This is most apparent when that infrastructure is weakened or disabled, as it is during a major blackout. In fact, an extended blackout would have a crippling effect on the fundamental structure of society.

Of course, modernizing the grid infrastructure requires an investment of considerable magnitude. But the resultant benefits, when viewed from a societal perspective, will return that investment many fold.

The quest for a self-healing grid will require a coalition of dedicated people determined to make a difference. If you agree with the vision and agree with the conclusions drawn in this paper, we urge you to get involved.

For more information

This document is part of a collection of documents prepared by The Modern Grid Initiative team. For a high-level overview of the modern grid, see "A Systems View of the Modern Grid."

For additional background on the motivating factors for the modern grid, see "The Modern Grid Initiative."

MGI has also prepared seven papers that support and supplement these overviews by detailing more specifics on each of the principal characteristics of the modern grid.

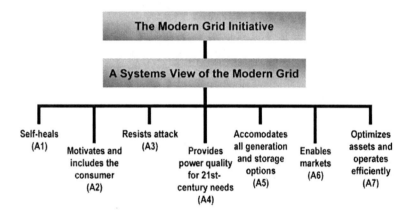

Documents are available for free download from the Modern Grid Web site.

The Modern Grid Initiative

Website: www.netl.doe.gov/moderngrid

Email: moderngrid@netl.doe.gov

(304) 599-4273 x101

BIBLIOGRAPHY

1. Amin, M. and B.F. Wollenberg "Toward a Smart Grid: Power Delivery for the 21st Century," IEEE Power and Energy Magazine, Vol 3, No 5, Sep/Oct 2005.

2. Amin, M. "Energy Infrastructure Defense Systems," Special Issue of Proceedings of the IEEE, Vol. 93, Number 5, pp. 861-875, May 2005

3. Amin, M. "Toward Self-Healing Energy Infrastructure Systems," cover feature in the IEEE Computer Applications in Power, pp. 20-28, Vol. 14, No. 1, January 2001

4. Amin, M. "Toward Self-Healing Infrastructure Systems," IEEE Computer Magazine, pp. 44-53, Vol. 33, No. 8, Aug. 2000

5. Apt, J., L. B. Lave, S. Talukdar, M. G. Morgan, and M. Ilic. 2004. Electrical blackouts: Repeating our mistakes. Working paper, Carnegie Mellon Electricity Industry Center, CE-04-01.

6. Dukart, J. R. 2003. The future of distribution. Transmission and Distribution World (January), http://tdworld.com/mag/power_future_distribution/.

7. Electric Power Research Institute. 2004. Electricity technology roadmap: Meeting the critical challenges of the 21st century. Summary report, product no. 1010929.

8. Electric Power Research Institute. 2003. The integrated energy and communication systems architecture: Volume II: Functional requirements. Palo Alto, CA: EPRI.

9. Federal Energy Regulatory Commission. 2006. Rules concerning certification of the electric reliability organization; and procedures for the establishment, approval, and enforcement of electric reliability standards. 18 CFR Part 39, Docket No. RM05-30-000, Order No. 672.

10. Glotfelty, J. 2004. Transforming the grid to revolutionize electric power in North America. Presentation to the U.S. Department of Energy Office of Electric Transmission and Distribution.

11. Huber, R. and R. Fanning. 2003. Distribution vision 2010. Transmission and Distribution World (January), http://tdworld.com/mag/power_future_distribution/.

12. Lee, S. T. and S. Hoffman. 2001. Power delivery reliability initiative bears fruit. IEEE Computer Applications in Power 14, part 3: 56–63.

13. Talukdar, S., J. Apt, M. Ilic, L. B. Lave, and M. G. Morgan. 2003. Cascading failures: Survival versus prevention. Electricity Journal 16 (November): 25–31.

14. U.S.-Canada Power System Outage Task Force. 2004. Final report on the August 14, 2003, blackout in the United States and Canada: Causes and recommendations.

15. U.S. Department of Energy Office of Electric Transmission and Distribution. 2003. "Grid 2030": A national vision for electricity's second 100 years.

16. Yeager, K. E. and C. W. Gellings. 2004. A bold vision for T&D. Paper presented at Carnegie Mellon University's Conference on Electricity Transmission in Deregulated Markets: Challenges, Opportunities, and Necessary R&D, Pittsburgh, PA.

Appendix A2:
A Systems View of the Modern Grid

MOTIVATES AND INCLUDES THE CONSUMER

Conducted by the National Energy Technology Laboratory
for the U.S. Department of Energy
Office of Electricity Delivery and Energy Reliability
January 2007

Office of Electricity
Delivery and Energy
Reliability

V2.0

TABLE OF CONTENTS

EXECUTIVE SUMMARY

The systems view of the modern grid features seven principal characteristics. (See Figure 1.) One of those characteristics is to provide the consumer with choices that benefit both consumers and the grid itself. We describe this as 'motivating and including the consumer'.

- Self-heals
- Motivates and includes the consumer
- Resists attack
- Provides power quality for 21st-century needs
- Accommodates all generation and storage options
- Enables markets
- Optimizes assets and operates efficiently

Figure 1: The Modern Grid Systems View provides an "ecosystem" perspective that considers all aspects and all stakeholders.

In the modern grid, consumers will be an integral part of the electric power system. They will help balance supply and demand and ensure reliability by modifying the way they use and purchase electricity. These modifications will come as a result of consumers having choices that will motivate different purchasing patterns and behavior. These choices will involve new technologies, new information about their electricity use, and new forms of electricity pricing and incentives.

From the modern grid's perspective, consumer demand, or *electric load*, is simply another manageable resource, similar to power generation, grid capacity and energy storage.

From the consumer's perspective, electric consumption is an economic *choice* that recognizes both the variable cost of electricity and its value to the consumer under a range of times, places, and circumstances.

Consumers with choices in how they purchase and use energy will be able to:

- Use price signals and other economic incentives (i.e. demand response or DR) to decide if and when to purchase electricity, and whether to produce or store it using a distributed energy resource (DER).

- Purchase "intelligent load" end-use devices that consume power wisely and that become integral parts of the grid to help optimize its operations and reliability.

Permitting consumers to face the underlying variability in electricity costs can improve economic efficiency, increase reliability and reduce the environmental impacts of electricity production. (*Hirst, E. and B. Kirby, Retail-load participation in competitive wholesale electricity markets, 2001*)

Each of these choices has already been demonstrated to provide numerous benefits to multiple parties. The technologies that enable each, such as advanced metering, smart thermostats and appliances, distributed generation, and energy storage, have been demonstrated to give utilities, system operators, retail marketers, electricity consumers and policy makers new tools for achieving their mutual and separate objectives. Much like the earliest personal computers and cellular phones, these technologies are poised for widespread adoption and deployment, as well as continual improvement.

The benefits of enabling the consumer to take a greater role are tangible and significant. For example, clipping the spikes of peak demand reduces the need to build new facilities, improves the utilization of existing plants and improves the environment by allowing the retirement or reduced use of inefficient generation.

Peak management activities also support a more efficient marketplace by acting as a dampening factor on wholesale electricity prices, which all consumers ultimately pay. In doing so, they help limit the amount of market power that electricity producers and sellers can exercise.

Environmental benefits also accrue because emissions, worse during peak demands, are substantially avoided.

As a result of these benefits being available, and customers taking actions to obtain them, the modern grid will be a more holistic and dynamic system where customers and their dynamic actions will be an integral part.

This document covers key elements of one of the seven principal characteristics of the modern grid — how to **motivate and include the consumer**. Although it can be read on its own, it supports and supplements "A Systems View of the Modern Grid," an overview prepared by the Modern Grid Initiative (MGI) team.

CURRENT AND FUTURE STATES

We begin by contrasting the current situation for consumers and their electricity purchases, with what it could be like in the future state with a modern grid.

CURRENT STATE

In today's environment, the vast majority of consumers are fully insulated from the volatility of wholesale electricity markets and the true underlying moment-to-moment cost to produce and deliver the electricity they consume. They purchase electricity under fixed, time-invariant prices that are set months or years ahead. The costs of generating that electricity, however, vary substantially from hour to hour, often by a factor of ten within a single day. (*Hirst and Kirby*)

Today there are new opportunities emerging that provide the consumer with better information on the actual cost of electricity. They also present a monetary incentive for consumers to modify their usage in response to that information.

These opportunities primarily fall under the new business and policy area known as demand response (DR). Examples of DR are time-based or dynamic pricing options where the price of the electricity purchased by a consumer varies by time-of-day – sometimes even hourly or more frequently. Other examples are programs offered by utilities or independent system operators (ISOs) where customers are paid to curtail or cut back their usage when electric system conditions would benefit.

Demand response offerings have been and can be made by utilities, systems operators or third parties such as retail marketers or companies that specialize in demand response technologies and services. These offerings can be made in either restructured markets or those that are still a traditional, vertically integrated model. Evidence shows that customers both large and small can be counted on to participate in demand response programs and market offerings.

Another area of opportunity is the use of distributed energy resources (DER). This refers to the use of generation systems that are on the customer side of the meter and which can be operated at times of the customer's choosing as an alternative to taking electricity off the grid. It also refers to the emerging storage technologies. (DER is covered extensively in Appendix A5, entitled "Accommodates All Generation and Storage Options," which is available from the Modern Grid Initiative at www.TheModernGrid.org.)

Retail consumers who modify their usage in response to price volatility help lower the size of price spikes. This demand-induced reduction in prices is a powerful way to discipline the market power that some generators would otherwise have when demand is high and supplies tight. And these price-spike reductions benefit all retail consumers, not just those who modify their consumption in response to changing prices. (*Hirst, E. and B. Kirby, Retail-load participation in competitive wholesale electricity markets, 2001*)

It should be noted that frequently a distributed energy resource is the technology that is used by a customer to participate in a demand response program.

The demonstrations of DR and DER to date illustrate the substantial benefits to reliability and economic stability attainable by motivating and including the consumer. Today, a number of technologies that support DR and DER are available, and many more are under development. The relevant information technologies and digital communications have become both more powerful and less expensive; they are ripe for deployment.

Ultimately, competitive electricity markets will feature two kinds of demand-response programs. First, some consumers will choose to face electricity prices that vary from hour to hour. Typically, these prices will be established in the day-ahead markets run by regional transmission organizations, such as those now operating in California, New York and the mid-Atlantic region. Second, some consumers will select fixed prices, as they have in the past, but voluntarily cut demand during periods of very high prices. In this second option, the consumer and the electricity supplier will share the savings associated with such load reductions. (*Hirst, E. and B. Kirby, Retail-load participation in competitive wholesale electricity markets, 2001)*

Future State

The future will see a robust and widespread link between energy consumers and the modern grid's operators. Creating this linkage will allow consumers to make informed consumption choices, which in turn will benefit both the consumer and utilities.

As technology improves and new policies allow or even encourage increased deployment, the number of customers actively participating will increase and costs will drop. DR and DER programs and market-based offerings will become even more attractive to consumers.

Achieving this customer participation means making it easy and understandable. And essential to this will be providing a user interface that successfully motivates and supports customer action. These interfaces can take a variety of forms, depending on the sophistication and desires of the consumer. They could range from a

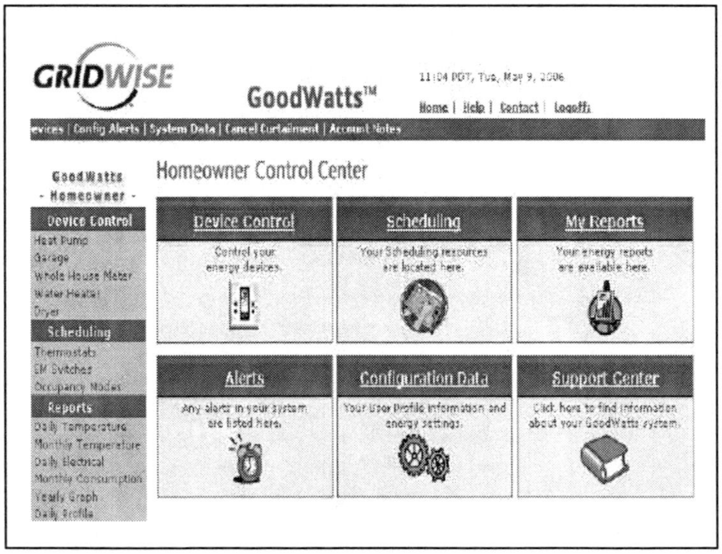

Figure 2: Consumers may be able to make their electricity choices via a simple point -and-Click Web interface, as in this example from a DOE GridWise demonstration project. Source: This graphic was produced at the Pacific Northwest National Laboratory under Contract DE-AC05-76RLO 1830 with the U.S. Department of Energy, and the GRIDWISE™ trademark is owned by Battelle Memorial Institute.

Page A2-5 Modern Grid Systems View: Appendix 2 v 2.0 *Motivates and Includes the Consumer* series of simple indicator/warning lights to detailed computer-generated displays of energy and pricing information. (See Figure 2.) Today's communications and electronic technologies create options that were just not viable in the past.

One example of a future architecture is shown in Figure 3, below. In this example, which visualizes the broad implementation of real time pricing, the mechanism to provide consumers with greater choice involves the insertion of a gateway unit between the energy company and the consumer's appliances. This gateway provides load control based on the consumer's pre-programmed price preferences. In a sense, the gateway acts here as the consumer's agent. New technologies such as computer agents to support consumer decisions and broadband over power lines (BPL) to communicate pricing and other information will enable more effective interaction between the energy company and the consumer.

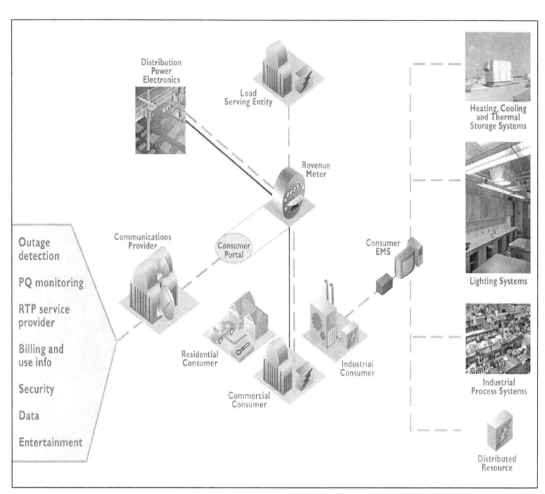

Figure 3. A consumer electricity portal, as envisioned by the Intelligrid research initiative, controls load based on the consumer's pre-programmed price preferences. Image courtesy of EPRI.

Motivated by economic incentives, consumers will adopt newly-available smart appliances. Such appliances will monitor electrical conditions such as voltage and frequency, and automatically turn on or off to support the stability of the grid. Some will also automatically respond to price signals. Distributed resources, another enabling technology, will further strengthen and expand consumers' support of grid operations. Today's digital revolution will seed the needed advances in metering, communications, and decision support. Innovation will be stimulated by competition and economics of scale.

REQUIREMENTS

Having described the current state and envisioned the future state, what are the requirements for achieving that future? This section describes the features and components of the modern grid that are required to offer informed choices to the consumer on how they use power.

FEATURES

Above all, the modern grid must address the consumer's primary objectives. Whether that is simply lowering the electrical bill at home or enhancing the productivity of manufacturing with cheaper, higher quality or more reliable energy, the overarching requirement is the same: choices and tools that are easy to understand and operate – even to the extent of being automatic.

Electricity customers, with the exception of some large businesses, do not want to become buyers and sellers of electricity – their main pursuit is economical and reliable electricity that can yield for them the true values they seek. An energy management program that operates in the background, quietly providing the quality, reliability and economics sought by the consumer is an example of solutions that can meet these customer desires and needs.

The modern grid must therefore strive to incorporate the consumer into grid operations in an automatic and cost-effective way. The system must have features that:

- Perform consistently within the rules, regulations and agreements between the utility and the consumer.
- Provide power and/or reduce load when needed or desired
- Deliver cost savings over time.

Consumer programs such as DR and DER must be cost effective to be successful. Lowering the costs of the required components — such as meters, communications and central support — is achievable only in mass quantity production. With sufficient consumer interest, the market itself would then support the required production scale needed to make these systems pay for themselves.

KEY COMPONENTS

There are a number of key components of the modern grid that are required to enable greater consumer choice in energy consumption and to link the consumer into the electricity practices of the grid:

- Consumer applications (such as DR systems) that are reliable, easy to use and tamper-resistant.
- Software applications for the consumer that respond to pricing signals from the utility — this agent software automatically

manages the consumer's usage based on price and within boundaries established by the consumer in concert with the utility.

- Smart communicating meters that measure both consumer usage and grid conditions to help the utility provide desired service at minimum cost.
- The communications infrastructure and control systems to support two-way information flow and load management.
- Processes, tariffs and incentive programs that serve both the utility and consumer.

Using these systems to the full benefit of consumers and the grid requires:

- Semi-autonomous processes and programs that enable both consumers and utilities to share the benefits of grid efficiencies.
- New pricing regimens, enabling consumer choice and planning as well as acceptable utility returns.
- Grid-friendly appliances that consumers can be encouraged to deploy.
- Multiple, affordable choices for consumer-usable DER.

BARRIERS

In the context of motivating and including the consumer, we have described the current state, explored the future state and identified the features and functions required to transport us from one to the other. This section discusses the barriers that must be overcome.

The technologies that enable DR and DER are now and will be for some time in a constant state of technological evolution. Research and development work remains to be done. For example, further development is required on:

- Cost-effective, secure, and reliable metering, communications, and information technologies.
- Systems and processes that more fully recognize and incorporate the active consumer role in grid management
- Extensive development in the design of user interfaces and the tariffs associated with them.

Much non-technological work also remains to be done.

- Consumer education will be necessary to promote the broad acceptance of voluntary programs.
- The design of innovative rate structures that benefit both the consumer and the power system's managers require more attention and testing.
- Many different stakeholders must agree upon a clear, auditable method to manage the various programs of rates and tariffs as consumer options multiply.
- Federal and state regulatory bodies need to set a clear direction, including reaching agreement on promoting DR programs. In the case of retail electricity pricing, state regulators have the authority to determine whether customers are provided with time-based pricing options or not.

Barriers in the way of accomplishing this work include:

- **Cultural views of electricity services** — The transition from a passive protected user to a proactive informed consumer should apply to electricity, just as it does to other products
- **Long-established regulations** — Current state legislatures and regulatory commissions' efforts exist largely to protect consumers from the risks of competition.
- **Lack of consumer education on electricity services** — The consumer must come to understand that the price of electricity should reflect its current cost of production and delivery so that more economic usage decisions can be taken. The vital role of electricity makes it important to recognize that everyone benefits from its optimal use.

Although the potential benefits of dynamic pricing are large, so too are the barriers to widespread adoption. State legislatures and regulatory commissions have inadvertently blocked consumer access to wholesale markets in their efforts to protect retail consumers, especially residential consumers, from the vagaries of competition. State regulators need to rethink their decisions on standard-offer rates that are set so low that new suppliers are unable to compete and consumers have no incentive to look elsewhere for a better deal. Although regulators should not force consumers to face dynamic pricing, neither should they make it difficult for them to do so. Ultimately, consumers will have to pay for prices that are set too low today. (*Hirst, E. and B. Kirby, Retail-load participation in competitive wholesale electricity markets, 2001*)

- **Slow process of technology deployment** — Advances in metering, communications, information processing and distributed resources are being deployed, but only gradually.

BENEFITS

Once we fill the gaps and overcome the barriers, the economic ripple effect of giving consumers informed choices will benefit all sectors of society. In fact, it is already happening, as this section reveals.

DEMAND RESPONSE

Demand Response has enjoyed considerable progress in overcoming the barriers of regulation and consumer education.

Diverse interest groups that include prominent regional and national stakeholders have formed the Demand Response Coordinating Committee.

Through the committee, the US is taking the lead on the International Energy Association's first DR project. Such issues as valuations, technology, coordination, barriers, and funding are on the agenda.

Recently, the chair of the Federal Energy Regulatory Commission described DR as "the silver bullet of market design." The US Congress has included strong DR provisions in the Energy Policy Act of 2005. This signals national recognition of DR's value in reshaping the industry.

The DOE Office of Electricity Delivery and Energy Reliability views DR as an essential element of the modern grid. At the same time, the Environmental Protection Agency believes DR can be a valuable driver of energy efficiency. This is because pricing signals will likely lead to more investment in efficient end-use devices. The integration of DR and energy efficiency is likely to provide additional synergies.

DR projects that included consumers have already produced very positive results. One program in Illinois was the first to clearly demonstrate just how effective DR can be. Independent evaluators found:

- Participants respond to peak period prices — Overall demand reduced by up to 20 percent with small changes in behavior.
- Participants saved money — Approximately 15 percent for the first two years of the program.
- Participants of all incomes benefited — Low-income households especially respond pro-actively to high prices.
- The meters were not expensive.
- Participants developed better understanding and attitudes about energy usage.

Independent System Operators (ISO) are increasing their focus on DR, both as a tool to mitigate emergencies and as a means to realize substantial economies. NY ISO has measured benefit ratios exceeding 5:1 with their emergency DR program.

PJM Interconnection, the world's largest electric grid operator, has stated that 20,000 MW of its load is served only 1 percent of the time. The huge value of shifting this load to lower-use periods is clear to PJM operators.

ISO New England has shown that DR programs can be very responsive, reaching committed reduction levels in less than 30 minutes

ISO New England and other cases prove that DR will be an important facet of another characteristic of the modern grid—enabling markets. In fact, each of the seven characteristics of a modern grid can be affected in one way or another by the broad deployment of systems and processes that include the consumer.

DISTRIBUTED ENERGY RESOURCES

Consumers may represent the largest market for DER well into the next decade as they use it to save money and improve reliability.

Deployment of DER will benefit the entire value chain of consumers — commercial, industrial and residential. Simple connections to the grid will accelerate consumer usage of small generation and storage devices and pave the way for larger ones.

Allowing the consumer to store and generate electricity in a coordinated way can support the grid by:

- Lowering the risk of load imbalances.
- Providing quality power for digital devices, regardless of local area fluctuations.
- Providing a wide range of economic and environmental benefits

RECOMMENDATIONS

With an understanding of the barriers to be overcome and the benefits that are attainable, this section summarizes the recommendations of the Modern Grid Initiative.

The transition to motivate and include the consumer will be gradual, with pockets of progress occurring first in those regions where regulators are most supportive. Over time, and with the emergence of increasingly effective programs and technology, consumer involvement will become the norm. Each step along the path will produce its own set of benefits.

Our recommended steps include:

- **Regulatory encouragement at federal and state levels.** Society would benefit greatly from clear directives that treat DR/DER programs as equal, or even preferred, solutions to the fundamental power system requirement of continuously balancing generation and load. Utilities would be far more likely to employ these tools if it was clear that their investment would not be questioned some time in the future. Regulatory clarity is also needed regarding the question of voluntary versus mandatory DR programs across the wide range of customer types.

- **Broader education regarding the opportunities to deploy DR and DER and the overall benefits they produce.** Education programs can reveal the many beneficiaries of well conceived programs that motivate and involve the consumer. The value of transitioning from passive protected users to proactive informed consumers must be conveyed clearly. In addition to higher quality, lower cost, greater customer choice and more reliable energy, there are substantial environmental benefits to be had — as well as national security and energy market stabilization gains. Such education, and the increased regulator and consumer receptivity it creates, can also encourage intermediaries (aggregators) to make wider use of these programs.

- **Continued improvement in the cost and performance of supporting technologies.** Considerable work is needed to develop and demonstrate the most effective, secure and reliable metering, communications and IT solutions. Available modern advances in digital communications can be applied to these utility applications — establishing reliable, secure low cost channels to each consumer. Today's digital electronics can enable accurate, sophisticated, low cost meters (actually consumer gateways that can incorporate decision-support agents) that allow intelligent interaction between consumers and grid operators. In addition, new smart appliances and enhanced distributed energy resources must be developed to help support the consumer's proactive interaction with the grid. Integration of such devices requires substantial engineering to insure compatibility with existing electric infrastructure.

- **Development of programs, tariffs and computer agents that satisfy both utility and consumer needs.** This includes the development of systems that enable timely *and effortless* interaction between the consumer and the power grid operator. Innovative rate structures that provide economic benefits to both the consumer and the utility are integral to these systems. New software programs are also needed to connect the new applications to the many utility legacy programs that will still be required for reliable and efficient overall operation of the grid.

Summary

In the modern grid, consumers become an integral active part of the overall electric power system. Consumer actions taken in their own self-interest will help balance electrical demand (loads) with electrical supply.

Motivating consumers to play that part means giving them the opportunity to make informed choices, profitably. Consumers with choices in their energy usage will be able to:

- Rely on price signals to decide when and if to purchase, store, or generate power.
- Employ smart devices that know the consumer's preferences about energy usage, and which respond automatically. Smart devices can also improve the stability of the power system.
- Relieve the grid's loads by reducing peak demand when necessary.

Demand response programs are examples of proven consumer receptivity to being included as a part of the power system. And DR has demonstrated measurable benefits to suppliers of electrical power.

Distributed Energy Resources are poised now for the same inevitable widespread deployment seen with personal computers and cell phones in years past. DER can also be an important component of consumer involvement, one that can complement large central generation and reduce the burden on the grid, while offering environmental, reliability and economic improvements.

For more information

This document is part of a collection of documents prepared by The Modern Grid Initiative team. For a high-level overview of the modern grid, see "A Systems View of the Modern Grid." For additional background on the motivating factors for the modern grid, see "The Modern Grid Initiative." MGI has also prepared seven papers that support and supplement these overviews by detailing more specifics on each of the principal characteristics of the modern grid. This paper describes the second principal characteristic: "Motivates and Includes the Consumer."

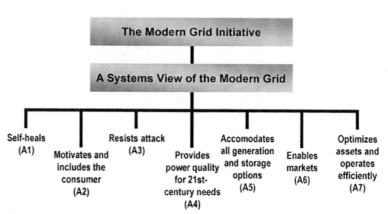

Documents are available for free download from the Modern Grid Web site.

The Modern Grid Initiative

Website: www.netl.doe.gov/moderngrid

Email: moderngrid@netl.doe.gov

(304) 599-4273 x101

BIBLIOGRAPHY

1. Borenstein, S., M. Jaske, and A. Rosenfeld. 2002. Dynamic pricing, advanced metering, and demand response in electricity markets. *Hewlett Foundation Energy Series*. University of California Energy Institute, Center for the Study of Energy Markets Working Paper Series (CSEM WP 105).

2. Delurey, D. 2005. National town meeting on demand response. Presentation given at the National Town Meeting on Demand Response, Washington, D.C.

3. Goldman, C. and R. Levy. 2005, Demand response in the U.S.: Opportunities, issues and challenges. Presentation given at the National Town Meeting on Demand Response, Washington, D.C.

4. Hirst, E. and B. Kirby. 2001. Retail-load participation in competitive wholesale electricity markets. Report prepared for Edison Electric Institute, Washington, D.C. and Project for Sustainable FERC Energy Policy, Alexandria, VA.

5. Hogan, K. 2005. Energy efficiency is a key part of demand response. Presentation given at the National Town Meeting on Demand Response, Washington, D.C.

6. Kolevar, K. 2005. Comments made at the National Town Meeting on Demand Response, Washington, D.C.

7. Morgan, R. 2005. Demand response: A regulator's perspective. Presentation given at the National Town Meeting on Demand Response, Washington, D.C.

8. Welch, T. 2005. Reflections on the role of demand response in electricity markets. Presentation given at the National Town Meeting on Demand Response, Washington, D.C.

9. Wood, P. 2005. Demand response: Making it work for customers. Presentation made at the National Town Meeting on Demand Response, Washington, D.C.

10. Electric Power Research Institute, Inc. (EPRI), Charlotte, NC. http://www.epri.com.

Appendix A4:
A Systems View of the Modern Grid

PROVIDES POWER QUALITY FOR 21ST CENTURY NEEDS

Conducted by the National Energy Technology Laboratory
for the U.S. Department of Energy
Office of Electricity Delivery and Energy Reliability
January 2007

Office of Electricity
Delivery and Energy
Reliability

v 2.0

TABLE OF CONTENTS

EXECUTIVE SUMMARY

Providing power quality for 21st century needs is one of the seven principal characteristics in a systems view of the modern grid. (See Figure 1.) Our future global competitiveness demands fault-free operation of the digital devices that power the productivity of our 21st century economy. And we need clean power to meet that demand.

- Self-heals
- Motivates and includes the consumer
- Resists attack
- Provides power quality for 21st-century needs
- Accommodates all generation and storage options
- Enables markets
- Optimizes assets and operates efficiently

Figure 1: The Modern Grid Systems View provides an "ecosystem" perspective that considers all aspects and all stakeholders.

When consumers think of power quality (PQ), they think of reliable power that is free of interruption, and clean power that is free of disturbances.

The focus of this paper is on the attribute of *clean* power. Issues about power *reliability* are treated by other papers in this collection of documents about principle characteristics of the modern grid. This paper provides a look at how the modern grid will help to **provide power quality for 21st century needs**. Although it can be read on its own, it supports and supplements "A Systems View of the Modern Grid," an overview prepared by the Modern Grid Initiative (MGI) team.

Power quality, or clean power, deserves such focus because of the importance of digital devices that have become the engines of so many industries in today's economy. There is hardly a commercial or industrial facility in the country that would not suffer lost productivity if a serious PQ event impacted its digital environment.

The level of delivered power quality can range from "standard" to "premium", depending on consumers' requirements. Not all commercial enterprises, and certainly not all residential customers, need the same quality of power.

The modern grid would supply varying grades of power and support variable pricing accordingly. The grade of delivered power is largely determined by the design of the electrical distribution facilities serving a given customer. Special attention can be devoted to minimizing the effect of perturbations. The cost of these premium features can be included in the electrical service contract.

The modern grid would support the mitigation of PQ events that originate in the transmission and distribution elements of the electrical power system. Its advanced control methods will monitor essential components, enabling rapid diagnosis and precise solutions to any PQ event. In addition, the grid's design will include a focus on the reduction of PQ disturbances arising from lightning, switching surges, line faults and harmonic sources. Its advanced components will apply the latest research in superconductivity, materials, energy storage, and power electronics to improve power quality.

Finally, the modern grid would help buffer the electrical system from irregularities caused by consumer electronic loads. Part of this will be achieved by monitoring and enforcing standards that limit the level of electrical current harmonics a consumer load is allowed to produce. Beyond this, the modern grid will employ appropriate filters to prevent harmonic pollution from feeding back into the grid.

Specific technologies and approaches the modern grid will bring to bear include:
- Power quality meters.
- System wide power quality monitoring.
- Grid-friendly appliances that control their high-load components, such as compressors and heating elements.
- Premium power programs that include dedicating office parks and neighborhoods to premium power usage.
- Various storage devices, such as Superconducting Magnetic Energy Storage (SMES) and advanced batteries, to improve power quality and stability, or to supply facilities needing ultra-clean power.
- A variety of power electronic devices that instantly correct waveform deformities.
- Monitoring of electric system health to identify and correct impending failures that could produce PQ problems.
- New distributed generation devices (e.g. fuel cells and micro-turbines) that can provide clean power to sensitive loads.

All this technology can be applied to the problem of power quality in the near future. However, to do so requires the coordinated efforts of government, utilities, regulators, and standards bodies such as IEEE. It also requires widespread education for all the modern grid's stakeholders.

The benefits of improved PQ could be tremendous in both cost avoidance and the resulting productivity gains. Clean, reliable power could also produce opportunities for economic growth to areas of the country previously denied the benefits of high-technology industry.

A PRIMER ON POWER QUALITY

Before we delve into the issues of power quality, we should understand the things that disrupt it, including sags, harmonics, spikes, and imbalances.

The power supplied by electric utilities starts out as a smooth sinusoidal waveform. This is the waveform produced at the power plant by electrical generators. (See Figure 2.)

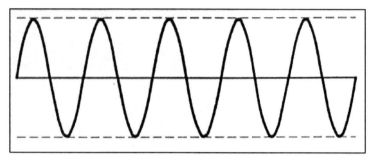

Figure 2: Normal power is supplied in smooth sinusoidal waveforms.

But as this power moves from the generator through the transmission and distribution systems and on to the customer's equipment, it can be affected by four kinds of perturbations that can distort its pure sine wave envelope:

1. Sags (undervoltages) — Voltage sags are the most common power disturbance. (See Figure 3.) They occur when very large loads start up, or as a result of a serious momentary overload or fault in the system. At a typical industrial site, several sags per year are not unusual at the service entrance. Many more occur at equipment terminals. Costs associated with sag events can range widely, from almost nothing to several million dollars per event (Primen Report. Power Quality Problems and Renewable Energy Solutions. September 2002.)

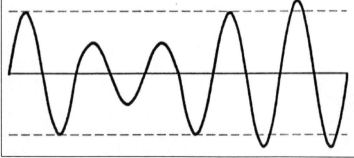

Figure 3: Voltage sags are the most common power disturbance, with associated costs ranging from zero to millions of dollars.

2. Harmonics — Harmonics are caused by "non-linear" loads, which include motor controls, computers, office equipment, compact fluorescent lamps, light dimmers, televisions and, in general, most electronic loads. High levels of harmonics increase line losses and decrease equipment lifetime. (See Figure 4.)

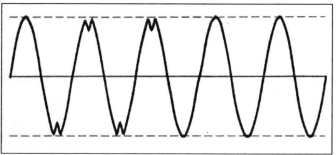

Figure 4: Non-linear loads create harmonic distortion, resulting in more line losses and reduced equipment life.

3. Spikes — Spikes are brief spurts of voltage (in the millisecond to microsecond range) when voltage can shoot up many times higher than normal. Spikes are caused by lightning and switching of large loads or sections of the power system network. They can disrupt the operation of data processing equipment and damage sensitive electronic equipment. (See Figure 5.)

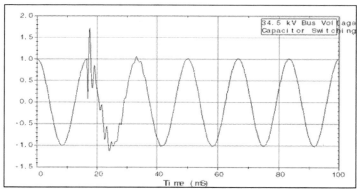

Figure 5: Spikes can damage sensitive electronic equipment.

4. Imbalances — Imbalances are steady-state problems caused by such things as defective transformers or uneven loading of grid phase wires. They are not as easy to identify as the other problems, which show up clearly in the waveforms, but they can gradually cause damage to equipment, especially to electric motors.

Note that in none of the above PQ events is power totally interrupted.

CURRENT AND FUTURE STATES

This section describes the current state of PQ and why problems persist. It then describes how power quality in the future would be enhanced by the modern grid.

CURRENT STATE

Voltage sags represent by far the largest PQ issue. Because voltage sags are mostly due to unforeseen and uncontrollable events, the number of voltage sags experienced in the power system varies from year to year. Several industry studies conducted in the last decade provide insight on the number of voltage sags at particular magnitudes and durations that may occur annually. (See Figure 6.)

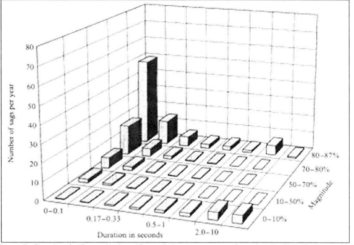

Figure 6: Results of a 5-year National Power Laboratory survey of sag density (duration, magnitude and frequency of occurrence) at 130 sites. Image courtesy of EPRI.

A modern grid should provide PQ that fully conforms to the customer's design criteria, as defined by industry standards.

The above diagram may be compared to the industry standard Semiconductor Equipment and Materials International (SEMI) F47 curve (green line in Figure 7.) While there is generally reasonable matching between what the utility supplies and what the customer designs to, there are also many points of non-compliance. This curve more clearly shows the points of non-compliance (points below the green line). Of particular note is the concentration of 30% to 60% magnitude voltages having between two and twenty cycle durations. None of these events conform to the Information Technology Industry Council (ITIC) curve criteria.

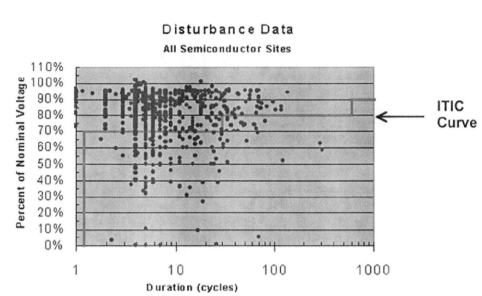

Figure 7: Sag Monitoring Data from 15 Semiconductor Sites show that power quality is not meeting customers' design criteria. Image courtesy of EPRI.

Power quality is an important issue for the information industry. For example, deciding where to locate power-sensitive server farms depends largely on the availability of clean, reliable power. Those criteria led to the selection of rural Grant County, Washington as the site for both Microsoft and Yahoo server farms.

Power quality is also a large issue for industrial and manufacturing facilities. Tiny power disturbances can wreak havoc with the increasingly complicated, computerized machinery found along assembly lines today. At the same time, customers must design their processes to conform to criteria such as the SEMI F47 curve. This graph of voltage sag events at a manufacturing facility is a good example of interruption levels in an industrial setting. (See Figure 8.) Events that caused process disruptions are circled. (Note that for the six circled events, the consumer's equipment did not meet the SEMI 47 standard.)

Based on both Figure 7 and Figure 8, it can be concluded that neither the power delivery supplier nor the industrial user has consistently conformed to the ITIC industry standard.

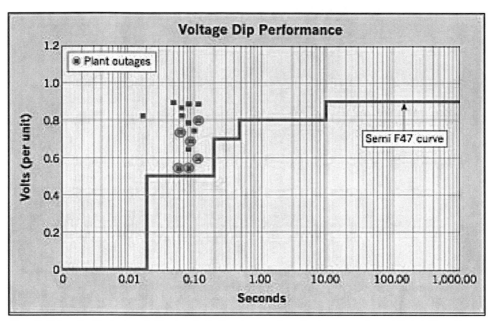

Figure 8: Voltage Sag Events at one facility show that both customer equipment and power supplies contribute to PQ problems. Image courtesy of EPRI.

The stakes are high: Work stoppages can cost a company up to $500,000 an hour, and power-related problems may cost U.S companies more than $100 billion a year. A 2001 Primen study concluded that power quality disturbances alone cost the US economy between $15-24 billion annually. The cost of a momentary disruption to various users in dollars per kilowatt is shown in Figure 9 below.

Category	Cost of Momentary Interruption ($/kW demand)	
	Minimum	**Maximum**
INDUSTRIAL		
Automobile manufacturing	$5.0	$7.5
Rubber and plastics	$3.0	$4.5
Textile	$2.0	$4.0
Paper	$1.5	$2.5
Printing (newspapers)	$1.0	$2.0
Petrochemical	$3.0	$5.0
Metal fabrication	$2.0	$4.0
Glass	$4.0	$6.0
Mining	$2.0	$4.0
Food processing	$3.0	$5.0
Pharmaceutical	$5.0	$50.0
Electronics	$8.0	$12.0
Semiconductor manufacturing	$20.0	$60.0
COMMERCIAL		
Communications, information processing	$1.0	$10.0
Hospitals, banks, civil services	$2.0	$3.0
Restaurants, bars, hotels	$0.5	$1.0
Commercial shops	$0.1	$0.5

Figure 9: Disruption Cost by Industry shows that some industries suffer more than others from power quality problems. Image courtesy of EPRI.

Problems persist with power quality. The number of sensitive loads continues to grow, while the costs to minimize PQ events remain relatively high. Clearly, the number of sensitive loads will only continue to grow with advances in communications and information technology.

There exists a large debate as to who should bear the costs of PQ improvement; the utility or the consumer. The development of new rate structures that offer premium quality has not been broadly adopted. Regulatory commissions have not placed a priority on resolving this dispute.

FUTURE STATE

Advanced technologies deployed by the modern grid will both mitigate power quality events in the power delivery system and protect end users' sensitive electronic equipment.

Because sensitive electronic loads represent an increasing portion of the total power system load, power quality will be of growing importance in the 21st century. Twenty years ago, the amount of electrical load associated with chips (computer systems, appliances, and equipment) and automated manufacturing was miniscule. The power system design was well suited to the type of loads that existed then. But ten years ago, the amount of load from chips and automated manufacturing had grown to about 10%. And in the future it can be expected to grow to more than half. The grid must change to accommodate this changing load characteristic.

The modern grid will be rich with technologies and devices that work at every level of power generation and delivery. All will contribute to clean and reliable power reaching the consumer. Included among these are:

- Power quality meters.
- System wide power quality monitoring.
- Grid-friendly appliances that control their high-load components, such as compressors and heating elements.
- Premium power programs that include dedicating office parks and neighborhoods to premium power usage.
- Various storage devices, such as Superconducting Magnetic Energy Storage (SMES) and advanced batteries, to improve power quality and stability or to supply facilities needing ultra-clean power.
- A variety of power electronic devices that instantly correct waveform deformities.
- Monitoring of electric system health to identify and correct impending failures that could produce PQ problems.
- New distributed generation devices (e.g. fuel cells and micro-turbines) that can provide clean local power to sensitive loads.

These applications are discussed further in the Requirements section.

Applying the advanced technologies that mitigate PQ events will require support and coordination among equipment makers, power providers, power users and standards bodies. The resulting design criteria and industry standards must be employed at every level of the electric system, including at the customer's load. This will ensure that the delivered power quality is consistent with the provider's capabilities and the needs of the consumer.

In the future, the modern grid will price power in accordance with the grade of power required by the user. The level of power quality required by consumers can vary, depending on the complexity of their equipment or criticality of their operations. As the data in Figure 9 clearly shows, a premium power offering holds greater appeal to a semiconductor manufacturer than to a newspaper printer, although both would benefit. Hence, customized premium power packages should be developed to meet these differing industry needs. Not all commercial enterprises, and certainly not all residential customers, need premium power.

REQUIREMENTS

We've broadly described the current state of power quality and discussed its future state. This section introduces the design requirements and solutions needed to make improved PQ an integral characteristic of the 21st century modern grid.

Commonly, 40% of power quality issues relate to the delivery of power from the utility, and 60% relate to the use of power within an industrial facility. (*Specification Guidelines to Improve Our Power Quality Immunity and Reduce Plant Opportunity Costs, RGLSolutions, IEEE PCIC 2002*)

The modern grid must apply power quality solutions wherever they're needed — where the power begins, where it gets distributed, or where it ends. Thus, power quality solutions, like the modern grid itself, must be autonomous and distributed. The devices that mitigate PQ events must be spread among transmission and distribution components of the modern grid, but also right at the sensitive load.

Distributing advanced power electronics at each level throughout the grid is key to solving many PQ problems. Many solutions fall under the broad heading of Flexible AC Transmission Systems (FACTS), even though some are actually deployed on distribution systems.

FACTS and related technologies, including Uninterruptible Power Supplies (UPS), are implemented and realized through the application of power semiconductor switches applied to high-speed controlled compensation devices. Examples include Static Compensators (STATCOM), Dynamic Voltage Restorers (DVR), and Thyristor Controlled Series Capacitors (TCSC). These FACTS devices may be connected in series and/or in shunt. While the STATCOM is connected at the load end in shunt, devices like the DVR and TCSC, having the capability of eliminating voltage sags and swells as well as rapid adjustment of network impedance, are connected in series with the line. Table 1 below illustrates the application of various power electronic devices.

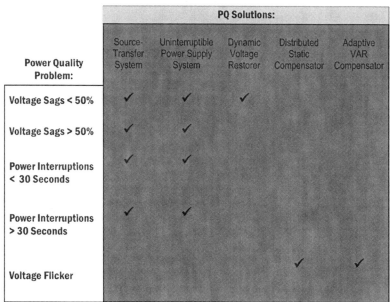

Power Quality Problem:	PQ Solutions:				
	Source-Transfer System	Uninterruptible Power Supply System	Dynamic Voltage Restorer	Distributed Static Compensator	Adaptive VAR Compensator
Voltage Sags < 50%	✓	✓	✓		
Voltage Sags > 50%	✓	✓			
Power Interruptions < 30 Seconds	✓	✓			
Power Interruptions > 30 Seconds	✓	✓			
Voltage Flicker				✓	✓

Table 1: Examples of Power Electronic Solutions that are key to providing power quality.

Transmission Level

At the transmission level, voltage sags are frequently the result of faults (short circuits), which can exist for many milliseconds. In the past, little could be done to reduce this effect. Today, high voltage static VAR compensators (SVCs) are fast enough to mitigate many of these events. However, these devices tend to be quite expensive, partly due to the small numbers deployed and also due to the cost of today's power electronic components. As component costs drop (discussed in MGI Advanced Components paper), these devices will become increasingly attractive to transmission system owners.

Looking toward the future, affordable current-limiting devices will be able to reduce the severity of voltage sags associated with faults. And eventually, lossless superconducting transmission lines will further reduce voltage sag concerns.

Distribution Level

At the distribution level, a variety of techniques are available to improve the quality of power delivered to the end customer. Since lightning is a major source of PQ problems, greater use of underground facilities can minimize this contribution.

Creating premium power quality business parks, where sensitive load customers can locate, can also be valuable. These parks can be directly connected by underground feeders from distribution substations. They can be fed by redundant feeders via high-speed source transfer switches, so that when one feeder is perturbed, the other can immediately take over.

The various power electronic devices shown in Table 1 can be deployed in many distribution applications. And distributed generation and storage resources located close to the load, including micro-grids and green power devices, can isolate the consumer from most grid disturbances.

Customer Level

Not all customers are equally impacted by poor quality of power. At one end of the spectrum, an integrated circuit manufacturer will likely incur very large losses if a PQ event shuts down or perturbs the process. At the other end of the spectrum, a homeowner may be only inconvenienced when a DVD player shuts down.

The power quality solution not only includes technologies that improve and maintain power quality, but also those that make customer loads more tolerant. Within the customer's facility, advanced devices will offer solutions to PQ sensitivity.

There are a number of ways customers can limit problems with transients in their facilities. It's best to start by selecting equipment that can withstand transients, and by using proper wiring/grounding practices. In addition, there are many spike suppression devices that can protect customer equipment.

Different sets of requirements must be specified to meet the needs for the different categories of customers: commercial/industrial and residential.

Commercial and industrial customers must be able to select the grade of power they need and then design their systems accordingly. Grid PQ mitigation techniques must then be coordinated with the customer's *load sensitivity characteristic* to prevent PQ events that can lead to plant outages.

Residential customers will also have varying power quality needs, depending on the sophistication of their home electronics. Here, much rests with the vendors of consumer products, which need to be designed to better tolerate common PQ events. In general, PQ events are more of an inconvenience than an economic burden to this class of customer. But with so many companies now based at home, the impact to the small business economy is not one to be ignored.

SPECIFIC SOLUTIONS FOR SPECIFIC PQ PROBLEMS

Each of the four PQ problem areas has its own technical solution, and all these solutions will be enabled by the advanced technologies of the modern grid. The modern grid will provide PQ that fully conforms to the customer's design criteria, as defined by industry standards. Standards such as SEMI 47 provide a basis for consumer load behavior and a realistic design target for service providers. Both consumers and service providers need a mutually acceptable standard in order to develop their respective designs.

Voltage Sag

Current-limiting and FACTS devices will help reduce the severity of voltage sags associated with power system faults. The most direct way to deal with voltage sags is by providing adequate buffering at the load. And for those customers who take advantage of the modern grid's distributed energy resources (DER), local generation could be provided in a variety of forms such as storage devices, micro-turbines and micro-grids.

Harmonics

Advanced filters will be very effective in the elimination of harmonic distortion. A series active filter, for example, presents a high impedance path to harmonic currents, thereby preventing them from flowing from the load to the source and vice versa.

In most cases, customer-owned equipment is the source of harmonics. Harmonics originating in customer equipment can also cause power quality problems for other utility customers, as well as to the power delivery system itself. Responsibility for controlling harmonics is twofold:

- **The customer is responsible for limiting harmonic currents that interfere with the power system.**
- **The utility is responsible for maintaining the quality of the voltage waveform.**

Since these responsibilities are highly interrelated, guidelines must establish harmonic limits for each party. Technical groups such as IEEE develop these guidelines and they must be enforced by utilities and state commissions.

Transients (Spikes)

Service providers will employ a number of system design strategies to minimize transients.

- Proper grounding and shielding, combined with the liberal application of lightning arresters, will minimize lightning-related spikes.
- Modern controlled switching techniques will minimize power system switching transients (e.g. capacitor bank switching).
- The use of the modern grid's advanced maintenance techniques — that prevent faults from occurring in the first place — will minimize transients related to power system faults.

While spikes on the grid can be reduced by methods described above, customers can also contribute to the solution. They can limit voltage spike problems in their facilities by selecting equipment that can withstand them and by employing proper wiring, grounding, and surge protection.

Voltage Imbalance

In the modern grid, voltage imbalance identification will happen quickly because modern communicating meters will report it to the service provider. Voltage imbalances can cause premature failure of motors and transformers due to overheating, and can cause electronic equipment to malfunction. The service provider will normally correct a severe voltage imbalance problem once it is identified.

KEY TECHNOLOGIES THAT OFFER SOLUTIONS

Carefully chosen and deployed, the key technologies of the modern grid will provide solutions that mitigate these power quality disturbances throughout the system:

- **Sensing and measurement technologies** — The broad deployment of modern meters will provide extensive information regarding the quality of power throughout the grid. This information will be valuable in both resolving problems as they are quickly identified, as well as in the design of grid enhancements and expansions. Also, new sensing techniques will monitor the health of equipment and predict potential failures that can create PQ problems.

- **Advanced components** — These will apply the latest research in superconductivity, fault tolerance, storage, and power electronics. Each of these components supports devices that improve power quality. Some examples include:
  - o   FACTS and related devices using power electronics.
  - o   Current Limiting Devices using superconductivity.
  - o   Superconducting devices such as synchronous condensers and SMES that improve voltage quality.
  - o   Intelligent switching devices that determine the integrity of a circuit before re-energizing it.
  - o   New clean power distributed resources employed at the local level that isolate loads from grid problems

- **Advanced control methods** — These will monitor essential components, enabling rapid diagnosis and precise solutions appropriate to any event. Advanced control methods are designed to maintain the grid in a stable state at all times and to provide extensive condition information. Proactive prevention of PQ events will be a result of this vast new data base.

- **Integrated communication** — This will support the new protection and control systems that make the grid more reliable and reduce the occurrence of perturbations that affect PQ. Near real time availability of data allows proactive actions that can prevent equipment deterioration, another source of PQ problems.

BARRIERS

We've described the modern grid requirements for providing higher power quality and we have noted potential solutions to meet those requirements. But there remain some barriers to the deployment of those solutions. We must address issues such as costs, government/regulatory policies, and industry standards.

The three primary issues that must be addressed are:

- Reducing the high costs of modern PQ-enhancing devices.
- Implementing policies and regulations to encourage investment in PQ programs, including those that provide pricing related to grades of power.
- Updating codes & standards.

HIGH COSTS OF DEVICES

The cost of PQ improvement devices needs to come down in order to encourage wide acceptance and usage. Greater use of these devices will reduce their costs as their supply increases and as new approaches to their design are developed. As with any product life cycle, more economical designs will be developed when it becomes clear that a significant market actually exists.

Power electronics has a key role in PQ, as described throughout this report. Power electronics also makes a significant contribution to a number of other modern grid characteristics, making it a key technology for grid modernization. Hence there is a huge potential market for advanced power electronic components. The reduction in cost of this important component will eliminate one barrier to achieving PQ that meets 21st century needs.

Advanced metering, with its ability to monitor a wide variety of PQ parameters, is an example of a PQ-related technology that has grown in sophistication and dropped in price due to an extensive emerging market (and resulting influx of players).

Technology advances that can be expected in the future (such as lower cost storage, distributed generation that delivers harmonic-free power; advanced monitoring that detects impending PQ events, and faster protective schemes) all contribute to reducing barriers to improved PQ.

POLICY AND REGULATION

The absence of a differentiated policy, regarding PQ delivery to customers with differing needs, is a large barrier. For those customers who suffer significant harm due to PQ events, a premium power product can be a solution for both buyer and seller. State regulatory commissions could do much to encourage PQ investment and pricing related to grades of power.

Only the regulator is in a position to encourage PQ solutions that represent the lowest overall cost to society, while providing a fair return to investors. Until increased investment in better PQ by regulated power delivery companies is encouraged by regulators, this missing incentive will remain another PQ barrier. In the meantime, debate will continue over who is responsible for making PQ better.

CODES AND STANDARDS

IEEE and other standards organizations have wide and deep influence on the design of consumer products, electrical system equipment, utilities, and power and communications systems.

Standards organizations have not created standards for categories of power quality that consumers choose from according to their needs. Standards for various grades of delivered power could serve as the basis for differentiated PQ pricing. Such standards will also help educate the many players involved in PQ issues, since this is a topic that is not well understood. This lack of understanding is itself a barrier.

BENEFITS

When barriers are overcome, the benefits will include both cost avoidance and new opportunities for economic growth.

Merely avoiding the productivity losses of poor quality power to commercial and industrial customers can shed billions of dollars of waste from the economy. The costs associated with power quality events at commercial facilities such as banks, data centers, and customer service centers can be tremendous, ranging from thousands to millions of dollars for a single event. The costs to manufacturing facilities can be even higher. Voltage dips that last less than 100 milliseconds can have the same effect on an industrial process as an outage that lasts several minutes or more (September 2002 Primen report).

The reduction of power quality problems will produce a proportional reduction in several categories of loss:

- **Scrapped materials** — This cost can be significant in industries where both the manufacturing process and product quality are extremely dependent on power reliability and quality.
- **Customer dissatisfaction**— Although difficult to quantify, this factor can create a negative perception that loses clients, revenue, and goodwill.
- **Lost productivity**— Even if the business shuts down, overhead costs continue and compound the resulting loss of revenue.
- **Consumer safety**— In some manufacturing processes, such as crane operation in steel production, power perturbations can create safety dangers.
- **Contractual violations**— Liquidated damage losses and litigation exposures can result from failing to meet specific deadlines.

Intelligently improving PQ in the nation's power system will offer opportunities to broaden and enrich the commercial bases of struggling communities and regions. Rural communities will be able to support clean, high-tech industries that demand high quality and reliable power. New jobs and higher tax bases will transform regions and communities that once depended solely on agriculture or single industries.

Since poor power quality leads to both shorter electrical equipment life and higher electrical (KWH) losses, economic and environmental benefits also accrue to the utility when PQ is improved.

RECOMMENDATIONS

Three broad actions would prepare the way for overcoming the barriers for PQ improvement and realizing its benefits.

1. PQ solutions must be tailored to the differing requirements of customers.

- Cost/benefit analyses should be conducted, taking into account the full range of benefits that improved PQ delivers. State utility commissioners, service providers and consumer representatives should work together to develop these studies. Those solutions with a favorable net present value to society should be adopted broadly. And such broad adoption will reduce the cost of solutions (such as those associated with power electronics devices) further advancing the spread of PQ enhancement programs.

- When the energy delivery company is the solution provider, the electric rates should include this associated cost and a fair return on the investment.

2. Government leadership is needed to hasten an answer to the question of who owns the PQ problem.

- PQ problems can originate in a wide variety of places along the electricity path. Therefore, federal agencies and state regulators need to become more involved in how to allocate costs of PQ solutions among transmission, distribution and consumer participants.

- Since customers have differing needs and priorities, a differentiated regulated approach makes the most sense.

3. Programs to provide PQ education should be developed and broadly publicized.

- Customers need to be better educated about the PQ issue so their facilities can be designed to accommodate today's PQ imperfections.

- For future planning by consumers, the emerging solutions should be widely publicized by the Modern Grid Initiative.

SUMMARY

We conclude with a summary of the key findings relevant to providing power quality for 21st century needs.

Clean electrical power is a necessity to commercial and industrial facilities that depend on sensitive digital control and communications systems to keep computing centers and manufacturing operations running productively.

Consumers, whether commercial or residential, have varying demands for power quality. With its advanced sensing and measuring technologies, the modern grid could deliver grades of power from standard to premium, and provide the ability to price the varying levels of PQ accordingly.

The consumer needs to recognize that *perfect power* is not a realistic goal. Therefore, consumer loads need to be designed to accommodate some power quality imperfections.

Almost half of PQ disturbances originate in the transmission and distribution elements of the electrical power system. The advanced monitoring and control functions of a modern grid, coupled with the wider application of conventional surge mitigation techniques, enable the diagnosis and solutions of many such PQ events.

The other major source of PQ events is the load from consumer electronics, which cause irregularities in voltage that feed back to the electrical power system. The advanced components being developed for the modern grid will also help address these problems.

By aligning the efforts of government, utilities, regulators, consumers and standards bodies, we can attain both improved PQ and choice in level of PQ in the near future.

The benefits to productivity in the economy could mean many billions of dollars in costs avoided. Harder to measure, but just as important, are the new opportunities for economic growth that emerge when new 21st century industries answer the call from regions that offer clean, reliable power.

For more information

Other principal characteristics of the modern grid, besides power quality, will be designed to provide reliable power to the consumer. These other characteristics (e.g. Self-heals, Resists Attack, Accommodates all Generation Options), are described in companion papers published by the Modern Grid Initiative.

This document is part of a collection of documents prepared by The Modern Grid Initiative (MGI) team. For a high-level overview of the modern grid, see "A Systems View of the Modern Grid." For additional background on the motivating factors for the modern grid, see "The Modern Grid Initiative." MGI has also prepared seven appendices that support and supplement these overviews by detailing more specifics on each of the principal characteristics of the modern grid. This paper describes the second principal characteristic: "Provides Power Quality for 21st Century Needs."

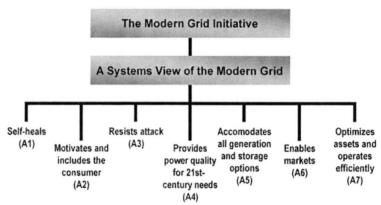

Documents are available for free download from the Modern Grid Web site.

The Modern Grid Initiative

Website: www.netl.doe.gov/moderngrid

Email: moderngrid@netl.doe.gov

(304) 599-4273 x101

BIBLIOGRAPHY

1. Bollen, M.H.J. 2000. Understanding power quality problems: Voltage sags and interruptions. New York: IEEE Press.

2. Falcon Electric. 2005. Power your customer's critical equipment reliably.

3. McGranaghan, M., M. Stephens, and B. Roettger. 2005. The economics of voltage sag ride-through capabilities. EC&M (May 1), http://www.ecmweb.com/mag/electric_economics_voltage_sag/index.html

4. Power Standards Testing Lab. 2000. Product announcement. April. http://powerstandards.com/whatsnew/MakeItWorse.txt

5. Southern California Edison Power Quality Department. *Power Quality Handbook.* http://www.sce.com/NR/rdonlyres/66BEEBD8-C9B9-4AE3-B2AC-473BEDCE21C0/0/PQhandbook.pdf

6. Whisenant, S., B. Rogers, and D. Dorr. 2005. Creating a business case to solve PQ problems. EC&M (May 1), http://www.ecmweb.com/mag/electric_creating_business_case/index.html

7. Primen Report. Power Quality Problems and Renewable Energy Solutions. September 2002

8. Primen Study: The Cost of Power Disturbances to Industrial and Digital Economy Companies. June 2001

9. Electric Power Research Institute, Inc. (EPRI), Charlotte, NC. http://www.epri.com.

Appendix A3:
A Systems View of the Modern Grid

RESISTS ATTACK

Conducted by the National Energy Technology Laboratory
for the U.S. Department of Energy
Office of Electricity Delivery and Energy Reliability
January 2007

Office of Electricity
Delivery and Energy
Reliability

v2.0

TABLE OF CONTENTS

EXECUTIVE SUMMARY

The systems view of the modern grid features seven principal characteristics. (See Figure 1.) The ability to resist attack is one of those characteristics and the subject of this paper.

- Self-heals
- Motivates and includes the consumer
- Resists attack
- Provides power quality for 21st-century needs
- Accommodates all generation and storage options
- Enables markets
- Optimizes assets and operates efficiently

Figure 1: The Modern Grid Systems View provides an "ecosystem" perspective that considers all aspects and all stakeholders.

The energy industry's assets and systems were not designed to handle extensive, well-organized acts of terrorism aimed at key elements.

The U.S. energy system is a huge network of electric generating facilities and transmission lines, natural gas pipelines, oil refineries and pipelines, and coal mines. Occasionally, these systems have been tested by large-scale natural disasters such as hurricanes and earthquakes. Generally, industries have restored energy relatively quickly. Sabotage of individual components has caused some problems, but the impacts have been managed. We've been lucky.

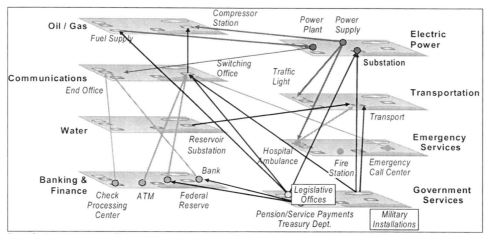

Figure 2: Interdependence of Energy and Other Sectors: Losing power in even one region damages the whole economy. Image courtesy of Science Applications International Corporation (SAIC).

The dependency of other key elements of the economy on the electrical power part of the energy system (see Figure 2) is apparent when millisecond outages disrupt sensitive digital processes, and outages extending days or weeks can deprive a community or region of running water. Telecommunications, financial, and health sectors try to ensure uninterrupted power by installing generators, batteries, or redundant systems. Even these, however, can be limited in their effectiveness. Generators, for example, are limited by the availability of fuel.

It is critical for the modern grid to address security from the outset, making security a requirement for all the elements of the grid and ensuring an integrated and balanced approach across the system.

Threats to the infrastructure are usually broken into two categories: physical (explosives, projectiles) attacks and cyber (computer launched) attacks. Whatever the specific nature of the threat, the designers of the modern grid should plan for a dedicated, well-planned, and simultaneous attack against several parts of the system.

The threat of both physical and cyber attack is growing and a widespread attack against the infrastructure cannot be ruled out. There is evidence that Al Qaeda has been tracking debates in the United States related to the cyber vulnerability of control systems in the energy infrastructure (Hamre, 2003).

- **Cyber attacks** — Computer security incidents are increasing at an alarming rate. According to the Government Accountability Office, in 2002, 70 percent of energy and power companies experienced some kind of severe cyber attack to their computing or energy management systems. (See Figure 3.)

With the introduction of digital technology throughout our society, the cost of outages (e.g., from equipment failure or weather-related incidents) has significantly increased – from $30 billion in 1995 to $119 billion in 2001 (National Research Council, 2002)

- **Physical attacks** — Physical attacks against key elements of the grid, or physical attacks combined with cyber attacks, cannot be discounted. From a terrorist viewpoint, damage from a physical attack may be more predictable than a cyber attack, and therefore promise more certainty in causing harm.

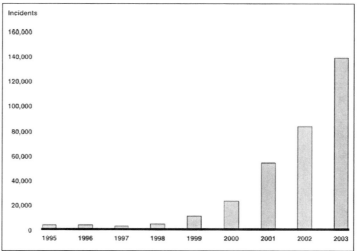

Figure 3: 70% of energy and power companies experienced some kind of severe cyber attack to either their IT or SCADA/EMS network. (GAO analysis based on Carnegie Mellon University's CERT Coordination Center data).

A study that surveyed over 170 security professionals and other executives concluded that, across industries, respondents believe that a large-scale cyber attack in the United States will be launched against their industry in the near future. (U.S. General Accounting Office, 2004)

Whether it is going to be a physical or cyber attack, the modern grid must resist two different attack strategies:

1. *Attacks on the power system*, in which the infrastructure itself is the primary target.
2. *Attacks through the power system*, in which attackers take advantage of power system networks to affect other infrastructure systems, such as telecommunications, financial, or government.

An emphasis on security throughout the development and implementation phases of the modern grid is critical. Such an emphasis would:

- Ensure lowest cost for system elements by addressing security concerns during the initial design and throughout the lifecycle.
- Demonstrate benefits of security enhancements to grid efficiency and vice-versa.
- Increase public and business confidence in the modern grid's resilience.

This paper covers four important topics:

1. The current and future states of grid security.
2. The requirements that, if met, would assure a disciplined systems approach and an industry/government partnership.
3. The benefits that accrue to developing a modern grid that resists attack.
4. Recommendations for moving forward.

Although it can be read on its own, this paper supports and supplements "A Systems View of the Modern Grid," an overview prepared by the Modern Grid Initiative (MGI) team.

CURRENT AND FUTURE STATES

Before we discuss how the modern grid might be developed to resist attack, we need to understand its current vulnerability and how to reduce that vulnerability in the future. This section explores the current state and the desired future state of grid security.

CURRENT STATE

The complexity of the current electrical power system and the reliance on critical nodes to operate without interruption create the potential for single-point failures that can result in widespread disruptions.

In today's grid, failure of a critical node may not be detected or corrected in time to avoid a major disruption. The weak link might become apparent on a hot summer day with specific power flow conditions as the potential catalyst for a widespread blackout, similar to one experienced on August 14, 2003.

The grid is aging, based largely on technology developed in the 1950s or earlier. This aging infrastructure is stressed by lack of adequate investment to meet the growing demand for electric power.

From 1988 to 1998, U.S. total electricity demand rose by nearly 30 percent but its transmission network's capacity grew by only 15 percent. From 1999 to 2009, analysts expect demand to grow by 20 percent, while planned transmission systems grow by only 3.5 percent (Amin 2003, 19–25).

Ironically, recent advances in technology and changes in the electricity sector, such as deregulation and dependence on 20th century technologies, may be adding to the security problem. Examples include:

- Increased reliance on unprotected telecommunications networks and on associated SCADA systems.
- The growth of independent power producers without the budget to address security.
- Outsourcing of maintenance and security by larger companies.

Attacks on the grid could be aided by today's easy accessibility to open sources of information. In the electric power industry, industry publications, maps and material are all available on the internet. These are sufficient to allow someone to identify the most heavily loaded transmission lines and the most critical substations in the power grid.

In general, deregulation has increased the number and types of industry players and interfaces, adding complexity and thereby increasing the potential for security gaps. (Committee on Science and Technology for Countering Terrorism, National Research Council, 178-179, 2002)

Hackers could gain access to "open" electric power control systems, crack passwords, and lower protective relay settings, causing circuit breakers to "trip" at normal current flow. They could raise, at the same time, the settings on neighboring circuit breakers so that diverted power would damage the infrastructure protected by those breakers (GAO 2004, 14–16).

Threats to the security of the grid's cyber backbone are increasing. Application of existing security technologies, such as encryption and the widespread use of routine security procedures could help somewhat. However, too many control devices in use on today's grid do not have the bandwidth and processing power to use even the current state of the art in cyber protection.

FUTURE STATE

The modern grid will address critical security issues from the outset, making security a requirement for all the elements of the grid and adopting a systems view that enables an integrated and balanced approach.

Planning for manmade threats will consider not only single, but also multiple points of failure. Parts of the system will need more risk reduction than others. With a systems view, security decisions will be based on prioritized options to reduce risk. Federal, state, and local officials will work with individual utilities to address acceptable risk, possibly with support from DOE and homeland security officials.

Federal, state and local policies and regulations will be developed that allow utilities and others in the electricity industry to recoup reasonable costs for security upgrades that are part of the overall system design.

Security will benefit from key modern grid technologies that include:
- Integrated Communications for real-time information & control.
- Sensing & Measurement.
- Advanced Components & Distributed Energy Resources (DER).
- Advanced Control Methods.
- Improved Interfaces & Decision Support.

Table 1 notes some of the security solutions offered by these technologies.

Modern Grid Key Technologies	Security Solutions
Integrated **Communications** for real-time information & control	▪ Use communication for prediction and decision support. ▪ Wide-area secure communications instead of internet monitoring. ▪ Monitor and respond to threat conditions instantaneously.
Sensing & **Measurement**	▪ Remote monitoring that detects problems anywhere in the grid. ▪ Events detected in time to respond.
Advanced **Components** & DER	▪ Tolerant and resilient devices. ▪ Fewer critical points of failure. ▪ Distributed, autonomous resources.
Advanced **Control Methods**	▪ Islanding to isolate vulnerable areas of the grid. ▪ Automated network "agents" for dynamic reconfiguring. ▪ Self-healing with preventive or corrective actions in real-time.
Improved **Interfaces** & **Decision Support**	▪ Operator training for response to attacks. ▪ System recommendations for best response. ▪ Simplification of operator interaction with the system.

Table 1: The key technologies of the modern grid contribute to solutions that resist attack.

A modern, more resilient grid will leverage technologies for rapid, wide-area communication of the status of grid components. New control technologies will quicken response to events and easily integrate DER.

Enterprises will focus people and processes on implementing and maintaining security. People with experience assessing risk and designing security in complex systems will help develop and operate the modern grid. Process improvements can provide a substantial benefit at low cost. Additionally, processes for resolution of inter-company and inter-regional issues will be put in place.

In the modern grid, implementing cost-effective options to enhance security will also have positive impacts on reliability and resilience. For example, the data required for computer simulations that provide operators with information to predict disruptions could also be used to identify and mitigate attacks against the grid.

Government and industry will jointly conduct exercises that will improve the security aspects of the modern grid, as well as its design and operation. Metrics will be used to gauge success and guide iterative improvements.

REQUIREMENTS

With a broad understanding of the current and future state of the electrical power system, we can now discuss some of the requirements that need to be met to move forward. This section explores system requirements, as well as requirements for policy and regulation, and codes and standards.

SYSTEM REQUIREMENTS

The systems approach to electric power security would identify key vulnerabilities, assess the likelihood of threats and determine consequences of an attack. The designers of the modern grid can draw on extensive experience developed by the Department of Defense in assessing threats and system vulnerabilities.

This approach would apply risk management methods to prioritize the allocation of resources for security, including R&D. Particular goals of security programs would include:

- Identification of critical sites and systems.
- Protection of selected sites using surveillance and barriers against physical attack.
- Protection of systems against cyber attack using information denial (masking).
- Dispersing sites that are high value targets.
- The ability to tolerate a disruption (self-healing characteristics).
- Integration of distributed energy sources and using automated distribution to speed recovery from attack.

Resilience must be built in to each element of the system, and the overall system must be designed to deter, detect, respond and recover from manmade disruptions.

For the modern grid to resist attack, it must reduce:

- The *threat* of attack by concealing, dispersing, eliminating or reducing single-point failures.
- The *vulnerability* of the grid to attack by protecting key assets from physical and cyber attack.
- The *consequences* of a successful attack by focusing resources on recovery.

Therefore, its system requirements must include those that:

- Implement self-healing capabilities.
- Enable "islanding" (the autonomous operation of selected grid elements).

- Provide greater automation, wide area monitoring, and remote control of electric distribution systems.
- Acquire and position spares for key assets.
- Use distributed energy resources.
- Ensure that added equipment and control systems do not create additional opportunities for attack.
- Rapidly respond to impending disruptions with the aid of predictive models and decision support tools.

A systems approach with government and industry teamwork will help the requirements and their costs to be allocated sensibly across the modern grid. Adopting a systems approach encourages balanced investment. Security investments must reinforce the weak links in the grid and avoid the costs of ineffective measures. For example, it does no good for a utility to build fences and hire guards to protect its power plant when an unscreened insider or an outside hacker exploiting unencrypted communications can disable the plant.

POLICY AND REGULATION REQUIREMENTS

Federal, state and local policies and regulations need to be developed to allow utilities and others in the electricity industry to recoup reasonable costs for security upgrades that are part of the overall system design.

For example, federal guidelines and regulations mandating the accommodation of distributed energy resources (DER) would require an investment from industry. The integration of distributed energy would enhance the reliability of the overall system, regardless of which entity owned and operated the DER.

Of course, the federal government could take on the role of funding selected security enhancements or pioneering development of certain advanced technologies that would support security of the modern grid. Nevertheless, the question remains as to what entities make the investment to satisfy requirements such as:

- Integrated DER.
- Real-time communications.
- Secure data transfer over wide areas.
- Integrated, standard fault detection and correction across systems involving multiple utilities.

Metrics to measure the results of security measures must also be used to allocate costs fairly. Utilities faced with investment costs to modernize the grid – even with possible government subsidies – will want a clear understanding of which security upgrades can be passed to the ratepayers. Coming up with the answers will require close coordination between federal and state regulatory authorities, DOE, and possibly homeland security officials.

CODES AND STANDARDS REQUIREMENTS

Grid owners and operators must take a systems view of security, applying industry best practices and standards.

The industry has already begun to establish best practices, primarily through the work of the North American Electric Reliability Council (NERC) Critical Infrastructure Protection Committee (CIPC). These practices are made available through the Information Sharing and Analysis Center for Electricity managed by NERC.

BARRIERS

The physical and cyber security of the electric industry is a growing concern. Evolving national security threats, increasing interoperability in the grid, and expanded use of open systems in the grid's architecture all contribute to serious vulnerabilities.

Most utilities have taken some action on security, but the question remains: Are we gaining ground or losing ground on security? Although we can't provide a definitive answer, we can pinpoint some of the specific barriers that must be overcome to achieve the Modern Grid vision of a system that resists intentional attack. These barriers include:

- **Incomplete understanding of threats, vulnerabilities and consequences.** Some utilities conduct vulnerability and risk assessments and a fraction of them apply the results to security upgrades. Industry as a whole lacks a standard approach to conducting these assessments, understanding consequences, and valuing security upgrades. Additionally, limited access to government-held threat information makes the case for security investments even more difficult to justify.

- **Perception that security improvements are prohibitively expensive.** When examined independently, the costs and benefits of security investments can seem unjustifiable. Approaching grid modernization from a systems perspective provides the significant leverage that can be used to improve security with related technology advances such as sensors, controls and communications.

- **Increasing use of open systems.** Open communication and operating systems are flexible and improve system performance, but are not as secure as proprietary systems. The increasing use of open systems must be met with industry approved and adopted standards and protocols that consider system security.

- **Increasing number of grid participants.** The growing number of participants in the electric system increases the complexity of physical and cyber security issues. Security measures must be built in to the functions that support DER owners, Independent Power Producers, and consumers active in demand response and automated metering programs.

- **Difficulty in recovering costs.** Utilities must be armed with sufficient knowledge and justification to make the case for security investments. Applying a cost-benefit analysis to the system as a whole will reflect the true value of security and system investments that support it.

BENEFITS

The modern grid will deliver substantial benefits if requirements to increase security are met. In this section, we focus on benefits unique to the characteristic of resisting attack.

Besides improving the modern grid's inherent resilience, there are some unique benefits of the modern grid's characteristic to resist attack. These benefits include:

- Deterring an attack in the first place, because it would have little effect.
- Improving the operational readiness of our defense forces by ensuring security-of-supply for electric power
- Reducing the social and economic impacts of a given disruption, for example:
    - Minimizing the costs of grid repair and costs associated with lost productivity.
    - Minimizing the loss of life associated with a loss of power for extended periods of time.
    - Reducing social disruptions.
    - Reducing the geographic extent of outages.
    - Improving the recovery time from outages.
- "Dual use" of security-related improvements to improve reliability such as:
    - Integration of DER.
    - Use of advanced modeling and simulation tools to prevent "normal" outages.
    - Use of spares to mitigate effects of equipment failure.
    - Application of demand response (DR) to increase system robustness.
    - Greater use of distribution automation.

Benefits attributable to other characteristics such as power quality and DER also satisfy security needs as well. Other papers in the MGI collection address these benefits.

RECOMMENDATIONS

To deploy a modern grid that resists attack, the recommendations of the Modern Grid Initiative rely on the coordinated efforts of planners, designers, developers, government, and industry.

Planners of the modern grid should:

- Leverage methods developed by DOD, DOE, and DHS to increase survivability of systems.
- Create a government-industry team, including state regulators, specifically to address issues of acceptable risk to the public from disruptions and ROI for industries' investments in security.

Designers and developers of the modern grid should:

- Consider security as a system requirement that could affect virtually every element and sub-system of the modern grid.
- Ensure that additional equipment and control systems added to the grid do not increase its likelihood of disruption and do not create additional opportunities for malevolent actions against it.
- Apply the ongoing work by industry, government, and academia on physical and cyber vulnerabilities.

Government and industry should:

- Define causes and consequences of outages — as required by the Department of Homeland Security in its development of the National Infrastructure Protection Plan.
- Share their concerns about the cost and expected benefits of security and ensure that the developers of the modern grid integrate security as an inherent characteristic — not as an optional feature.

Once key elements of the modern grid with security-related aspects are identified, the cost/benefit analysis and the risk analysis can be undertaken to determine the benefit of incorporating security upgrades. Then we will be well on our way to a modern grid that resists attack.

Summary

The threat of both physical and cyber attack is growing and a widespread attack against the infrastructure cannot be ruled out.

The 20th century electrical power system is aging and its 1950s infrastructure was never designed to handle well-organized acts of terrorism.

In the 21st century, it is critical for the modern grid to address security from the outset, making security a requirement for all the elements of the grid and ensuring an integrated and balanced approach across the system.

Whether the threat is physical or cyber, the modern grid must resist attacks that employ two different strategies:

1. *Attacks on the power system*, in which the infrastructure itself is the primary target.
2. *Attacks through the power system*, in which attackers exploit power system, networks to affect other economic sectors such as telecommunications, financial, or government.

The complexity of the current electrical power system and the reliance on critical nodes to operate without interruption create the potential for single-point failures that can result in widespread disruptions. To resist organized attacks, the modern grid must consider the risk of multiple points of failure, not just single ones.

The systems approach to grid security would identify key vulnerabilities, assess the likelihood of threats that could exploit those vulnerabilities, and determine the probability and the possible consequences of a successful attack. Technologies and solutions to address security issues will complement communications, computing, decision-making support, self-healing aspects, and equipment improvements being developed for the modern grid.

Implementing cost-effective technologies to enhance the security of the grid will have positive impacts on reliability and resilience. Resilience must be built in to each element of the system. The modern grid (if designed in a way to deter, detect, respond, and recover from manmade disruptions) will also achieve other desired characteristics.

The electrical power industry, partnered with government, must maintain a systems view of security, applying industry best practices and standards. They must establish metrics to measure security effectiveness and allocate costs fairly. Federal, state, and local policies and regulations need to be developed to allow utilities and others in the electricity industry to recoup reasonable costs for security upgrades that are part of the overall system design.

Addressing the grid as a system will require unparalleled government and industry cooperation. To deploy a modern grid that resists attack, the recommendations of the Modern Grid Initiative rely on the coordinated efforts of developers, government and industry to define and adhere to a total systems approach.

For more information

This document is part of a collection of documents prepared by The Modern Grid Initiative team. For a high-level overview of the modern grid, see "A Systems View of the Modern Grid." For additional background on the motivating factors for the modern grid, see "The Modern Grid Initiative." MGI has also prepared seven papers that support and supplement these overviews by detailing more specifics on each of the principal characteristics of the modern grid. This paper describes the third principal characteristic: "Resists Attack."

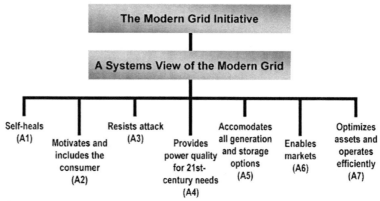

Documents are available for free download from the Modern Grid Web site.

The Modern Grid Initiative

Website: www.netl.doe.gov/moderngrid

Email: moderngrid@netl.doe.gov

(304) 599-4273 x101

BIBLIOGRAPHY

1. Amin, M. 2003. North America's electricity infrastructure: Are we ready for more perfect storms? *IEEE Security and Privacy* 1 (September/October): 19–25.

2. Amin, M. 2001. Toward self-healing infrastructure systems. *IEEE Computer Applications in Power, pgs 20-28.*

3. Apt, J. Causes of Major Disturbances. Carnegie Mellon University.

4. Committee on Science and Technology for Countering Terrorism, National Research Council. 2002. *Making the nation safer: The role of science and technology in countering terrorism.* Washington, D.C.: National Academies Press.

5. Hamre, J. 2003. "Cyberwar! Interview with John Hamre." PBS Frontline, February 18.

6. Idaho National Laboratory. Access denied: Defending the network against hackers. Fact sheet.

7. Sandia National Laboratory. Electric power network surety: Critical infrastructure surety. Fact sheet.

8. U.S. General Accounting Office. 2004. Critical infrastructure protection: Challenges and efforts to secure control systems. Report to congressional requestors. GAO-04-354.

9. U.S. Department of Homeland Security. 2004. National Response Plan.

10. Yeager, K. E. and C. W. Gellings. 2004. A bold vision for T&D. Paper presented at the Carnegie Mellon University Conference on Electricity Transmission in Deregulated Markets, Pittsburgh, PA.

Appendix A5:
A Systems View of the Modern Grid

ACCOMMODATES
ALL GENERATION AND
STORAGE OPTIONS

Conducted by the National Energy Technology Laboratory
for the U.S. Department of Energy
Office of Electricity Delivery and Energy Reliability
January 2007

Office of Electricity
Delivery and Energy
Reliability

v2.0

TABLE OF CONTENTS

EXECUTIVE SUMMARY

The systems view of the modern grid features seven principal characteristics of a modern grid. (See Figure 1.) Accommodating alternatives to generating and storing electrical power is one of those characteristics. Our economy depends too much on large, centralized generation facilities and not enough on distributed energy resources.

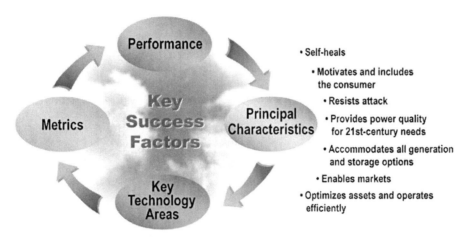

- Self-heals
- Motivates and includes the consumer
- Resists attack
- Provides power quality for 21st-century needs
- Accommodates all generation and storage options
- Enables markets
- Optimizes assets and operates efficiently

Figure 1: The Modern Grid Systems View provides an "ecosystem" perspective that considers all aspects and all stakeholders.

This document provides detail on one of those important qualities: The ability to safely and seamlessly accommodate a wide variety of generation, from massive centralized plants to small solar panels and everything in between.

The modern grid must accommodate not only large, centralized power plants, but also the growing array of distributed energy resources (DER). Today, grid-connected distributed generation supplies only 3% of our total. Going forward, DER will increase rapidly all along the value chain, from suppliers to marketers to customers. Those distributed resources will be diverse and widespread, including renewables, distributed generation and energy storage. Our goal should be widespread adoption of DER, similar to what has occurred with computers, cell phones, and the Internet. (See Figure 2.)

Coping with that diversity will require a host of new and improved functions. Achieving a modern grid will require additional developments in real-time pricing, in smart sensors and controls, in a broadly accepted communications platform, and in advanced tools for planning and operation. It will also require clear standards for interconnection, performance and metrics.

507

Barriers exist to accommodating a wide variety of generation. Perhaps the biggest constraint is the slow development of the new functions described above. In addition, the interaction of DER with different distribution networks is not well understood. The total cost of DER is still too high, and consumers have little motivation to invest, limiting deployment to the electric industry itself. Because stakeholders have conflicting agendas – and because nobody wants to bear the costs of the "societal" benefits – vitally important projects are not being funded.

Integrating multiple generation alternatives will provide significant benefits. The result will be a more reliable, secure, efficient power grid. That grid will also be safer, less expensive and friendlier to the environment.

There is a path forward. Accelerating the deployment of DER – and the ability of the modern grid to accommodate DER and all other kinds of generation – will require a clear, consistent vision with buy-in from all stakeholder groups. It will need greater research and development, financial incentives to customers and utilities, and stimulation of customer demand. Regional transmission operators and utilities should explore DER as a solution to transmission congestion. Regional pilot programs to demonstrate the value of DER are also needed to bring increased visibility to these important resources.

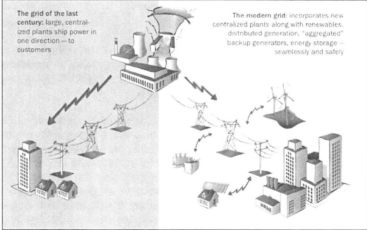

The grid of the last century: large, centralized plants ship power in one direction — to customers

The modern grid: incorporates new centralized plants along with renewables, distributed generation, "aggregated" backup generators, energy storage — seamlessly and safely

Figure 2: The modern grid accommodates all generation and storage options

This document covers:

- The current state and likely future of generation alternatives.

- The requirements for accommodating diverse generation sources (and the gap between what we have and what we need).

- The barriers to meeting those requirements.

- The benefits of success.

- The recommendations for accelerating adoption of this important characteristic of the modern grid.

CURRENT AND FUTURE STATES

Before we discuss *how* to support a wide variety of generation options, we need to understand where we are today and what kinds of options will be accessible tomorrow. This section explores the current state and the probable future state of generation.

CURRENT STATE

A substantial gap exists between existing and desired amounts of DER, particularly in the renewables category. Most of our electricity comes from centralized plants. Of the distributed generation we do have, most is "dirty" (from internal combustion engines) and disconnected from the grid. The U.S. is dominated by big, centralized generating facilities. Large generators (coal, nuclear, and hydro) made up over 75% of net generation in 2004. (See Figure 3.)

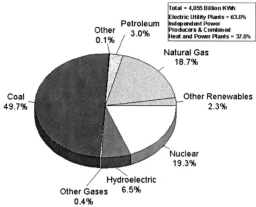

Figure 3: U.S. electric generation fuel diversity (Energy Information Administration, Form EIA-860, "Electric Power Annual" 2006 www.eia.doe.gov/cneaf/electricity/epa/figes1.html)

Small, widely dispersed plants account for an estimated 234GW of distributed generation (DG). Small reciprocating engines, used to supply emergency or standby power, account for about 81% of that total. Combined heat and power (CHP) is second at 9%. Combustion turbines are third with 7%. However, very little of that DG is connected to the grid – only 30GW or just over 3% of our 953GW total.

FUTURE STATE

The future offers several growth pathways for DER, depending on how technologies and markets evolve. A modernized grid should be prepared for five likely scenarios, itemized below.

1. DER will increase dramatically. The modern grid must expect and enable a substantial increase in new energy sources. California's Renewable Portfolio Standard (RPS) program requires investor-owned utilities to provide 20% of their electricity from renewable sources by 2010, and 33% by 2020. DER is likely to grow rapidly because of cost, regulations, environment and speed to market.

2. DER will be everywhere. Deployment will occur along the entire value chain. Suppliers will install it. Power marketers will embrace it. And all type of customers – commercial, industrial, residential -- will adopt it. The goal should be to expect and enable the same widespread deployment that occurred with personal computers and cell phones.

3. DER will be grid-connected. Standalone generation will continue to be common. But in the future, much more DER will be connected to the grid at many different points – at transmission voltages, at distribution voltages, and at AC and DC networks and micro-grids.

4. DER will be aggregated. Both the sources of power and the users of that power will often be aggregated. For instance, wind and solar units may be aggregated into energy "farms" and scattered backup generators into "peaking plants."

5. DER will be diverse. DER will not be dominated by any one size or type of generation. Instead, it will include a wider variety, from those already available to those not yet invented and popularized.

The diversity of distributed energy resources will include sources with a relatively small capacity such as photovoltaic (PV), wind, fuel cells, plug-in hybrid vehicles and energy storage. These devices will typically be connected to low voltage distribution lines or through a DC micro-grid. Their benefits and affordability will lead to a significant increase in the deployment of DER by consumers. In fact, consumers may represent the largest market well into the next decade as they use distributed generation to save money and improve reliability.

But that diversity will include large plants, too. Big power customers and marketers will invest in combined heat and power (CHP) and non-utility generation facilities. Large combustion turbines will be built at a rate consistent with fuel costs and will be located closer to load centers than conventional, centralized power stations.

As we now turn our attention to what is required to reach our DER goals, it is important to remember that the modern grid must also accommodate new centralized plants. We will need conventional, centralized power stations -- coal, oil, gas, nuclear – to meet the increase in demand.

REQUIREMENTS

Accommodating a variety of generation options will require a host of new or improved grid functions. This section discusses the importance of generation alternatives and the most essential functions needed to implement those alternatives.

THE IMPORTANCE OF GENERATION ALTERNATIVES

It is crucial that the United States move away from its current over-reliance on big, centralized generation. Our electric system must accommodate a wider variety of options, often lumped together as *distributed energy resources (DER)*. The main options include:

- **Distributed generation (DG)** – small, widely dispersed plants
- **Renewables** – wind, solar, biomass, etc.
- **Energy storage** – in essence, giant "batteries" and "capacitors"
- **Demand response (DR)** – decreasing demand instead of increasing supply in response to peak loads,

Renewables such as wind and solar can be either distributed or centralized—from individual, isolated wind turbines, or centralized giant wind farms.

DER complements rather than displaces other generation. The 21st-century power system will need a diverse portfolio of generation options. Table 1 maps DER technologies to market applications.

Technology	End user	Grid support	Energy supply
	CHP Premium power Backup power Peak shaving	Asset management Reliability Power quality	In-city generation Renewable
Internal combustion engine	✓	✓	
Combustion turbine	✓	✓	✓
Microturbine	✓		
Fuel cell	✓		
Photovoltaic	✓		✓
Wind	✓	✓	✓
Energy storage	✓	✓	✓
Biomass and waste	✓		✓

Table 1 – DER Options and Market Applications

Figure 4: Renewable generation sources are an important option: The modern grid must accommodate electricity from renewable sources such as wind, solar, and biomass.

DER offers advantages that accomplish many of the key success factors of the modern grid. DER increases our safety factors. For instance, allowing different plant types and fuel types makes the modern grid more resilient because a single failure or unavailability of a specific fuel type will not affect all plants, so there is no common impact to many. And decentralizing generation increases geographical dispersion.

DER improves the ability to manage the system. For instance, diversifying sizes provides flexibility in operations, while diversifying startup (ramp) rates improves response times. Energy storage devices improve power quality and reliability.

On top of all that, DER can save money. It is often faster and easier to install DER at the point of pain rather than siting, permitting, and financing a giant plant and then transmitting the power long distances. This speed and "right-size, right-place" advantage can translate to millions of dollars in reduced risk, lowered interest costs and faster solutions to costly problems. What's more, accommodating renewables and other low-emission sources reduces environmental impact.

ESSENTIAL FUNCTIONS

Real-time pricing

When gasoline prices rise significantly, consumers get very clear "price signals" posted on signs outside the filling station. In response, they may look for alternatives, such as conservation, ethanol-based gasoline and fuel-efficient vehicles. But residential electric customers are not billed on a real-time basis. They receive a monthly bill that charges the same amount whether the electricity was used at expensive (peak) times or low-cost times. Until electric customers get price signals, they will not be motivated to pursue DER.

Tariff features are usually allowed and approved by Public Utility Commissions (PUC), so those entities must re-calculate rate designs with price as a function of time. Then technologies need to get the price signals to consumers, so they can make a decision.

Providing those signals requires smart meters, information gateways, and technologies that allow transmission and distribution operators to send pricing information. The real-time pricing information will tell suppliers, marketers, DER vendors and consumers when it makes sense to buy more DER. That investment will in turn spur the development of next-generation DER devices, making them even more cost effective.

Smart sensors and controls

Integrating DER into the system requires advances in the research and commercialization of smart sensors, protective relays and control devices. Lower cost sensors and controls will reduce DER installation costs, ensure stable operation of interconnected DER units and safeguard line crews and the public during maintenance and restoration. These devices will be needed even more as autonomous operations increase.

On the customer side of the meter, we need energy-management systems to monitor and control DER operations and demand-response requests from the utility.

Communications infrastructure

We need a standard, ubiquitous, integrated communications platform to enable all power system components to intercommunicate. Smart sensors and controls must communicate, but today's grid lacks communications integration and standardization. In most cases, communication does not yet reach to the consumer level. For system operators to integrate new generation sources, communication systems must be able to handle energy price signals and commands.

The lack of a standard platform causes hesitation. Buyers fear their investment will be stranded by technology change (as happened with beta versus VHS standards in the video tape industry). To prevent this, the communications platform of the future must have an open architecture acceptable to vendors, consumers and utilities. Such a platform will reduce the concern for stranded investments and will stimulate DER deployment.

Controls and tools for operation and planning

The modern grid will incorporate generation sources that are smaller, decentralized and often intermittent. But today's operating models cannot reliably operate this new configuration. We require several new tools and technologies:

- **New operating models and algorithms** to address the transient and steady-state behavior of the modern grid, and the integration of large amounts of DER.
- **Improved operator visualization techniques and new training methodologies** to enable system operators (both distribution and transmission) to work together to manage systems in both routine and emergency operations.
- **Advanced simulation tools** that can provide a more complete understanding of grid behavior, especially where a large number of diverse DER units are deployed. These tools are also needed to assist system planners in designing reliable power systems in this new environment.

- **Methods for resolving the unique maintenance and operational challenges** created by DER, demand response, and other new generation sources.
- **Advanced system-planning tools** that assess the benefits (and consider the uniqueness) of DER to locate optimal sites for power stations.

Interconnection codes and standards

Interconnection and operation codes and standards need to be more quickly adopted across the industry to support DER implementation. Efforts are underway to develop fair and uniform interconnection standards at the federal level and in individual states (California, Michigan, Minnesota, New York, Texas, Wisconsin and others).

The development of these standards will enable DER to be easily integrated with the modern grid—called "plug and play"—to connect any power generation into the grid and communicate fully.

Performance standards

Performance standards and metrics must also be developed. We must ensure that DER owners continuously meet their obligations to grid operators. Regulatory groups should perform periodic audits and enforce compliance when needed. Each owner should perform self-assessments. (For instance, does the unit follow dispatch instructions within tolerances?)

Metrics

Key metrics need to be developed and promulgated to provide the transparency needed to most effectively support the safe operation of the modern grid. Some areas where metrics might be established include:

- DER percentage of system-wide capacity, energy, and ancillary services
- Improvements in system and customer reliability
- Improvements in power quality
- Improvements in transmission congestion
- Energy prices, with and without congestion
- Capital investments and deferred investments
- Reduction in emissions and other environmental impacts
- Reduction in system losses

Other

The new generating sources of DER must be able to do the following:

- Auto start, load, and shut down in response to price signals and commands from system operators.
- Represent a significant amount of capacity, energy and voltage support on an aggregated, system-wide basis.

- Integrate safely and reliably with legacy distribution topologies (e.g., long radial feeders) and operations.

BARRIERS

Although **DER** adoption is occurring today, significant barriers remain to meeting the requirements of the previous section and proceeding to full-scale deployment.

To achieve the level needed to support modern grid operations, DER must occur at three levels: 1) electric system, 2) marketer and 3) customer level (residential, commercial, and industrial). The goal should be widespread adoption similar to what has occurred with computers, cell phones, and the Internet.

Several constraints prevent this level of deployment. Prominent among these are the slow development of the elements discussed earlier in the "Requirements" section. In addition, several other factors are holding DER back:

- **Distribution system behavior is not well understood.** We need to further study how various distribution systems interact when DER of many types and designs are broadly deployed (particularly their behavior during upset conditions).

- **Total cost of ownership is high.** The lifetime cost is too high for existing DER devices to compete with traditional alternatives (investment, operation, maintenance, fuel, etc.). Advances in R&D and commercialization are needed to make it more competitive with conventional generation. Already over 225GW of backup generation currently exists on consumer premises. However very little of this capacity is connected to the grid and dispatched by system operators because of environmental and cost considerations.

- **Consumers are not motivated to invest.** Getting varied generation options depends on motivating marketers and residential, commercial and industrial consumers to invest in DER. Until this occurs, DER investment will primarily be funded by the electric industry, limiting its deployment.

- **Conflicting agendas exist among stakeholders.** For example, deployment of DER by consumers negatively affects utility revenues. Societal benefits desired by government are normally not considered in the business plans of marketers and utilities. As a result, some projects are not being funded. They are often the very projects most important for achieving the modern grid.

BENEFITS

As we overcome the barriers to the modern grid, the seamless integration of diverse generation and storage options will deliver substantial benefits.

RELIABILITY

Combining power generation and storage options with a modern grid's advanced communication and control systems results in better reliability and power quality:

- Reduces dependency on the transmission system by strengthening the distribution system.
- Increases operational flexibility during routine, emergency and restoration activities.
- Improves power quality during times of system stress (i.e., peak load, storms, etc.) and reduces system restoration time following major events.
- Reduces transmission losses and congestion by locating generation closer to loads.
- Increases "ride-through" capability and momentary voltage support.
- Reduces the chances for a common mode failure to affect overall operation of the entire grid.

Improvements such as these put us on the road to a "self-healing" grid. In response to signals from system operators and smart sensors, DER will respond in real time with preventive and corrective actions so that reliability issues are avoided or at least mitigated.

SECURITY

The ability of the modern grid to accommodate a wide variety of options can reduce its vulnerability to security attacks and improve its security during major events. Some specific security enhancing features include:

- Decentralization to the distribution level reduces the grid's vulnerability to a single attack.
- Large quantities of smaller DER, coupled with smaller quantities of large centralized generation, reduce the impact of a unit's failure on overall grid operation.
- Diversity in DER gives operators more choices in response to a security emergency.
- Diversity in a geographic location provides alternate means to restore the grid following a major event.
- Diversity of fuels at central generating stations (coal, oil, gas, nuclear, hydro) coupled with diversity of fuels at decentralized

DER (wind, solar, gas, hydrogen for fuel cells, etc.) increases the probability that adequate fuel supplies will be available.

ECONOMIC

Accommodating a variety of generation and storage options adds to the modern grid's economic advantages:

- Eliminates or defers some large capital investments in centralized generating plants, substations, transmission and distribution lines, reducing overall costs by tens of billions of dollars over a 20-year period. *[Pacific Northwest National Laboratory, 2003]*
- Enables consumers to participate in the electricity market (and partially fund new generation).
- Reduces peak demand, transmission congestion and peak prices.
- Increases the grid's robustness and efficiency, leading to cost savings and eventual lower rates.
- Encourages retail electricity markets (capacity, energy, ancillary services) and, potentially, emissions markets.

EFFICIENCY

The modern grid is made more efficient by accommodating many generation alternatives:

- Increases options for system planners to address future demand issues.
- Increases options for system operators to improve the utilization of grid assets.
- Improves asset utilization since plants located near load centers reduce transmission losses.

ENVIRONMENTAL QUALITY

Accommodating generation alternatives is environmentally friendly:

- Encourages the deployment of smaller DER sources including those based on clean technology.
- Encourages greater use of hydro, solar, and nuclear power that produce zero emissions.
- Reduces the need for new centralized generating stations and transmission lines.

SAFETY

Finally, the modern grid improves safety, protecting workers and the public. It mitigates the hazards of interconnecting large numbers of diverse generating sources and energy storage devices.

RECOMMENDATIONS

Considering the barriers to meeting modern grid requirements and the benefits to be attained in overcoming these barriers, what are some of the steps we can take right now?

Table 2- Recommended Steps
Establish a clear vision
Increase R&D
Generate financial incentives
Expand customer side options
Explore DER as a congestion solution
Add demonstration programs

Specific actions are needed to rapidly accelerate deployment of generation options, including:

- **Establish a clear vision:** Stakeholders need a clear, consistent vision for the modern grid that identifies the role of generation, DER and DR. Then research, policies, statutes and reforms may be put in place that endorse this vision, remove barriers and align regulations with the goal of achieving the modern grid.

- **Increase R&D.** Additional research, development, and standards are needed to realize the elements cited in the "Requirements" section.

- **Generate financial incentives:** National energy legislation and regulations are needed to provide financial incentives for investing in DER. These programs should consider total societal benefits. Investment incentives should be made available to consumers. And financial relief should be given to corporations who make generation investments to help offset possible stranded investments.

- **Expand options for customers:** Although great strides have been made in DER research and commercialization, much remains to be done. Stimulation for further improvements should come from the customer side. That is, the value proposition for the consumers and marketers needs to be real and needs to depend on price signals (or perhaps legislation). For example, fuel cells for the home and business, energy storage devices, and hydrogen fuel technologies are expected to be components in the future DER portfolio.

- **Explore DER as a congestion solution.** Regional transmission operators (RTOs) need to put a higher priority on the use of DER as a solution to transmission line congestion.

- **Add demonstration programs.** State-level and regional demonstrations are needed to prove the value and uses of DER, coupled with innovations in grid design and automation. These demonstrations would establish a baseline to encourage follow-on investments in DER-enabled grid infrastructure.

SUMMARY

The ability to accommodate a wide variety of generation options is essential to realize the full promise of a modern grid. Generation will increasingly include renewables and distributed generation, alongside energy storage and other "non-traditional" sources.

Coping with that diversity will require a wide range of new and improved functions, including real-time pricing, smart sensors, integrated communications, advanced decision support tools and more.

If we successfully integrate large and small generation sources, then we will gain a grid that is more reliable, secure, safe, efficient and environmentally friendly. At the same time, it will cost less to operate and maintain.

Barriers exist that may slow our progress. Our understanding of DER and its interactions is still limited, and prices remain high. Development is slow and haphazard. Important projects remain unfunded.

Despite these challenges, there is a path forward.

The Modern Grid Initiative is working with a wide range of stakeholders. The MGI objectives are to:
- Further define the vision of a grid that accommodates many different generation sources.
- Pinpoint the research and technology gaps.
- Better understand how DER fits into the larger, integrated whole.

This document is part of a collection of documents prepared by The Modern Grid Initiative (MGI) team.

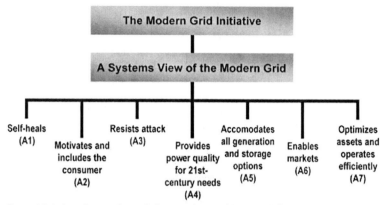

For a high-level overview of the modern grid, see "A Systems View of the Modern Grid." For additional background on the motivating factors for the modern grid, see "The Modern Grid Initiative." Seven appendices support and supplement these overviews by detailing more specifics on each of the principal characteristics of the modern grid. This paper describes the fifth principal characteristic: "Accommodates all Generation and Storage Options."

Documents are available for free download from the Modern Grid Web site.

The Modern Grid Initiative

Website: www.netl.doe.gov/moderngrid

Email: moderngrid@netl.doe.gov

(304) 599-4273 x101

BIBLIOGRAPHY

1. K.E. Yeager and C.W. Gellings, December 2004, *A Bold Vision for T&D,* Carnegie Mellon University Conference on Electricity Transmission in Deregulated Markets.

2. U.S. Department of Energy, January 2004, National Electric Delivery Technologies Roadmap.

3. D. Rastler, EPRI, December 2004, Distributed Energy Resources: Current Landscape and a Roadmap for the Future.

4. EPRI document 1008415, Distributed Energy Resources: Current Landscape and a Roadmap for the Future, December 2004.

5. New Rules Project-Democratic Energy website, Communities and Government Working on our Energy Future, Interconnection Standards Efforts Underway, http://www.newrules.org/electricity/producers.html.

6. Energy Information Administration website, Electricity Generation, Electricity Statistics, http://www.eia.doe.gov/cneaf/electricity/page/at_a_glance/gen_tabs.html.

Appendix A6:
A Systems View of the Modern Grid

ENABLES MARKETS

Conducted by the National Energy Technology Laboratory
for the U.S. Department of Energy
Office of Electricity Delivery and Energy Reliability
January 2007

Office of Electricity
Delivery and Energy
Reliability

v2.0

TABLE OF CONTENTS

EXECUTIVE SUMMARY

The systems view of the modern grid features seven principal characteristics, one of which is the characteristic of fully enabling markets for electrical power. (See Figure 1.)

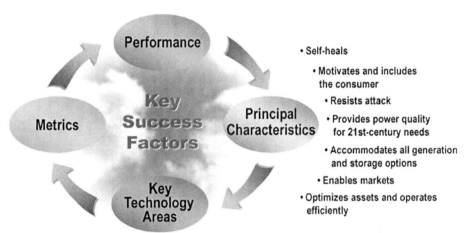

• Self-heals

• Motivates and includes the consumer

• Resists attack

• Provides power quality for 21st-century needs

• Accommodates all generation and storage options

• Enables markets

• Optimizes assets and operates efficiently

Figure 1: The Modern Grid Systems View provides an "ecosystem" perspective that considers all aspects and all stakeholders.

Correctly designed and operated markets efficiently reveal cost-benefit tradeoffs to consumers by creating an opportunity for competing services to bid. In general, the fully functioning modern grid will account for all of the fundamental dynamics of the value/cost relationship. Some of the independent grid variables that must be explicitly managed are energy, capacity, location, time, form (e.g., high voltage vs. low voltage; AC vs. DC), rate of change of capacity (e.g., ramp rates), resiliency (e.g., ability to accommodate perturbations), and energy quality/value (similar to return on investment). Markets can play a major role in the management of these variables.

The challenge for the modern grid is to allow, as much as possible, regulators, owners and operators, and consumers to modify the rules of business to suit operating and market conditions. Markets can enable efficient operation under both low stress and high stress conditions. Markets can enable automatic reconfiguration of facilities and equipment as needed to operate reliably and efficiently. Figure 2 shows the differing time frames for market operations and the supporting infrastructure required for those operations.

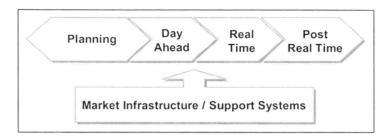

Figure 2: Time frames for market operations and their supporting infrastructure define the concept of enabling markets.

"...better and cheaper technologies will be invented once retail energy is subject to free entry and exit. No one knows what combination of technology, cost and consumer preferences will be selected. And that is why the process must be exposed to the trial-and-error experiment called free entry, exit and pricing. As in other industries, investors will risk their own capital – not your tax dollars or a charge on your utility bill – for investments that fail. Also, as in other industries with dynamically changing product demand, competition will force prices to be slashed off-peak, and increased on-peak to better utilize capacity." (Vernon Smith – 2002 Nobel laureate in economics, *Wall Street Journal*, 2003)

- **Planning**—This part of the market provides the long-term and intermediate-term regional infrastructure (generation, transmission and demand) planning activities that forecast load and congestion, develop capacity and adequacy, and schedule outages.
- **Day ahead**—This part of the market provides the short-term planned capacity requirements, MW injections, MW withdrawals, financial transmission rights (FTR) and ancillary services.
- **Real time**—This part of the market provides the real-time generation dispatch, management of injections and withdrawals, congestion management, ancillary services, and real time reliability management.
- **Post-real time**—This part of the market provides the settlement of the energy dispatch and financial transactions, as well as analysis and auditing of the Day Ahead and Real Time market operations.

Market infrastructure and support systems are critical factors in the success of enabling electricity markets in the modern grid. Advanced components and widespread communication in the modern grid will support market operations in every time frame above and provide full visibility of data to the market participants.

This document covers:
- The current and future states of electrical markets.
- The requirements for moving to a modern grid that enable these markets.
- The barriers to be overcome.
- The benefits of fully enabled markets.
- Recommendations to move forward.

Although it can be read on its own, this paper supports and supplements "A Systems View of the Modern Grid," an overview prepared by the Modern Grid Initiative (MGI) team.

CURRENT AND FUTURE STATES

Before we detail the requirements to realize the modern grid's "enable markets" characteristic, we need to understand the difference between the current state of electrical markets and their potential future state.

CURRENT STATE

The majority of the nation's electrical power system operates in accordance with rate structures established by state utility regulators. Rates are based on expenses plus a reasonable return on investment. Some expenses are passed directly to the consumer. Examples include fuel cost adjustments and power line losses.

Retail markets operate in several regions of the nation, governed by state requirements. Retail markets typically separate production costs, i.e. the cost of generating electricity, and transportation costs, i.e. transmission plus distribution expenses. This transportation expense is sometimes referred to as the wires cost.

However, the consumer is served by the same electrical infrastructure (wires) used before the retail choice was enacted, so the cost structure of the wires is the same and is billed as a fixed price-per-unit of energy. There are no savings to the consumer from retail choice as far as the wires cost is concerned.

In retail markets, the consumer may choose from a list of producers of electricity. But the production of the energy is also billed as a fixed price-per-unit of energy. Since the competing producers of electricity operate within the same wholesale competitive market, their price variations are minor or the market would not choose them to produce electricity in the wholesale market in the first place.

Wholesale markets select producers of electricity on an economic merit order, so the least expensive units are selected before more costly units. A constant fixed wires charge plus a production charge which only varies slightly has resulted in a small change in the total consumer energy bill.

The resulting minimal cost savings to consumers is the reason for low participation rates in the retail choice programs. There are only two rate options in the Time-of-Day (TOD) program, on-peak rates and off-peak rates. Hours of use for on-peak and off-peak time periods are recorded with electric meters, and a consumer bill is normally produced monthly. The ratio of peak to off-peak rates ranges only from about two to three. There is a much higher degree of variability in the wholesale market hourly prices where a daily high to low hourly price ratio of ten to one is a common occurrence.

There is a win-win scenario where the consumer takes a more active role, joins with other consumers into a large coalition representing a large load, service or generation supply and sells those products in the marketplace at a fair price. The power of the consumer has reduced prices in other markets and if organized, can do the same for electricity. The operators of the market also win in that they get a product that was previously not available to reduce the cost of delivered energy or increase reliability at a minimum cost.

Even though retail choice and wholesale markets have been around for decades, they have not become the normal way we conduct energy business. The Federal Energy Regulatory Commission (FERC) State of the Markets Report, a staff report by the Office of Market Oversight and Investigation published in January of 2004 states that operating or forming markets serve 70% of the U.S. population. While that may be true, participation in many of those markets is low. The graph and table below show greater detail. (See Figure 3 and Table 1.)

Figure 3: States Participating in Retail Choice, Status of State Electric Industry Restructuring Activity. (DOE EIA 2003)

State	Penetration	State	Penetration
Arizona	Not Reported	New Hampshire	Not Reported
California	16%	New Jersey	2%
Connecticut	Not Reported	New York	21%
Delaware	Not Reported	Ohio	18%
DC	13%	Oregon	11%
Illinois	12%	Pennsylvania	4%
Maine	32%	Rhode Island	Not Reported
Maryland	4%	Texas	Not Reported
Massachusetts	25%	Virginia	Not Reported
Michigan	11%		

Table 1: For those states that have tried retail choice, consumer participation is low. ("Status of State Electric Industry Restructuring Activity", DOE EIA, Feb 2003; "Status of Electric Competition in Michigan", Michigan Public Service Commission, January 2005)

Wholesale market operations are also operating in several regions of the nation, governed by the FERC in coordination with state utility regulators. The process has four steps:

1. Generators initiate offers to sell their energy to the market and load-serving entities submit bids to purchase it.

2. When a balance is reached between sellers and purchasers, then all loads are served and the market is declared to be 'cleared.'

3. Market participants are advised of the 'cleared' results to include injection MW, withdrawal MW, hours and prices for each hour thereby initiating their market responsibilities.

4. Settlements occur based upon the bids, offers and actual injections and withdrawals of energy per hour.

States participating in wholesale markets are shown in Figure 4.

Figure 4: States participating in wholesale market competition. (DOE EIA 2005)

In the United States, the extent of the electricity market today can be described as follows:

- The majority of wholesale transactions are bilateral and long term
- A small portion of the country is using real-time wholesale energy markets
- A smaller portion of the country is using day-ahead wholesale energy markets
- A small portion of the country is using wholesale ancillary services markets
- A portion of the country is using zonal pricing models in wholesale markets
- A small portion of the country is using nodal (locational) pricing models in wholesale markets
- Retail choice (retail electricity markets) represents less than 5% of the electric load in the nation

The DOE National Transmission Grid Study, published in May 2002, states that the benefits of the existing wholesale electricity markets in the United States are $13 billion annually. A study of the Eastern Interconnect 1999–2003, which was released in 2005, determined that consumers realized $15.1 billion in value from wholesale electric competition.

Yet another study, conducted by the Fraser Institute in late 2004, determined that the United States could have dropped electricity prices 7%–9% between 1997 and 2002, following the completion of industry restructuring and the offering of a wholesale market to all consumers. This corresponds to an annual national benefit of $10–$13 billion, based on $0.04/kwh and an average of 3,600 billion kwh consumed annually.

FUTURE STATE

In the future, the scheduling and use of electricity will be fully commoditized by creating open-access markets across the country based on wholesale and retail models. These economically constrained market operations will drive reliability in the grid and open utilities and consumers to new service models that better fit the needs of all grid participants. Electricity markets in the future will integrate many diverse technologies and control functions to include the following:

- Suppliers in the various markets will be seamlessly integrated across all generating unit sizes (from 10kw to 1,300MW)
- The vast majority of all types of consumers (industrial, commercial, and residential) will participate in the market seamlessly through various forms of decision-assistant software
- All loads will have some measure of intelligent control, enabling new demand-response (DR) markets

Regional differences in the transmission level of the electricity network affect the enabling of markets. In the Northeast, load is concentrated, so the network is compact. In the Northeast, it may take only 100 miles of transmission line to "touch" one million consumers. In the West, except for major metropolitan areas, load is spread over an expansive geography, so the network must travel long distances between major loads with few lines supporting it. Varied circumstances create varied reliability and economic limitations. In the West, it may take over 1000 miles of transmission line to "touch" one million customers.

This would suggest that the cost of physically enabling a wholesale market in the West might be ten times that in the Northeast. However, if underground is required, the cost of transmission construction in the Northeast is significantly more costly per mile. While challenges differ in the West and Northeast, it is equally true for both regions that curbing peak loads and making loads more predictable are common elements of wholesale markets, and all regions will improve in grid reliability through advanced control and protection.

The competitive wholesale market has steadily made more services available to participants, including bilateral transactions, scheduling, day-ahead markets, and reserve sharing.

If present trends hold, generation resources of the future will be dispersed throughout the load areas mostly to minimize fuel transportation costs and they will be much smaller in electrical size. This change in generator size-mix over time requires new thought on how to control this vastly distributed resource as well as the impacts in the marketplace. Studies show that the nation can expect to add an additional 120,000 distributed generating units under 50MW over the next 20 years.

The future will reveal the need for new market elements. The experiences of the various Regional Transmission Organizations (RTO) and Independent System Operators (ISO) show that participants in the market are active in requesting new services and market forums:

- Expansion of the ancillary services market offerings.
- Introduction of renewables, carbon trading, and other specialty markets.
- Inclusion of DER market operations and other consumer-rich markets at the wholesale and retail market levels.

REQUIREMENTS

Having examined the current state of electrical markets and their desired future state, what requirements must the modern grid meet to fully enable markets?

The basic requirements of electricity markets are the existence of an adequate physical and informational infrastructure, sound market rules, vigilant oversight, and fair and equitable access.

- **Adequate infrastructure**—Markets can affect load and load can affect network reliability, therefore proper enabling of electricity markets supports a more reliable grid. A properly enabled marketplace assumes that the information and control architecture is adequate to provide needed information to appropriate decision makers.

- **Sound market rules**—Markets are based on proven first principles of physics and economics.

- **Vigilant oversight**—In each phase of the electricity market there is independent monitoring and review of operations and participants to assure fairness in the market and reliability of the grid.

- **Fair and equitable access**—The foundation of the market is the idea that "it is the same electricity market for all who qualify." Those who "qualify" have learned how to function in the market both financially and operationally.

DESIGN CONCEPT

To establish new markets and market services requires the development of new tariffs, systems, information flows in real-time and training for market participants. Therefore, new markets and services must roll out in logical pieces. For example, a market may open with a real time market only and as the market gains experience, later it will open a day-ahead market. An ancillary services market would follow the real time and day-ahead markets. These have been the actions at both the Midwest Independent System Operator (MISO) RTO and PJM RTO.

The seamless architecture of the modern grid can and must extend the electricity market into the electric distribution level. The distribution consumer may participate in demand response (DR) or distributed energy resource (DER) programs which when aggregated, become a market commodity in the wholesale market. While nearly all the distribution systems are operated as state-chartered monopolies, this is the area where the greatest variability in customer needs resides.

In order to apply DR and DER as market commodities, the modern grid needs to expand current electricity market thinking to include

designing for open-access market participation. This expansion of market thinking may take place in the wholesale market, the retail market, a new intermediate market, or some combination of these markets.

Today, FERC regulates interstate wholesale markets and state and local agencies regulate retail markets. For the modern grid to provide seamlessly integrated markets, it must include interstate wholesale markets, regionally based retail markets, and a new intermediate market that joins them at the distribution level.

Summarized, the design concept must include:
- Fully effective wholesale markets
- Selective expansion into retail markets
- New, presently unidentified markets that may not fit the traditional wholesale/retail model

DESIGN FEATURES AND FUNCTIONS

The design of the modern grid must be consistent enough to enable the electricity market to operate coast-to-coast and deliver economic benefits. In addition, the modern grid requires more sophisticated models to analyze options, refine market performance and design new markets. The basic design can be described in the context of the market's time horizons and infrastructure shown previously in Figure 2.

Planning

- **Power systems coordination and planning**—Long-term development of strategies that improve overall reliability of the grid and accommodation of future loads
- **Load forecasting**—Long- and intermediate-term forecasting of future loads and load profiles across the grid in sufficient detail for power systems coordination and planning
- **Facility and operational data**—Developing the necessary data integration and acquiring the necessary grid facilities information (attributes and behaviors) to support accurate modeling of the grid for planning and operational purposes
- **Long-start resource commitment**—Planning that incorporates the long start-up sequence for large, central generation plants into the market operations and reliability coordination of the grid
- **Congestion management**—Factoring congested transmission pathways into the planning for current operations, new grid assets, and upgrades
- **Reliability planning and coordination**—Establishing operating strategies that implement the performance and planning goals of the grid
- **Generation and transmission outage coordination**—Determination of outage impacts and scheduling for the purpose of minimizing challenges to reliability and fair market operations

- **Financial transmission rights** (FTR)—Running the advanced (forward contracts) allocation of FTRs to asset holders and establishing the simultaneous feasibility requirements of the upcoming auction of rights in the market

Day-Ahead

- **Generation supply offers**—The introduction of offers by generators to supply the grid with a specified amount (and profile) of MW for a specified period of time at a specified start time, based on an estimated price at a specific grid node
- **Demand bids**—The submission of bids to take power from the grid and serve loads of a specified amount (and profile) of MW for a specified period of time at a specified start time, based on an estimated price at a specific grid node
- **Physical bilateral transactions**—A specific agreement between one seller of generation and one buyer of power for serving a load of a specific amount of MW for a specified period of time at a specified start time for a specified price
- **Financial transactions (bilateral, FTRs, virtuals)**—A financial hedging function where one party takes control of a specified amount of transmission capacity at a specified grid node for a price; as the transmission service is used by the party or other parties, the difference between the agreed price and the eventual real-time market price is settled
- **Ancillary services offers and bids**—The introduction of offers and submission of bids for spinning reserve requirements, volt/VAR support needs, demand response needs, renewable energy credits, etc.
- **Market results (clear day ahead)**—The process by which the electricity market settles the financial commitments made and accepted during the day-ahead market period
- **Re-offer period**—A short period at the end of the clearing of the day-ahead market where unfulfilled non–real-time offers and bids can be reintroduced to the electricity market

Real-Time

- **Supply offer instructions**—The continuous process of sending generating unit production targets to market participants to balance supply and demand at least cost while recognizing current operating conditions.
- **Security Constrained Economic Dispatch (SCED)**—An algorithm-based continuous process to simultaneously balance injections and withdrawals at least cost, manage congestion, and produce ex-ante LMP to establish resource baselines
- **Physical bilateral transactions**—The process of executing previously planned bilateral agreements
- **Prices**—The continuous process of determining and publishing real-time pricing every 5-minute interval at every commercial node in the electricity grid

- **Re-dispatch**—The process of dispatching previously uncommitted (but available) generation to fulfill an emerging demand in real-time at previously set prices or market prices
- **Emergency ancillary services**—The process of dispatching available reserves to manage congestion or fulfill a demand (load) that another supply (generator) has failed to serve

Post-Real Time

- **Metering/Meter Data Management Agent (MDMA)**—The system where real-time metering data is submitted on behalf of each market participant to the grid operator through a real-time portal
- **Settlement calculations**—The performance of a series of computations on received metering data (real-time metered MW, cleared day-ahead MW, and real-time prices) at different post-market day intervals (operating day, 7 days, 14 days, etc.) to reconcile supply, demand, and associated pricing in the Day-Ahead, Real-Time, and FTR markets
- **Accounting and billing**—The transference of market settlements into individual market participant accounts and creating the appropriate payments and invoices for receivables
- **Settlement disputes**—The correction of differences and disagreements in the market results before payments and invoices are made
- **Market auditing**—The on-going, independent review of market operations and settlements to assure market fairness and openness

All present day markets ensure that the uplift cost for these services is kept at an acceptable level.

MARKET INFRASTRUCTURE AND SUPPORT SYSTEMS

Market infrastructure and support systems for the modern grid must be complete, robust, and of high quality. The following functions and processes are required for a market infrastructure.

- **Systems functions**—Tagging and scheduling, Organization for the Advancement of Structured Information Standards (OASIS), day-ahead/real-time market systems, power system simulation of the network and the real-time market, accounting system, configuration control, reserve sharing applications, settlements application, algorithms for clearing market prices, etc.
- **Business processes**—Structure methods for converting tariffs to market processes, consensus market rules, continuity plans, published grid operator and control area functions, emergency operations and processes to engage, etc.
- **Market participant readiness functions**—Open participant portal, market participant training and certification, client relationship management, credit management system, etc.
- **Independent market monitor (IMM) functions**—IMM interface/connectivity, real-time market data access, etc.

- **FTR functions**—FTR applications to include an allocation system, auction system and simultaneous feasibility test
- **LMP and state estimation (SE) functions**—Fundamental nodal model of the grid, high-speed network model, real-time telemetry, state estimator applications, reliability assessment and commitment (RAC) application, commercial node application, energy node application, LMP model, adequate resolution and precision in each system, etc.
- **Contingency functions and processes**—Back-up market and control operations systems and facilities, a process to accommodate RAC failure, systems for LMP/settlement with SE failure, unit dispatch system with SE failure, congestion relief process/system with SE Failure, Net Scheduled Interchange system with SE failure, etc.
- **Control area functions and processes**—Operator routine and emergency operations training, market participant interface (message and response) via communication and telemetry, systems for network visibility, etc.
- **Joint operating agreement functions and processes**—Seams agreements, data communication with the rest of the nation's regional markets, handling of marginal losses and congestion across seams, etc.

OTHER REQUIREMENTS

Common Information Model

The Common Information Model (CIM) architecture is a critical element for standardizing data shared by the various elements of a modern grid.

"...the CIM facilitates the integration of Energy Management System (EMS) applications developed independently by different vendors, between entire EMS systems developed independently, or between an EMS system and other systems concerned with different aspects of power system operations, such as generation or distribution management." (IEC 61970-301)

CIM is a necessary foundational element of successful market development because the real-time information important to proper operation comes from hundreds of sources and dozens of entities (transmission owners, generators, market participants, etc.). For example, at MISO, state estimation, location marginal pricing, and network topology depend on commonly used information across more than 30 EMS/SCADA and GIS systems at utilities.

Communications

For electricity markets to function properly, near real-time information communication must flow seamlessly between market systems and monitoring and control systems throughout the region. This will require a much dispersed, highly reliable, multi-variant communications infrastructure.

Policy and Regulation

In a June 2005 testimony to the House Government Reform Subcommittee on Energy and Resources, FERC Chairman Pat Wood made three important points about enabling markets in a modern grid:

1. "The industry and its regulators (state and federal) must find ways to accelerate investment in transmission, if customers are to receive the many benefits achievable with competitive wholesale markets . . ."

2. "...Most traditional, vertically integrated utilities with retail service obligations must go before their state commissions to seek retail rate recovery for any investment they make in new transmission. This can involve opening up all of their costs as well as their entire rate structure for reevaluation, a step few utilities desire . . . "

3. ". . . Finally, development of a robust inter-utility transmission grid may come into conflict with an individual utility's fiduciary responsibility to its shareholders if such a grid will allow competing generators to more economically serve the transmission-owning utility's wholesale customers."

Codes and Standards

Voluntary standards must be adopted by federal and state authorities as new law regulates performance of the grid and markets. For example, the Organization for the Advancement of Structured Information Standards (OASIS) has developed standards for business transactions, legal, education and biometric uses. There are many new standards in development including tax, voting and medical. A trusted professional organization such as OASIS must emerge to manage the data protocols of all the energy-related transactions of the new energy market.

Quality

Like the real-time operating systems that manage the modern grid, market infrastructure support systems must utilize ISO-certified or Capability Maturity Model Integration (CMMI)-based software solutions as standards to improve the openness, scalability, and maintainability of electricity markets.

User Interface

Traditionally, the electric grid has been managed by a select few individuals, with little interface between systems and consumers. Energy management systems use interfaces based on the specialized knowledge of grid operators.

With the introduction of electricity markets, new users will bring a wide variety of needs, skills, and levels of knowledge. The user interface will require an easier, more socialized interface to accommodate this constituency.

This trend will gradually expand as the electricity marketplace is demystified and user interfaces become easy to use in an open-access environment. In time, accessing the modern grid electricity market will be as easy as logging on to eBay or Amazon.com.

BARRIERS

Meeting some of the basic requirements of enabling markets in the modern grid means overcoming some significant barriers.

Parochial regulations, limited skills of market participants, complex information infrastructure, and burdensome capital investments comprise some of the more common barriers.

- **Regulations** – Both federal and state regulations are required to support full scale integrated markets to fulfill the needs of all consumers. The regulators of low energy cost states are naturally reluctant to have their low cost energy sent out of state to the detriment of their consumers. The regulators of high energy cost states feel pressure from their base constituency to block market entry and keep out low cost energy providers. This preserves the profits of the high cost resident energy providers, who are also high tax payers and influential in the political process through lobbying efforts. In the change to fully enabled markets, there will be winners and losers.

- **Market Participant Skills** – An educated consumer is required to effectively operate in a market. All market participants should try through legal means to make as much money in a market as possible. That is true of all markets and should be expected in a new energy market. The best way to maximize profit is to understand the detailed mechanisms of the market tariffs and billing procedures. This is no small effort and every mistake will likely cost the market participant money.

- **Information Infrastructure** – A vast amount of data is required to operate and run a market. In the startup of the Midwest Independent System Operator Market, over one hundred thousand telemetry data points of system components were required to assure accurate pricing and operating signals. The collection and transmission of this data requires an extensive communications network. If multiple control areas are involved in making up a market, then common information protocols, security provisions and timing add to the complexity.

- **Capital Investment** – It takes a large capital investment to set up, operate and monitor a market. The market participants are the only source of funds. Generators must be convinced that they will be able to sell at higher prices and load serving entities representing the consumers must be convinced that they will be able to purchase power at cheaper prices. It takes a great deal of optimism to open a market with such risks.

BENEFITS

As barriers are overcome, the infrastructure to enable markets will gradually be implemented and important benefits will accelerate the evolution of the modern grid.

As consumers respond to market data about increases in price, demand will be mitigated. Plus, consumers become more engaged in determining alternate lower cost solutions, which spurs new technology and process development. As consumers suffer interruptions, the load profile and generation profile shift as alternate load management and distributed generation schemes become more prevalent in the industrial, commercial, and residential sectors. These drivers and changes result in fewer and briefer interruptions.

From a marketplace looking for alternate lower costs solutions, the modern grid will be able to offer a wide array of load-management strategies. Distributed generation, energy storage, demand-response strategies and new ways to effectively manage voltage will emerge to cope with a more volatile operating environment. In addition, such marketplaces spawn new ways to improve performance. The addition of a residential fleet of distributed energy resources (DER) would also provide service for the self-healing feature of the modern grid.

Fully enabled electricity markets will also drive smarter decisions about where to locate grid resources. Examples include the locational marginal price (LMP) as an added input to independent power producer (IPP) generation siting in Wisconsin, and added input to siting distributed generation in Connecticut. From a systems view of the modern grid, it is important for generation siting to have all the related information available to make the best decisions possible.

The modern grid's fully enabled market would open the electricity infrastructure to all consumers, not just transmission owners (TO) and independent power producers (IPP). Extending electricity market participation to a wider stakeholder group (e.g., distribution companies, distributed generation owners, and consumers) can greatly increase the performance and reliability benefits of a market, whether wholesale or retail. For example, the open access of the cable television industry has greatly expanded services, now providing telephone and internet along with programming selections. The open access to the electricity market will likely result in a similar expansion of commerce.

RECOMMENDATIONS

Four broad actions taken in parallel would ensure that the modern grid's design and implementation would support fully enabling markets for electrical power.

1. Modify existing and create new polices and regulations that remove the economic and political barriers to integrated markets. It will take a systems view and, most likely, federal directives to align policies toward integrated markets that benefit all consumers. There are complex issues to be undertaken and, as in any political process, no one wants to end up with a disadvantage.

2. Provide widespread market education to all stakeholders in the modern grid, especially distribution level consumers. Thousands of consumers may join a demand response group that may sell its DR product in the real time market. Knowing the potential gain and potential liability of such a venture is a must for an informed consumer.

3. Standardize the communication of market information throughout the design of the modern grid with equipment, software processes and protocols. A broker of market information is to the advantage of all market participants, like the reporting of stock values. Any consumer should have access through non-proprietary equipment.

4. Incentivize capital investment. Regulators wish to have access to competitive markets to lower prices, yet are reluctant to authorize investments that spread benefits outside their regulatory jurisdiction. Options for resolution include cost sharing, benefit sharing or some combination of the two.

SUMMARY

"Enables Markets' is the name we give to one of the seven principal characteristics featured in the systems view of the modern grid. It is an important characteristic because well-designed and operated markets efficiently reveal cost-benefit tradeoffs by creating opportunities for competing services to bid.

Such a market for the modern grid would be attained by a market infrastructure that supports the four time frames of market operations:

1. **Planning** – For intermediate and long-term regional operations.
2. **Day ahead** – For short-term capacity requirements.
3. **Real time** – For managing generation dispatch, congestion relief and reliability.
4. **Post-real time** – For settlement of transactions, analysis and auditing.

Attaining fully enabled market infrastructure for the modern grid requires action in four broad areas:

1. **Regulations** – Both federal and state regulation must be overhauled to remove parochial barriers to a seamless integration of electrical markets.
2. **Market education** – To maximize their profits, all market participants must gain a detailed understanding of the operations and economics of electrical power services.
3. **Information infrastructure** – The collection and distribution of the vast amount of market data requires an extensive network and common information protocols.
4. **Capital investment** – Market participants must be reasonably certain their investment will be profitable, given the large commitment needed to set up, operate, and monitor the market.

Even partial successes in these four broad areas would provide substantial benefits.

- As consumers respond to market data about increases in price, demand will be mitigated and consumers will seek alternate lower cost solutions in new technologies and products.
- Consumer demand for solutions will stimulate a market for distributed generation, energy storage, and other distributed energy resources that, in turn, improve the reliability of the modern grid.
- Fully enabled markets will provide data for smarter decisions about where to locate grid resources.

- Access by all consumers, wholesale and retail, to the electrical market will expand commerce for future services and products that support their needs for lower cost energy.

For more information

This document is part of a collection of documents prepared by The Modern Grid Initiative team. For a high-level overview of the modern grid, see "A Systems View of the Modern Grid." For additional background on the motivating factors for the modern grid, see "The Modern Grid Initiative." MGI has also prepared seven papers that support and supplement these overviews by detailing more specifics on each of the principal characteristics of the modern grid. This paper describes the sixth principal characteristic: "Enables markets."

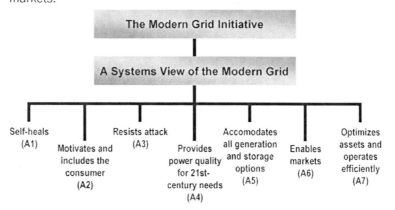

Documents are available for free download from the Modern Grid Web site.

The Modern Grid Initiative

Website: www.netl.doe.gov/moderngrid

Email: moderngrid@netl.doe.gov

(304) 599-4273 x101

BIBLIOGRAPHY

Baer, W., B. Fulton, and S. Mahnovski. 2004. Estimating the benefits of the GridWise initiative: Phase I report. Rand Corporation technical report, document no. TR-160-PNNL.

1. California Independent System Operator. 2006. Market redesign and technology upgrade tariff.

2. Widergren, S. 2005. GridWise™ architecture council interoperability path forward whitepaper v1.0.

3. Federal Energy Regulatory Commission. 2006. Rules concerning certification of the electric reliability organization; and procedures for the establishment, approval, and enforcement of electric reliability standards. 18 CFR Part 39, Docket No. RM05-30-000, Order No. 672.

4. Kelliher, J. T. 2006. Opening statement on Energy Policy Act (EPAct) of 2005. FERC, http://www.ferc.gov/press-room/statements/kelliher/2006/02-02-06-kelliher-epact.asp.

5. Midwest Independent Transmission System Operator. 2004. Midwest market fundamentals. Course presented at Midwest ISO's Market Participant Training, Carmel, IN.

6. North American Electric Reliability Council Control Area Criteria Task Force. 2001. The NERC functional model: Functions and relationships for interconnected systems.

7. PJM Interconnection. 2006. Markets. http://www.pjm.com/markets/markets.html

8. U.S. Department of Energy, Office of Electric Transmission and Distribution. 2003. "Grid 2030": A national vision for electricity's second 100 years.

9. Sutherland, R. 2003. Estimating the benefits of restructuring electricity markets: An application to the PJM region. Center for the Advancement of Energy Markets, Version 1.1.

10. Yeager, K. E. and C. W. Gellings. 2004. A bold vision for T&D. Paper presented at the Carnegie Mellon University Conference on Electricity Transmission in Deregulated Markets, Pittsburgh, PA.

11. "Wholesale Market Operations Combined Information from RTOs and ISOs." Energy Information Administration Official Energy Statistics from the U.S. Government. September 2005. <http://www.eia.doe.gov/cnef/electricity/wholesale/wholesale.html>.

Appendix A7:
A Systems View of the Modern Grid

OPTIMIZES ASSETS AND OPERATES EFFICIENTLY

Conducted by the National Energy Technology Laboratory
for the U.S. Department of Energy
Office of Electricity Delivery and Energy Reliability
January 2007

Office of Electricity
Delivery and Energy
Reliability

v2.0

TABLE OF CONTENTS

EXECUTIVE SUMMARY

The systems view of the modern grid features seven principal characteristics needed to achieve a modern grid. (See Figure 1.) One of those characteristics is *optimizes assets and operates efficiently.* How we might attain this characteristic and contribute to positive returns on investment is the subject of this paper.

- Self-heals
- Motivates and includes the consumer
- Resists attack
- Provides power quality for 21st-century needs
- Accommodates all generation and storage options
- Enables markets
- Optimizes assets and operates efficiently

Figure 1: The Modern Grid Systems View provides an "ecosystem" perspective that considers all aspects and all stakeholders.

Unlike today's grid, the *modern grid* will apply the latest technologies to optimize the use of its assets. For example, optimized capacity can be attainable with *dynamic ratings*, which allow assets to be used at greater loads by continuously sensing and rating their capacities.

Maintenance efficiency involves attaining a reliable state of equipment or "optimized condition". This state is attainable with *condition-based maintenance*, which signals the need for equipment maintenance at precisely the right time.

System control devices can be adjusted to lower losses and eliminate congestion. Operating efficiency is increased by selecting the least cost energy delivery system available through these adjustments of system control devices. Optimized capacity, optimized condition and optimized operations will result in substantial cost reductions.

In the modern grid, asset optimization does not mean that each asset will reach its maximum operating limit. Rather, it means that each asset will integrate well with all other assets to

maximize function while reducing cost. For example, load-sharing would routinely adjust the loads of transformers or lighten loads of transmission line sections.

Optimized maintenance will be possible when, for example, equipment monitors send a "wear" signal as part of a *predictive maintenance* regime or a direct malfunction signal in a *condition-based maintenance* regime.

Key technologies to be applied by the modern grid will provide the infrastructure, processes and devices to support both these examples of optimized asset utilization and maintenance, plus many more.

This paper explores how the modern grid would make such optimization possible. We address these important topics:

- The present and future of asset utilization and maintenance.
- Requirements of its implementation in the modern grid.
- Barriers between today's reality and modern grid requirements.
- Expected benefits.
- Recommendations to move forward.

Although it can be read on its own, this paper supports and supplements "A Systems View of the Modern Grid," an overview prepared by the Modern Grid Initiative (MGI) team.

CURRENT AND FUTURE STATES

Before we discuss how the modern grid will optimize assets and operate more efficiently, we need to understand this characteristic's current state and its future possibilities.

CURRENT STATE

In today's grid, the data systems to ascertain real-time asset utilization are not typically available. The current utilization rate of assets within a utility, if it is measured at all, is mostly limited to transformers and transmission lines. Both assets use a load divided by capacity calculation for percent loading. A larger percentage is considered better.

At a few larger utilities, the average utilization of transformers at distribution substations was measured at about 40%. At transmission substations, utilization of transformers was measured at about 50%. Using a dynamic rating to measure and adjust loading improved the utilization of both groups by about 6%.

Operators only know the condition of equipment when they perform maintenance or when failures occur. For example, after a maintenance overhaul, they assume that the equipment has been refurbished to an almost new condition. Unfortunately, data systems to support better assessments of equipment conditions are not common, and data mining tools and wear algorithms are rare.

Predictive maintenance, which applies condition-based algorithms to predict future failures and signal the need for maintenance, is not commonly used by utilities.

More commonly, equipment maintenance occurs on a regular time interval or sometimes by diagnostic testing. Regular time interval maintenance is called preventative maintenance, whereas diagnostic maintenance involves performing a health checkup with limited testing on a regular basis.

Diagnostic testing requires taking equipment out of service. Key components are tested and, if all tests show positive results, the equipment goes back in service. If problems are found, then corrective maintenance occurs prior to restoring the equipment.

Critical technology tools to optimize assets and their maintenance are not widely used. Such technologies include a Common Information Model (CIM) and Substation Automation (SA). Likewise,

Intelligent Electronic Devices (IED) may exist at Extra High Voltage (EHV) Substations, but presently have no widespread use.

Complicating optimized asset management and efficient operations is the lack of standard communication systems in the industry. Widespread communication is limited because common communication systems are not available throughout the utility service area. Common systems were originally not required and the expense to upgrade is often prohibitive.

For larger utilities, there is no consensus on how to measure cost of maintenance. Most have done some benchmarking, but the performance of maintenance is usually based on the previous year's actual expenditures. The data and systems do not exist to predict future maintenance expenses.

A great number of sensors are in the marketplace, usually targeted at the transformers and circuit breakers. Nevertheless, many equipment types remain without the sensors to gather needed data for wear algorithms to process.

Probabilistic Risk Assessments (PRA) are presently only used in the nuclear sector of the electric utility industry. Current research offers ways to show a PRA presentation of overloads, voltage violations, and voltage stability warnings for Regional Transmission Organizations (RTO). Grid operations would benefit greatly from this kind of information, but widespread acceptance and implementation has yet to occur.

FUTURE STATE

The future state of optimizing assets and operating efficiently would include the widespread installation of sensors to provide equipment condition in real-time. This information may be gathered as a direct reading, as with a vibration monitor or as a derived estimation using a wear algorithm. Automated analysis, such as comparing the wear to a threshold value, would signal an exceeded threshold to the asset manager. The asset manager would then perform maintenance, no sooner than necessary.

Using Common Information Model (CIM), Substation Automation (SA), and sensors with widespread communications enables "just-in-time" maintenance. (See Table 1.) These key technology tools help to accurately gather and transmit the required data to a processing center to develop an *equipment maintenance condition status*. Only those equipment units in immediate need of maintenance would have maintenance crews dispatched.

In operating the modern grid, optimization can occur when generation resources identify untapped capacity, thus avoiding the startup of more costly generator resources. Dynamic real-time data reveals when and where unused capacity is available. Finding and using that excess capacity avoids the cost of starting up more costly generation. The use of excess capacity also applies to transformers, transmission lines, and distribution lines. For example, avoiding the startup of a residentially placed Distributed Energy Resource (DER) on a cold winter night could be possible if the distribution system were capable of carrying the heavy load from the substation.

As modern grid sensors provide more data, asset planning is also optimized. The optimization in planning occurs in the selection and timing of the installation of new assets. Using the data from all grid sensors, planners can decide more economically when, where, what, and how to invest in modern grid improvements.

Technology Tools	Functions
Sensors/IEDs	Detect and measure conditions in near real--time to assess equipment.
Common Information Model (CIM)	Provides system-wide commonality of data used to measure the condition of equipment.
Widespread Communications	Provides system-wide exchange of data between equipment and asset managers.
Substation Automation (SA)	■ Provides the internal substation communications path as well as many of the IEDs required to effectively monitor the equipment. ■ Provides both a local and remote human-machine interface. ■ Provides monitoring (such as infrared imaging) of equipment for visible signs of health. ■ Scales up from substations to transmission and distribution facilities.

Table 1: Key technology tools provide important functions needed to optimize the use and maintenance of modern grid assets.

REQUIREMENTS

Moving from our current state to the future state of optimizing assets and operating efficiently requires employing technologies at several levels in the modern grid.

GATHERING AND DISTRIBUTING DATA

In the modern grid, the approach to asset utilization requires:

- **Gathering the data for processing by the optimization applications.**
- **Distributing the data widely in real-time.**

To accomplish these two functions, sensors must first be installed at the equipment and messages must be sent when defects or trends are discovered.

Asset managers would analyze high-risk areas and even individual pieces of equipment for immediate or contingency action. As new real-time information updates the model, operators of the grid will have advanced visualizations depicting congestion areas, probable equipment failures, and the potential consequences. Asset managers will be able to foresee the consequences of maintenance tasks (or lack of maintenance) and make informed business decisions.

Gathering useful data and distributing it where and when needed requires the use of these specific technology tools:

- **Sensors/Intelligent Electronic Devices**
- **Common Information Model**
- **Widespread Communications**
- **Substation Automation**

Sensors/Intelligent Electronic Devices (IED)

Wide varieties of sensors already exist for many of the equipment types employed by the modern grid. From a functional basis, the processing can be done locally or remotely, but the preference is to trend results and then to query in near real-time to assess equipment under investigation. Additional sensors will be required to fill in the gaps between what is known about the equipment and what needs to be known about the equipment. Such sensors may include monitors of vibration, chemical analysis, acoustics, temperature or any of the electrical parameters used in the delivery of electricity.

Common Information Model (CIM)

The CIM is a vital ingredient to the data collection from equipment. CIM will be the single most important data validation methodology in place because, by definition, it associates the equipment with the performance criteria to be measured. It thus enables validation of the equipment's quality parameters.

Widespread Communications

A communications path is required to get data and information from the equipment to the asset manager. Even Supervisory Control and Data Acquisition (SCADA) only covers about half of the substations, so large-scale communications will be required to implement this characteristic of the modern grid.

Substation Automation (SA)

Substation automation functionality must be extended to the distribution level. Predictive maintenance routines are greatly enabled by the implementation of a substation automation scheme. Applying SA technology more widely would allow monitoring more equipment and expand the base of power quality data.

SA would also provide both a local and remote human-machine interface. In addition, it will provide the ability to consolidate and prescreen data, thus reducing the data load on the communications system between the substation and the operations center.

Substation automation technologies would support wider use of remote cameras to help other elements of the modern grid resist attack. These same cameras can be used to view equipment for visible signs of health (such as infrared imaging and thermography). As asset optimization methods become more widely used, substation data will be sent to more control areas. As a result, the increased observations will reduce the time to estimate status of equipment throughout the grid.

LEVELS OF ASSET OPTIMIZATION

Asset optimization technologies must satisfy requirements at several levels of the electrical power system, namely:

- **Distribution level**
- **Operations level**
- **Regional Transmission Organization (RTO) and Control Area level**
- **Planning level**

Distribution

Asset optimization at the *distribution level* requires configuring circuits and operating capacities to minimize losses. Such software solutions already exist but are not common in the US.

Operations

Asset optimization in operations requires real-time dynamic ratings for both lines and transformers. While lines may have inconsistent temperature environments, average values may apply for the short term. Technologies are required that embed temperature sensors inside the conductor at regular intervals for a complete temperature profile of the line. Transformers should have a temperature sensor in the substation. Operating a transformer closer to its limit could save a re-dispatch of generation, thus increasing efficiency while increasing the utilization of the asset.

RTO and Control Areas

Asset optimization at the RTO and control area level requires data to be integrated to show the big picture. Loop flows, or the passing of energy through an area for use by others outside that area, may be avoided by an operating configuration only visible at the RTO level where multiple control area schemes are presented. That configuration may save costly upgrades, enable more economic dispatch, and enable greater asset utilization for all parties.

Planning

Planners need to know the options available to optimize asset loading. Maximum demand might be rarely needed. Conversely, it might be urgently needed during an emergency procedure. Having all the data showing different ways to optimize loads would provide planners with the details to manage assets more flexibly and effectively.

APPLICATIONS AND DEVICE TECHNOLOGY REQUIREMENTS

Key technologies applied by the modern grid will help close the gaps between the grid's current and future states. The modern grid infrastructure will be required to integrate both applications and device technologies.

Applications Technologies

Real-time dynamic rating - Real-time dynamic rating applications will allow existing assets to be used at greater loads under certain conditions. Since heat is a limiting factor in the operation of electrical equipment, heat-mitigating conditions such as cold weather or elevated wind may offer increased capacity during windows of opportunity.

Probabilistic Risk Assessment (PRA) - Probabilistic Risk Assessment (PRA) assists in *operations and maintenance* decisions because asset managers can know the *probability* of failure of their assets. However, unless they understand the *consequences* of that failure, the true risk is hidden. A PRA combines the probability of failure and the consequences of that failure to arrive at a real-risk factor.

PRA assists in *planning* decisions because planning managers often know the consequences of an asset failure, but few may know the probability of failure of their assets. As with maintenance and operations, a PRA combines the probability of failure and the consequences of that failure to arrive at a real-risk factor. Using true risk probabilities, the planner will have more information to arrive at more informed decisions.

Common Information Model (CIM) - The CIM ties the identification of equipment to its measurements. There are six different CIM categories: wires, SCADA, load, energy scheduling, generation, and finance. Each category has specialized formats to encompass the information being transmitted. CIM has been around for over a decade and its implementation in one or more categories is ongoing in most states. The requirement is that CIM be adopted as the industry standard and fully integrated into the modern grid.

Failure rate analysis - Failure rate analyses collect data about known equipment failures. Failure rate data includes the asset nameplate information, the failed component, the cause of failed component, utilization history, operating cycle, percent loading, environment, and location. The analyses by themselves can become a performance indicator and improvement tool.

Root-cause analysis of failure rates may provide insights to solutions to eliminate failure altogether. It requires a great deal of information to perform failure rate analyses. Filling the gaps between sensors, communications, and condition-based monitoring systems will enable this analysis to be automated and thus cost-effective.

Condition-based and predictive maintenance - Both condition-based maintenance and predictive-maintenance methodologies are dependent on meeting requirements for sensors and communications technologies. Software presently exists for detailed analysis and presentation applications. The solution lies in the implementation of monitoring technology, which will enable a just-in-time maintenance regime.

Application technologies and the solutions they provide are summarized in Table 2 below.

Application Technologies	Solutions
Real-time dynamic rating	Allows existing assets to be used at greater loads.
Probabilistic Risk Assessment (PRA)	Integrates Real-Time Contingency Analysis (RTCA) with the probability of failure determined by the condition-based monitoring system.
Advanced monitoring	Extension of SA functions to other grid elements.
Real-time visual observation	Remote-operated cameras in substations and other critical grid locations.
Sensors and IED	Usage expanded to equipment throughout the grid.
Communications	Wide communication connecting all substations and distribution switch locations.
Common Information Model (CIM)	Adoption as an industry standard.
Failure rate analysis	Applied throughout the grid by widespread sensors.
Condition-based and predictive maintenance	Monitoring technology and widespread communication will allow just-in-time maintenance.

Table 2: Filling the gaps between current and future states with modern grid solutions.

Device Technologies

Advanced monitoring technologies - Advanced monitoring, as applied in substation automation, could be integrated into other levels of the grid as well. Equipment state and parameters could then be viewed in real-time by other control elements of the grid such as central utility headquarters, distribution centers or even other substations.

Advanced monitoring technologies may also provide a needed solution to identifying the precursors to underground cable failure. This identification may lead to the execution of operating guides to reduce the effect of such failures on the grid. Advancements in Phasor Measurement Unit (PMU) technologies may offer real-time assessment of grid flows and help to determine whether system stability requires the opening of tie lines.

Real-time visual observation – Remotely operated cameras in substations and other grid critical locations can deter or warn of attack. Infrared scans can also offer real-time assessment of local heating of grid elements such as risers, connectors, and bushing terminals, as well as visual observations of transformer-cooling radiator operation.

Sensors & Intelligent Electronic Devices (IED) - The installation of IEDs in the substations provides extended protection as their primary function. They also provide a wealth of information available by the simple connection to a communications port. IEDs are common in EHV substations, but they have not yet been used at lower voltage class levels.

The grid can use a variety of existing sensors to implement many of the applications required to enable a full asset optimization program. These sensors include relay IEDs, oil pump monitors, vibration monitors, thermometers, pressure gauges, and specific gas detectors, to name just a few. There are some missing sensors in the sensor family, mostly due to cost or choice of maintenance policy. A low cost combustible gas analyzer would be an example of a sensor to yield real-time data.

Communications - Widespread communications infrastructure to all substations and distribution switch locations has not occurred. Communication types include telephone, fiber optics, microwave, cellular, and broadband over power line carrier (BPL).

PERFORMANCE STANDARDS

Asset Optimization has many ways to define performance. The most common ones measure reliability and economics. Some examples include:

Reliability

- CAIDI — A weighted CAIDI average of differing construction types (i.e. overhead, underground, network) and various penetration levels of modern grid elements to arrive at a uniform measure of local grid reliability.
- Forced Outage Rate — How often a power plant is out of service due to non-planned events.
- Equipment Failure Rate — How often equipment fails in service.

Economics

- Cost per delivered MW — O&M dollars spent to serve a megawatt.
- Cost per installed MW — Capital cost to serve a megawatt.
- Cost per MW transformed — O&M cost of transformers per MW.

BARRIERS

There are examples of making asset optimization and operating efficiencies work, yet they have not spread throughout the industry. The near-term payback and ROI do not compete well with alternate investment opportunities.

Barriers to full-scale implementation of needed applications and device technologies include risky investments, the existing outdated equipment base, and reluctance to change.

- **Weak Business Cases** – The quantity of required sensors and their attendant communications infrastructure demands large financial investment. By itself, the business case for achieving asset optimization is weak. On the bright side, there have been many pilot programs implementing partial solutions confined to small demonstration areas. When the sensors and communications infrastructure are integrated with the other characteristics of the modern grid, the benefits increase from the addition of the self healing, power quality and enabling market benefits, resulting in a viable business case.

- **Incompatible Equipment** – Some of the older equipment types do not have the embedded sensors to enable a predictive methodology, and in some cases, it may not be cost-effective to install it. With a mix of some automated maintenance equipment and some manual maintenance equipment, the benefits are diluted, further weakening any business case.

- **Reluctance to Change Processes** – It will be a hard sell to change long standing maintenance practices and beliefs. Some equipment types may require equipment modeling research and development while a great many equipment types will surely require improvement and refining of the wear models, predictive analytics and dynamic rating algorithms currently in use. While automation will replace many human labor tasks, it will still require process change to achieve success.

BENEFITS

The benefits of optimizing assets and operating efficiently can be viewed in the context of the modern grid's key success factors.

Optimizing asset utilization and introducing efficiencies in operation and maintenance will have a positive impact on key success factors of the modern grid. (See Table 3.)

Key Success Factors	Impact of Optimizing Assets & Operating Efficiently
Reliable	Advanced Monitoring Components provide data for: • Dynamic equipment ratings. • Level-loading assets. • Predictive maintenance.
Secure	Substation Automation provides remotely monitored real-time surveillance devices that can also be used in other parts of the grid.
Economic	Advanced sensors and software provide: • Failure rate analysis and Root Cause Analysis for performance improvement. • Opportunities for greater power densities in existing assets.
Efficient	The combined improvements in asset utilization, operations and maintenance help to: • Retire inefficient equipment. • Put more energy through current assets. • Obtain higher reliability of equipment with a lower cost of maintenance
Environmentally Friendly	Decrease pollutants through the prediction and removal of potential sources of toxic leaks and spills
Safe	Reducing the exposure to accident through reduced maintenance effort

Table 3: Asset optimization and reduced maintenance will have a positive impact on each key success factor of the modern grid.

RELIABLE

The use of advanced monitoring technologies will provide the information for asset management programs to realize substantial cost savings through improvements in reliability. The detailed awareness of component and equipment condition reduces human errors in performing maintenance, and helps to avoid outages for unnecessary maintenance.

Advanced monitoring technologies will:

- Allow for dynamic (continuous) equipment ratings, enabling greater use of existing assets.
- Enable better level-loading of assets by removing overloads and high stress conditions through the increased use of other assets.
- Attain some improvements in CAIDI because those monitoring technologies integrate with a communications path and process algorithms to enable the condition-based and predictive maintenance regimes.

SECURE

A security benefit resides in Substation Automation with the installation of low cost remote cameras that can monitor equipment. This function can be scaled to other grid elements.

ECONOMIC

Asset optimization has the opportunity for gaining economic benefits because greater power densities can be attained using the same existing assets. In an energy market, this increase in utilization increases revenue for the asset owner, and at the same time lowers energy costs for load-serving entities. This win-win scenario offers economic incentive to both asset owners and users of energy.

EFFICIENT

Modern grid assets will remain in service longer and have a lower maintenance profile, thus attaining a higher asset utilization. Asset utilization measures performance, loads carried, and maintenance costs. Heavily loaded assets that perform for long times with little maintenance are best utilized.

While there is no known goal for standard cost per equipment item, an obvious benchmark is the "best-in-class" performer. Drill-down analyses of all other performers will further aid in the asset optimization process because economic analyses can identify previously undiscovered improvement opportunities.

ENVIRONMENTALLY FRIENDLY

As asset utilization optimizes, environmental quality becomes easier to maintain and improve. For example, leaks from sulphur hexafluoride equipment decrease as do leaks from oil-filled equipment. Avoiding spills and harmful emissions from inefficient equipment contributes to a cleaner environment.

SAFE

Optimizing maintenance results in performing less maintenance. Less maintenance work equates to less exposure to accident and thus increased safety to maintenance personnel.

RECOMMENDATIONS

The optimization of modern grid assets, like other principal characteristics of the modern grid, will depend on motivating investors, thorough planning, and changing the way the industry performs its operations.

Successful pilot programs and demonstrations in the past indicate some fundamental actions needed to pave the way for the long-term benefits of asset optimization and efficient operations.

1. **Develop investment criteria that acknowledge the systems view of the modern grid and its long-term benefits to the national economy.** Business cases that are based on returns to local providers or distributors are too narrow in scope and often too short term in duration.
2. **Provide incentives to replace older equipment with equipment designed to operate in a 21st century communication and information system infrastructure.** Too many necessary upgrades are proven non-cost effective due to high costs of purchase.
3. **Replace Equipment with 'no maintenance' versions.** Where equipment replacement is warranted because of depreciation and obsolescence, asset managers should consider 'maintenance elimination'. In this final maintenance process, high maintenance equipment is replaced by equipment that requires little to no maintenance at all.
4. **Install Sensors to Assess Equipment.** Until the utility environment is populated with self-maintained equipment, there will be a need to develop a myriad of low cost sensors that monitor equipment/system health. While vendors may invest in the research to develop such devices, utilities might also invest in common research with regulator approval.
5. **Begin training in the industry that will prepare operators to use the new maintenance tools available.** Business-as-usual operations will not exploit the technologies needed to optimize and efficiently maintain modern grid assets.
6. **Adopt the Common Information Model (CIM)** - The CIM ties the identification of equipment to its measurements. This validates the data source thus eliminating the source of many present day errors. The recommendation is that CIM be adopted as an industry standard and fully integrated into the modern grid.

SUMMARY

Whether optimizing assets or operating efficiently, information is the key. Gathering that information is one challenge for the modern grid, communicating it widely is a second, and processing it usefully is a third.

Gathering the information is the function of sensors and Intelligent Electronic Devices (IED) comprising advanced monitoring technologies. Sensing and measurement technologies of the type employed in substation automation must extend their reach in the modern grid to other control elements of the electrical power system, including transmission and distribution.

Communicating the information requires both a common language such as the Common Information Model (CIM) and a widespread communications infrastructure. A common suite of advanced communications components and protocols will characterize the modern grid solution to this need.

Processing information usefully is the domain of software applications for predictive and condition-based maintenance as well as dynamic load ratings for efficient asset utilization and operations. Though these applications exist today, they are effectively starved for usable information from most elements of the electrical power system.

To meet these three challenges, the modern grid will apply elements of its five key technology areas as presented in the systems view:

1. Integrated communications.
2. Sensing and measurement technologies.
3. Advanced components.
4. Advanced control methods.
5. Improved interfaces and decision support.

By integrating these types of current and future technology solutions for asset utilization and maintenance, information will be better gathered, better communicated, and better processed to achieve this important characteristic of the modern grid.

For more information

This document is part of a collection of documents prepared by The Modern Grid Initiative team. For a high-level overview of the modern grid, see "A Systems View of the Modern Grid." For additional background on the motivating factors for the modern grid, see "The

Modern Grid Initiative." MGI has also prepared seven papers that support and supplement these overviews by detailing more specifics on each of the principal characteristics of the modern grid. This paper describes the seventh principal characteristic: "Optimizes assets and operates efficiently."

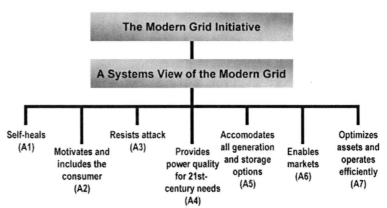

Documents are available for free download from the Modern Grid Web site.

The Modern Grid Initiative

Website: www.netl.doe.gov/moderngrid

Email: moderngrid@netl.doe.gov

(304) 599-4273 x101

BIBLIOGRAPHY

1. Clark, H. K. 2004. It's time to challenge conventional wisdom. *Transmission & Distribution World* (October 1). http://tdworld.com/issue_20041001/.

2. Electric Power Research Institute. 2002. Performance-focused maintenance. Palo Alto, CA: EPRI. Product ID: 100825.

3. Federal Energy Regulatory Commission. 2006. Rules concerning certification of the electric reliability organization; and procedures for the establishment, approval, and enforcement of electric reliability standards. 18 CFR Part 39, Docket No. RM05-30-000, Order No. 672.

4. Houbaer, R. and K. Gray. 2000. Power transformer asset management. Paper presented at APWA International Public Works Congress NRCC/CPWA "Innovations in Urban Infrastructure."

5. Humphrey, B. 2003. Asset management, in theory and practice. *Energy Pulse* (June 26), http://www.energypulse.net/centers/article/article_email.cfm?a_id=386.

6. McDonald, J. 2003. Successful integration and automation relies on a strategic plan: Automation requires integration. *Electric Energy T&D* (January/February), http://www.electricenergyonline.com/article.asp?m=9&mag=11&article=76.

7. Robinson, C. J. and A. P. Grinder. 1995. *Implementing TPM: The North American experience.* University Park, IL: Productivity Press.

8. Terez, T. J. 1990. Managing change in the 1990s: Strategies for the operations manager. Arrow Associates.

9. Vanier, D. J. 2000. Advanced asset management: Tools and techniques. Paper presented at APWA International Public Works Congress NRCC/CPWA "Innovations in Urban Infrastructure."

10. Weaver, C. N. 1995. Managing the four stages of TQM: How to achieve excellent performance.

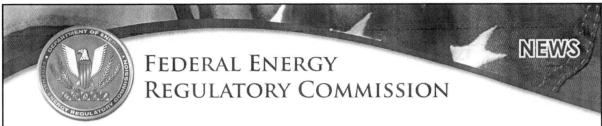

FEDERAL ENERGY REGULATORY COMMISSION

July 16, 2009
Docket No. PL09-4-000

NEWS MEDIA CONTACT
Mary O'Driscoll - 202.502.8680

FERC Adopts Policy to Accelerate Development of Smart Grid

The Federal Energy Regulatory Commission (FERC) took a major step today to accelerate the development of a smart electric transmission system that could provide long-term savings for consumers by improving the efficiency and operation of the grid. The Smart Grid Policy Statement sets priorities for work on development of standards crucial to a reliable and smart grid.

Smart grid advancements will apply digital technologies to the grid, enabling two-way communications and real-time coordination of information from both generating plants and demand-side resources. This will improve the efficiency of the bulk-power system with the goal of achieving long-term savings for consumers. And it will help promote wider use of demand response and other activities that give consumers the tools they need to control electricity costs.

The final policy issued today closely tracks the proposed policy issued March 19. It sets priorities to guide industry in development of smart grid standards for achieving interoperability and functionality of smart grid systems and devices. It also sets out FERC policy for recovery of costs by utilities that act early to adopt smart grid technologies. More than 70 sets of comments from interested groups indicated broad support for the proposed policy.

"Changes in how we produce, deliver and consume electricity will require 'smarter' bulk power systems with secure, reliable communications capabilities to deliver long-term savings for consumers," FERC Chairman Jon Wellinghoff said. "Our new smart grid policy looks at the big picture by establishing priorities for development of smart grid standards, while giving utilities that take the crucial early steps to invest in smart grid technologies needed assurance about cost recovery."

"The smart grid policy provides a roadmap that will guide the transformation of the old grid into the grid of the future, while providing for fair regulatory treatment to consumers and utilities," Commissioner Suedeen G. Kelly said.

"It's our responsibility to help protect the security and reliability of the nation's electric grid by adopting effective cyber-security standards for the smart grid," Commissioner Philip D. Moeller said. "If we do that right, consumers can look forward to exciting new products and services from a smarter, safer and more efficient grid."

"The policy statement provides important guidance to focus and expedite ongoing industry efforts to develop interoperability standards – this will enable entrepreneurs to deploy new market based technologies to improve efficiency and reliability," Commissioner Marc Spitzer said. "Equally important, this policy statement is a step toward smarter rates that will enable customers to control their personal use of electricity."

The new policy adopts as a Commission priority the early development by industry of smart grid standards to:

 FEDERAL ENERGY REGULATORY COMMISSION
WASHINGTON, DC 20426

WWW.FERC.GOV

- Ensure the cybersecurity of the grid;
- Provide two-way communications among regional market operators, utilities, service providers and consumers;
- Ensure that power system operators have equipment that allows them to operate reliably by monitoring their own systems as well as neighboring systems that affect them; and
- Coordinate the integration into the power system of emerging technologies such as renewable resources, demand response resources, electricity storage facilities and electric transportation systems.

The policy also provides for early adopters of smart grid technologies to recover smart grid costs if they demonstrate that those costs serve to protect cybersecurity and reliability of the electric system, and have the ability to be upgraded, among other requirements.

Importantly, the policy statement also explains that by adopting these standards for smart grid technologies, FERC will not interfere with any state's ability to adopt whatever advanced metering or demand response program it chooses. In adopting this policy, FERC continues to abide by the Federal Power Act's jurisdictional boundaries between federal and state regulation of rates, terms and conditions of transmission service and sales of electricity.

The policy will take effect 60 days after publication in the *Federal Register*.

-30-

R-09-25

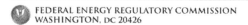

FEDERAL ENERGY REGULATORY COMMISSION
WASHINGTON, DC 20426

WWW.FERC.GOV

March 19, 2009
Docket No. PL09-4-000

MEDIA CONTACT
Mary O'Driscoll – 202.502.8680

P□□□□□□□□□□□□□0□Statement and Action Plan

The Federal Energy Regulatory Commission's proposed policy statement and action plan provides guidance to inform the development of a smarter grid for the nation's electric transmission system. It focuses on:

- Prioritizing the development of key standards for interoperability of Smart Grid devices and systems;
- A proposed rate policy for the interim before the standards are developed for Smart Grid investments that demonstrate system security and compliance with Reliability Standards, the ability to be upgraded, and other criteria.

Smart Grid advancements will apply digital technologies to the grid, and enable real-time coordination of information from both generating plants and demand-side resources. This will improve the efficiency of the bulk-power system, with the goal of achieving long-term consumer savings, and will enable demand response and other consumer transactions and activities that give consumers the tools to control their electricity costs.

The Energy Independence and Security Act of 2007 (EISA) requires FERC, once sufficient consensus has been achieved through a process coordinated by the National Institute of Standards and Technology (NIST), to adopt standards and protocols necessary to ensure smart-grid functionality and interoperability in the interstate transmission of electric power and in regional and wholesale markets.

The proposed policy statement and action plan seeks to provide guidance on standards for four important elements of the Smart Grid:

1. **Cyber Security:** Advise NIST to assure each standard and protocol is consistent with the overarching cyber security and reliability requirements of EISA and FERC Reliability Standards.
2. **Inter-System Communications:** Identify standards for common information models for communication among all elements of the bulk power system – regional market operators, utilities, demand response aggregators and customers.
3. **"Wide-Area Situational Awareness":** Ensure that operators of the nation's bulk power system have the equipment that gives them a complete view of their systems so they can monitor and operate their systems.
4. **Coordination of the bulk power systems with new and emerging technologies:** Identify standards development that would help accommodate the introduction and expansion of renewable resources, demand response and electricity storage to help address several bulk power system challenges. Also identify standards development that could help accommodate another emerging technology, electric transportation.

For more information, go to http://www.ferc.gov/industries/electric/indus-act/smart-grid.asp. Comments on the proposed policy statement and action plan are due 45 days after publication in the *Federal Register*.

FEDERAL ENERGY REGULATORY COMMISSION
WASHINGTON, DC 20426

WWW.FERC.GOV

126 FERC ¶ 61,253
UNITED STATES OF AMERICA
FEDERAL ENERGY REGULATORY COMMISSION

18 CFR Part Chapter I

[Docket No. PL09-4-000]

SMART GRID POLICY

(Issued March 19, 2009)

AGENCY: Federal Energy Regulatory Commission.

ACTION: Proposed Policy Statement and Action Plan.

SUMMARY: This proposed policy statement and action plan provides guidance to inform the development of a smarter grid for the Nation's electric transmission system focusing on the development of key standards to achieve interoperability of smart grid devices and systems. The Commission also proposes a rate policy for the interim period until interoperability standards are adopted. Smart grid investments that demonstrate system security and compliance with Commission-approved Reliability Standards, the ability to be upgraded, and other specified criteria will be eligible for timely rate recovery and other rate treatments. This rate policy will encourage development of smart grid systems.

DATES: Comments on the proposed policy statement and action plan are due

[Insert_Date 45 days after publication in the **FEDERAL REGISTER**]

Docket No. PL09-4-000 - 2 -

FOR FURTHER INFORMATION CONTACT:

David Andrejcak
Office of Electric Reliability
888 First Street, N.E.
Washington, D.C. 20426
(202) 502-6721
david.andrejcak@ferc.gov

Elizabeth H. Arnold
Office of General Counsel
888 First Street, N.E.
Washington, D.C. 20426
(202) 502-8818
elizabeth.arnold@ferc.gov

Ray Palmer
Office of Energy Market Regulation
888 First Street, N.E.
Washington, D.C. 20426
(202) 502-6569
ray.palmer@ferc.gov

SUPPLEMENTARY INFORMATION:

126 FERC ¶ 61,253
UNITED STATES OF AMERICA
FEDERAL ENERGY REGULATORY COMMISSION

Before Commissioners: Jon Wellinghoff, Acting Chairman;
Suedeen G. Kelly, Marc Spitzer,
and Philip D. Moeller.

Smart Grid Policy   Docket No.   PL09-4-000

PROPOSED POLICY STATEMENT AND ACTION PLAN

(Issued March 19, 2009)

1.   The Commission is issuing this proposed policy statement to articulate its policies

and near-term priorities to help achieve the modernization of the Nation's electric

transmission system, one aspect of which is "Smart Grid" development. Smart Grid

advancements will apply digital technologies to the grid, and enable real-time

coordination of information from generation supply resources, demand resources,[1] and

distributed energy resources (DER).[2] This will bring new efficiencies to the electric

system through improved communication and coordination between utilities and with the

grid, which will translate into savings in the provision of electric service. Ultimately the

---

[1] For purposes of this proposed policy statement, "demand resources" refers to the set of demand response resources and energy efficiency resources and programs that can be used to reduce demand or reduce electricity demand growth.

[2] DER comprises dispersed generation devices and dispersed storage devices, including reciprocating engines, fuel cells, microturbines, photovoltaics, combined heat and power, and energy storage. See International Electrotechnical Commission, International Standards IEC 61850-7-420.

Docket No. PL09-4-000   - 2 -

smart grid will facilitate consumer transactions and allow consumers to better manage

their electric energy costs. These technologies will also enhance the ability to ensure the

reliability of the bulk-power system. The Commission's interest and responsibilities in

this area derive from its authority over the rates, terms and conditions of transmission and

wholesale sales in interstate commerce, its responsibility for approving and enforcing

mandatory reliability standards for the bulk-power system in the United States, and a

recently enacted law[3] requiring the Commission to adopt interoperability standards and

protocols necessary to ensure smart-grid functionality and interoperability in the

interstate transmission of electric power and in regional and wholesale electricity

markets. The development and implementation of these interoperability standards is a

challenging task, which requires the efforts of industry, the states and other federal

agencies, in addition to the Commission. The Commission intends to use its authority, in

coordination and cooperation with other governmental entities, to help achieve

interoperability in a timely manner. Achievement of interoperability will not only

increase the efficiency of the bulk-power system, with the goal of achieving long-term

consumer savings, but will also enable demand response and other consumer transactions

and activities that give consumers the tools to better control their electric energy costs.

Reaching this goal will also help promote the integration of significant new renewable

---

[3] Energy Independence and Security Act of 2007, Pub. L. No. 110-140, 121 Stat. 1492 (2007) (EISA).

Docket No. PL09-4-000   - 3 -

power into the transmission system and help state and federal initiatives to promote

greater reliance on renewable power and meet future demand growth to satisfy the

Nation's energy needs.

2.   The purpose of the policy statement the Commission ultimately adopts will be to

prioritize the development of key interoperability standards, provide guidance to the

electric industry regarding the need for full cybersecurity for Smart Grid projects, and

provide an interim rate policy under which jurisdictional public utilities may seek to

recover the costs of Smart Grid deployments before relevant standards are adopted

through a Commission rulemaking. Specifically, development of interoperability

standards for inter-system communication, system security, wide-area situational

awareness, demand response, electric storage, and electric transportation should be

prioritized and accelerated. The work done on certain standards will provide a

foundation for development of many other standards.

3.   In addition, as further explained below, for the near term we propose certain rate

treatments to encourage investment in Smart Grid technologies that advance efficiency,

security, reliability and interoperability in order to address potential challenges to the

bulk-power system. We recognize that a key consideration of public utilities in deciding

whether to invest in Smart Grid technologies may involve the potential for stranded costs

associated with legacy systems that are replaced by Smart Grid equipment. Additionally,

as the electric system may require several of the new capabilities of the Smart Grid

Docket No. PL09-4-000   - 4 -

before interoperability standards have been developed, we recognize the need for guidance for jurisdictional entities. Thus, to offer some rate certainty and guidance regarding cost recovery issues, the Commission is proposing a rate policy for the interim period until final interoperability standards are adopted. The Commission also proposes that smart grid investments that demonstrate system security and compliance with Commission-approved Reliability Standards, the ability to be upgraded, and other specified criteria will be eligible for timely rate recovery and other rate treatments. For now, we propose as an interim rate policy to accept single-issue rate filings submitted under FPA section 205 by public utilities to recover the costs of Smart Grid deployments involving jurisdictional facilities provided that certain showings are made. In other words, we propose to consider Smart Grid devices and equipment, including those used in a Smart Grid pilot program or demonstration project, to be used and useful for purposes of cost recovery if an applicant makes the certain showings, as described below.

4.   We seek comments from the industry on these and other steps the Commission can take to encourage and expedite the development of interoperability standards and implementation of Smart Grid projects. In the near future, we may convene a technical conference for further public input on these issues.

I.   <u>Background</u>

5.   Under the Federal Power Act (FPA), the Commission has jurisdiction over the transmission of electric energy in interstate commerce by public utilities, and over the

Docket No. PL09-4-000   - 5 -

reliable operation of the bulk-power system in most of the Nation.[4] The Commission

also was given a new responsibility under the EISA, discussed further below, to issue a

rulemaking to adopt standards and protocols to ensure Smart Grid functionality and

interoperability in interstate transmission of electric power and in regional and wholesale

electric markets.

6.   Section 1301 of the EISA states that it is the policy of the United States to support

the modernization of the Nation's electricity transmission and distribution system to

maintain a reliable and secure electricity infrastructure that can meet future demand

growth and to achieve each of several goals and characteristics, which together

characterize a Smart Grid.[5]   These goals and characteristics are:

> (1) Increased use of digital information and controls
> technology to improve reliability, security, and efficiency of
> the electric grid. (2) Dynamic optimization of grid operations
> and resources, with full cyber-security. (3) Deployment and
> integration of distributed resources and generation, including
> renewable resources. (4) Development and incorporation of
> demand response, demand-side resources, and energy
> efficiency resources. (5) Deployment of "smart" technologies
> (real-time, automated, interactive technologies that optimize
> the physical operation of appliances and consumer devices)
> for metering, communications concerning grid operations and
> status, and distribution automation. (6) Integration of "smart"
> appliances and consumer devices. (7) Deployment and
> integration of advanced electricity storage and peak-shaving

---

[4] 16 U.S.C. 824, 824o.

[5] EISA sec. 1301, to be codified at 15 U.S.C. 17381.

Docket No. PL09-4-000 - 6 -

technologies, including plug-in electric and hybrid electric vehicles, and thermal storage air conditioning. (8) Provision to consumers of timely information and control options. (9) Development of standards for communication and interoperability of appliances and equipment connected to the electric grid, including the infrastructure serving the grid. (10) Identification and lowering of unreasonable or unnecessary barriers to adoption of smart grid technologies, practices, and services.[6]

7.   Section 1305(a) of EISA directs the National Institute of Standards and

Technology (the Institute) ". . . to coordinate the development of a framework that

includes protocols and model standards for information management to achieve

interoperability of smart grid devices and systems."[7]   A helpful description of

interoperability is "the ability of a system or a product to work with other systems or

products without special effort on the part of the customer . . ."[8]   In order to achieve the

---

[6] Id.

[7] EISA sec. 1305(a), to be codified at 15 U.S.C. 17385(a).

[8] Testimony of Patrick D. Gallagher, Ph.D., Deputy Director, National Institute of Standards and Technology, before the Committee on Energy and Natural Resources, United States Senate, March 3, 2009, available at: http://www.nist.gov/director/ocla/nist%20pgallagher%20smart%20grid%20testimony%20senate%20e&nr%203-3-09.pdf. According to the GridWise Architecture Council, the term "interoperability" refers to the ability to: (1) exchange meaningful, actionable information between two or more systems across organizational boundaries; (2) assure a shared meaning of the exchanged information; (3) achieve an agreed expectation for the response to the information exchange; and (4) maintain the requisite quality of service in information exchange (i.e., reliability, accuracy, security). See GridWise Architecture Council, Interoperability Path Forward Whitepaper at 1-2, 2005, available at: http://www.gridwiseac.org/pdfs/interoperability_path_whitepaper_v1_0.pdf. The GridWise Architecture Council was formed by the U.S. Department of Energy to

(continued...)

Docket No. PL09-4-000   - 7 -

Smart Grid characteristics and functions listed in EISA section 1301, interoperability of Smart Grid equipment will be essential.

8.   Finally, pursuant to the EISA, once the Commission is satisfied that the Institute's work has led to "sufficient consensus" on interoperability standards, we are directed to "institute a rulemaking proceeding to adopt such standards and protocols as may be necessary to insure smart-grid functionality and interoperability in interstate transmission of electric power, and regional and wholesale electricity markets."[9]

9.   The Commission appreciates the Institute's work to assess current Smart Grid standards and infrastructure to identify gaps, and is aware of its plans to create a knowledge base to enable effective communication among stakeholders and a roadmap to lay out a recommended course toward a highly interoperable grid.[10] In general, we expect that the Institute will recommend standards to the Commission that have resulted from the Institute's coordination with standards development organizations and technical experts. The Commission will initiate rulemakings as individual or suites of standards[11] achieve sufficient consensus. The Commission will consider the most effective and

---

promote and enable interoperability among the many entities that interact with the Nation's electric power system. See http://www.gridwiseac.org/about/mission.aspx.

[9] EISA sec. 1305(d), to be codified at 15 U.S.C. 17385(d).

[10] See Testimony of Patrick D. Gallagher, Ph.D., infra n.8.

[11] A suite of standards would consist of a group of related standards.

Docket No. PL09-4-000   - 8 -

efficient ways to interact with the Institute and standards development organizations

between the issuance of a notice of proposed rulemaking on submitted standards and a

final rule adopting standards. We invite comment on this proposed approach.

10.   The Commission will continue to take an active role in helping to ensure that the

participants in the Institute's process effectively prioritize and sequence future standards

development efforts. We invite comments on what factors the Commission should

consider in determining when the Institute's work has led to "sufficient consensus" on

interoperability standards to warrant instituting a rulemaking proceeding. We also seek

comment and ideas on how to identify and stage the adoption of successive waves of

interoperability standards. Finally, we seek comment as to whether there should be some

formal process for parties to seek Commission guidance if negotiations on certain

interoperability standards reach an impasse.

II.   Discussion

A.   Urgency of Achieving Certain Smart Grid Functionalities

11.   As noted above, rather than directing the Institute to develop interoperability

standards of its own, Congress charged the Institute with coordinating such development.

The EISA specifically requires the Institute to solicit input from, among others, a range of

Docket No. PL09-4-000   -9-

existing standards development organizations that rely on extensive negotiation in order

to achieve broad industry consensus on proposed standards.[12]

12.   The EISA contains no specific deadline for the creation of interoperability

standards; instead, it provides for a consensus-based process. However, there is a sense

of urgency within industry and government for the development of standards for and

deployment of smart grid technologies generally. The Commission is particularly

interested in the development of Smart Grid functions and characteristics that can help

address challenges to the Commission-jurisdictional bulk-power system. These include

the cross cutting issues of cybersecurity and the further development of common

information models to allow useful exchange of electric system information (e.g.,

standard definitions).   Broad policy goals also need to be addressed such as optimizing

the transmission system to reduce congestion and improve reliability, security and

efficiency; encouraging increased reliance on demand response; state and possibly

national climate change initiatives such as Renewable Portfolio Standards and other

efforts that result in increased reliance on variable renewable resources; and the potential

---

[12] The EISA specifically names the IEEE (formerly known as the Institute of
Electrical and Electronics Engineers), and the National Electrical Manufacturers
Association. Other relevant existing standards development organizations could include
the International Electrotechnical Commission (IEC), the American National Standards
Institute (ANSI), the German Standards Institute (actually Deutsches Institut für
Normung), the International Organization for Standardization, and the International
Telecommunication Union.

Docket No. PL09-4-000   - 10 -

for increased and variable electricity loads from the transportation sector. We discuss in

turn the importance of each of these in driving the need for Smart Grid capabilities and

the standards to achieve interoperability of smart grid devices with the electric grid and

its associated users and infrastructure.

Cybersecurity and reliability

13.   Absent any consideration of the Smart Grid concept, other activities and events

currently taking place in various regions raise physical and cybersecurity concerns for the

electric industry. For example, utilities have already taken advantage of the existing

communications infrastructure and capabilities of the Internet to aid their marketing

operations. While typically not connecting their more sensitive control center systems

directly to the Internet, many entities have nevertheless upgraded those systems to use

Internet-based protocols and technologies. This, coupled with the fact that the non-

Internet-connected control center operations may be connected to the same corporate

network as the Internet-connected marketing systems, means that there may be an

indirect Internet vulnerability to those sensitive control systems. Accordingly, without

adequate protections, these preexisting utility efforts potentially increase the exposure of

the bulk-power system to cybersecurity threats. Cybersecurity and physical security have

been ongoing concerns for the Commission and the electric industry with the advent of

the mandatory and enforceable federal bulk-power system reliability regime in place in

most of the United States under the oversight of the Commission pursuant to FPA section

Docket No. PL09-4-000   - 11 -

215.[13] Pursuant to this section 215 authority, the Commission recently approved eight

cyber and physical protection related reliability standards.[14]

14.   The fact that a smarter grid would permit two-way communication between the

electric system and a much larger number of devices located outside of controlled utility

environments commands that even more attention be given to the development of

cybersecurity standards. Therefore, the Commission proposes to advise the Institute to

undertake the necessary steps to assure that each standard and protocol that is developed

as part of the Institute's interoperability framework is consistent with the overarching

cybersecurity and reliability mandates of the EISA as well as existing reliability standards

approved by the Commission pursuant to section 215 of the FPA. The Commission

proposes to make consistency with cybersecurity and reliability standards a precondition

to its adoption of Smart Grid standards. We seek comment on these proposals.

15.   In order to fully incorporate measures to protect against cyber and physical

security threats, we also propose to advise the Institute to take the necessary steps to

assure that its process for the development of any interoperability standards and protocols

---

[13] 16 U.S.C. 824o.

[14] See Mandatory Reliability Standards for Critical Infrastructure Protection, Order
No. 706, 73 FR 7368 (Feb. 7, 2008), 122 FERC ¶ 61,040, reh'g denied and clarification
granted, Order No. 706-A, 123 FERC ¶ 61,174 (2008). Notably, section 215(a) of the
FPA, 16 U.S.C. 824o(a), defines the terms "reliability standard," "reliable operation," and
"cybersecurity incident."

Docket No. PL09-4-000 - 12 -

leaves no gaps in cyber or physical security unfilled. We are concerned that this could be a particular problem where separate groups of interested industry members independently develop and advocate select standards or protocols for the Institute's consideration. We seek comment on this proposal.

Inter-system communication and coordination

16. There is an urgent need to further develop a common semantic framework (i.e., agreement as to meaning) and software models for enabling effective communication and coordination across inter-system interfaces. Such standards could play an important role in the movement to a smarter grid that is capable of addressing challenges to the operation of the bulk-power system. The bulk-power system can be thought of as a system of systems.[15] In order to enable a smarter grid, particularly one capable of addressing the bulk-power system challenges discussed below, effective interfaces must be developed between and among all of these systems (i.e., inter-system interfaces) and common information model standards appear to be powerful tools to enable such inter-system interfaces. The Commission proposes to identify standards for common information models for inter-system interfaces as a high priority for accelerated development. We seek comment on this proposal.

Integrating renewable resources into the electric grid

---

[15] See Appendix A for a graphic representation of the various systems.

Docket No. PL09-4-000   - 13 -

17.   Several groups of states have been working on aggressive regional carbon control

measures,[16] and one regional effort has already begun operation in the form of the

Regional Greenhouse Gas Initiative.[17] Federal legislation addressing carbon control and

other environmental and climate related matters may follow. These initiatives point

toward a shift in the mix of fuels that will be used to generate electricity, and an

associated shift in where new generation resources are located. Additional transmission

capacity to ensure deliverability of those new generating resources will be needed in the

form of new transmission lines and more efficient use of existing infrastructure. Also,

additional demand resources, generation resources, and DER will be needed to reliably

integrate variable generation into the electric grid. Efforts to address these challenges

could benefit from the enhanced capabilities associated with certain aspects of the Smart

Grid; among them, the ability to maximize the capability and use of existing and new

transmission capacity,[18] and foster the deployment and integration of demand resources,

generation resources and DER.

---

[16] See, e.g., the Western Climate Initiative
(http://www.westernclimateinitiative.org/) and the Midwestern Greenhouse Gas
Reduction Accord (http://www.midwesternaccord.org/).

[17] See http://www.rggi.org/home.

[18] For example, a smarter grid could enable an increase in transmission capacity
through a switch from static to dynamic transmission line ratings enabled by the
advanced sensor, communications, and information technology capabilities associated
with a smarter grid.

Docket No. PL09-4-000   - 14 -

18.   As of December 2008, the Nation had 25,170 MW of wind generation based on

nameplate capacity.[19] According to the 2008 Long-Term Reliability Assessment by the

North American Electric Reliability Corporation (NERC), an additional 145,000 MW of

wind power projects are planned or proposed over the next ten years.[20] Accordingly, it is

evident that in a relatively short period of time, some parts of the bulk-power system may

face the need to effectively integrate unprecedented amounts of variable generation

resources. This is significant because operators of variable generation have less control

over when the resource is available to produce electricity, in contrast with more

conventional fossil and nuclear generation.

19.   Large amounts of variable generation raise several important operational and

planning issues, including: (1) resource adequacy (potential loss and unavailability of

variable resources at peak periods and other critical times such as loss of other generators

or transmission lines); (2) resource management (potential for over-generation by

variable resources during off-peak periods when there is insufficient load to

accommodate such generation); and (3) reduced system inertia (potential loss of system

---

[19] Source, American Wind Energy Association's website:
http://www.awea.org/projects/.

[20] North American Electric Reliability Corporation, 2008 Long-Term Reliability
Assessment at 12.

Docket No. PL09-4-000   - 15 -

stability due to the high penetration of variable resources with low inertia properties).[21]

Given sufficient time and resources, a variety of solutions to these concerns may be

feasible. For example, investment in large amounts of electricity storage could ultimately

address both the resource adequacy and resource management concerns,[22] although

technical and economic issues remain to be addressed before such investment is likely to

become significant. In the meantime, Smart Grid-enabled demand response capabilities[23]

---

[21] Inertia is the physical property which allows an object in motion to continue to stay in motion, absent other forces. Traditional dispatchable generating units (such as thermal and hydro power plants) utilize large rotary generators which have large amounts of inertia. This property has a tendency to stabilize the bulk-power system with an output response in the event of a disturbance. Variable resources, such as wind and solar, have less or no inertia and, as such, cut back more quickly in response to disturbances (e.g., frequency excursions), which may contribute to power system instabilities.

[22] The Electricity Advisory Committee, which was formed by the Department of Energy to provide it with advice on a number of electricity issues, recently issued a report, Bottling Electricity: Storage as a Strategic Tool for Managing Variability and Capacity Concerns in the Modern Grid, December 2008. This report asserts that there are many benefits to deploying energy storage technologies into the Nation's grid: (1) a means to improve grid optimization for bulk power production; (2) a way to facilitate power system balancing in systems that have variable renewable energy sources; (3) facilitation of integration of plug-in hybrid electric vehicle power demands with the grid; (4) a way to defer investments in transmission and distribution infrastructure to meet peak loads (especially during outage conditions) for a time; and (5) a resource providing ancillary services directly to grid/market operators.

[23] The Smart Grid concept envisions a power system architecture that permits two-way communication between the grid and essentially all devices that connect to it, ultimately all the way down to large consumer appliances. Efforts at realizing this concept focus on standardization to enable all of this new equipment to be manufactured economically in support of widespread adoption by consumers. Once that is achieved, a significant proportion of electric load could become an important resource to the electric

(continued...)

Docket No. PL09-4-000   - 16 -

could add important new tools to deal with both resource adequacy and resource

management concerns.[24] Demand response reductions in load can help address the

resource adequacy concerns surrounding unexpected loss of variable generation, and

EISA envisions, among other things, the development of large new pools of demand

response resources.[25]

20.   With respect to the resource management concerns surrounding potential over-

generation, this situation tends to arise during off-peak periods when load is at its lowest

and system operators have already turned off all traditional generation except their large

conventional units that, for primarily operational reasons, must be operated in a nearly

steady state around the clock.[26] If large amounts of variable generation begin producing

---

system, able to respond automatically to customer-selected price or dispatch signals
delivered over the Smart Grid infrastructure without significant degradation of service
quality. For purposes of this proposed policy statement we will refer to such new
demand response capability as Smart Grid-enabled demand response capability.

[24] A recent NERC Draft Special Report recognizes that "[d]emand response has
already been shown in some balancing areas to be a flexible tool for operators to use with
wind generation [footnote omitted] and is a potential source of flexibility equal to supply-
side options." NERC, Special Report Accommodating High Levels of Variable
Generation at 45; available at
http://www.nerc.com/docs/pc/ivgtf/IVGTF_Reporta_17Nov08.pdf.

[25] See, e.g., EISA sec. 1301(4), (5), (6), (8), and (9), to be codified at 15 U.S.C.
17381(4), (5), (6), (8), and (9).

[26] There can also be an economic justification for around-the-clock operation
because large conventional units tend to have relatively higher capital costs and lower
running costs. However, their generally slow and difficult start-up and cool-down
sequences are the main reason why they cannot be started and stopped easily to address
                                                                        (continued...)

Docket No. PL09-4-000   - 17 -

power during such periods, then the supply of electricity would exceed the demand for

electricity and risk unbalancing the bulk-power system. In order to bring the system back

into balance in a situation where easily dispatchable generation or demand resources are

not available, system operators may have to require variable generation to reduce output.

However, at such times this variable generation may be producing the lowest priced

energy on the system, so reducing or eliminating its output would not be economically

efficient. If a system existed whereby entities[27] could receive a timely signal to

temporarily shift their demand from peak to off-peak, and if such load shifts could be

controlled by the system operator, then such "dispatchable" demand response could

alleviate to some degree the resource management concerns associated with over-

generation from the other side of the supply/demand equation. Again, the urgency to

develop and implement those aspects of a smarter grid that can enable such demand

response capability is clear.[28]

21.   The future potential for a large and variable new class of electric load, specifically

electricity-powered vehicles, also presents challenges that may deserve special attention

and priority in the consensus-based interoperability process being coordinated by the

---

over-generation situations.

[27] Such entities would need to have invested in the equipment necessary to reliably
measure and control either their own load or the load of clients that they manage under
contract.

[28] See, e.g., EISA sec. 1301(4), (5), (6), (8), and (9).

Docket No. PL09-4-000   - 18 -

Institute. In addition to the plans of major automobile manufacturers to roll out plug-in

hybrid vehicles starting in 2010, it is possible that large numbers of pure electric vehicles,

sometimes known as neighborhood electric vehicles, could be purchased as second cars

for short-haul daily commuting or for other purposes.[29] Judging by the observed intensity

of electric utility and state government interest in this area,[30] the potential for a

significant shift in personal transportation technology to electric power in the near future

cannot be discounted.

22.   The timing of vehicle charging activities is an illustration of the effect electric

vehicles can have on the operation of the electric grid. If charging takes place during

peak periods it could require a large investment in new generation, demand response

resources and/or transmission capacity to meet the resulting higher peak loads. However,

charging off-peak could actually improve the operation of the electric system, for

example by improving existing generation asset utilization or by providing an electricity

storage solution to address the potential for over-generation by variable resources in off-

---

[29] See, e.g., Kris Osborn, Services Plan to Buy Electric Cars, Federal Times, November 17, 2008, at 3 (noting that Army, Navy, and Air Force plan to purchase a total of 30,000 neighborhood electric vehicles for use on military bases).

[30] See, e.g., John S. Adams, Bill benefits 'medium-speed' electric cars, Great Falls Tribune, January 9, 2009 (reporting on efforts in the Montana legislature to ease restrictions and ownership and use requirements on "medium speed" electric vehicles, which could include electric vehicles of up to 5,000 pounds gross vehicle weight), available at http://www.greatfallstribune.com/article/20090109/NEWS01/901090337.

Docket No. PL09-4-000   - 19 -

peak periods. Ultimately, large numbers of plug-in electric vehicles have the potential to

provide some ancillary services like distributed energy storage or, when aggregated,

regulation service. In all cases, however, the enhanced information processing and high-

speed communications and control capabilities of the Smart Grid would be extremely

helpful, perhaps necessary, in dealing with the challenges and opportunities associated

with large numbers of new electric vehicles on the bulk-power system.

23.   Additionally, these and other changing patterns of electricity generation and use

are increasing the frequency with which congestion on transmission facilities becomes

binding and raises costs for consumers. The Smart Grid concept includes the deployment

of advanced sensors and controls throughout the electric system that should maximize the

capability and use of existing and new transmission capacity.

24.   For all of the reasons discussed above, which may represent direct challenges to

the reliable operation of the bulk-power system and wholesale power markets, the fact

that many utilities are already beginning to deploy Smart Grid related systems, and the

substantial funding for Smart Grid in the American Recovery and Reinvestment Act,[31]

---

[31] See American Recovery and Reinvestment Act, Pub. L. No. 111-5, Title IV, Subpart A, __ Stat. ___, ___ (2009) (ARRA).

Docket No. PL09-4-000   - 20 -

the Commission herein proposes a targeted acceleration of certain aspects of the

interoperability standards process as described further below.[32]

B.   Development of Key Interoperability Standards

25.   As discussed above, several important trends indicate a strong national interest in

expediting the development and deployment of the types of technologies and capabilities

associated with a smarter grid. To achieve these types of capabilities, Smart Grid

technologies must be interoperable.[33] The Commission understands that a consensus-

based interoperability standards development process typically requires time to reach

consensus, but also recognizes that recent efforts by the Institute and several industry

groups, including the OpenSG Subcommittee of the Utility Communication Architecture

International User Group (OpenSG Subcommittee) and the GridWise Architecture

Council, have developed concepts to prioritize the large set of potential standards, and

have suggested principles for expediting development of a set of transmission and

distribution systems standards that will facilitate many other important standards

development activities. The Commission is committed to identifying these key

transmission and distribution standards and working with the Institute to expedite their

---

[32] This is consistent with the Institute's approach of prioritizing standards and functionalities that may impact reliability. See NIST Smart Grid Issues Summary, March 10, 2009, available at: http://www.nist.gov/smartgrid/ (in case link is temporarily unavailable at this website, please request it via e-mail at: smartgrid@nist.gov).

[33] See Gridwise Architecture Council, Interoperability Path Forward Whitepaper, infra n.8.

Docket No. PL09-4-000   - 21 -

adoption. The Commission believes that focusing on the priorities identified below will

help to remove uncertainty for developers of standards applicable to all levels of the grid.

26.   The Institute has issued for comment a "Smart Grid Issues Summary" that will act

as an interim roadmap, starting with high priority standards that are largely based on

existing broadly accepted standards.[34] Leveraging existing standards to the greatest

extent practical should shorten the time required to finalize needed interoperability

standards.

27.   The Commission proposes to prioritize the development of standards for two

cross-cutting issues and four key grid functionalities involving interfaces between utilities

(e.g., regional transmission organizations (RTO) to utilities outside the RTO), utilities

and customers, and utilities and other systems (e.g., energy management systems). These

cross-cutting issues and key functionalities are proposed as the first level of work to be

accomplished in the interoperability standards-setting process. Swift progress on

adopting standards for these cross-cutting issues and key functionalities is necessary for

the transmission operator/RTO to address the bulk-power system challenges identified

above.

28.   The two cross cutting issues are first, cybersecurity (and physical security to

protect equipment that can give access to Smart Grid operations) and second, a common

---

[34] See infra n.32.

Docket No. PL09-4-000   - 22 -

semantic framework and software models for enabling effective communication and

coordination at the boundaries of utility systems where these interface with customer and

other systems (and hence provide "inter-system" functionality).[35] The four key grid

functionalities are wide-area situational awareness, demand response, electric storage,

and electric transportation.

<u>System Security</u>

29.   We propose two initial overarching principles regarding security that Smart Grid

applications must address in order to comply with the need for full cybersecurity and with

the Commission's bulk-power system concerns, consistent with our authority under

section 215 of the FPA.[36] First, we believe that a responsible entity subject to

Commission-approved reliability standards, such as the Critical Infrastructure Protection

Reliability Standards, must ensure that it maintains compliance with those standards

during and after the installation of Smart Grid technologies. Indeed, many Smart Grid

installations will need to be included on a responsible entity's list of critical assets to be

---

[35]The concept of the Smart Grid as a "system of systems" and the importance of
the need of first focusing on the inter-system interfaces are presented in a paper by the
OpenSG Subcommittee and Smart Grid Executive Working Group entitled <u>Smart Grid
Standards Adoption: Utility Industry Perspective</u> (Utility Perspective Paper), <u>available
at</u>: <u>http://osgug.ucaiug.org/Shared%20Documents/Forms/AllItems.aspx</u>. A graphic that
illustrates these concepts is found in Appendix A.

[36] 16 U.S.C 824o.

Docket No. PL09-4-000   - 23 -

protected under the Commission-approved NERC Critical Infrastructure Protection

Reliability Standards.

30.   Second, to the extent that they could affect the reliability of the bulk-power

system, Smart Grid technologies must address, the following considerations: (1) the

integrity of data communicated (whether the data is correct); (2) the authentication of the

communications (whether the communication is between the intended Smart Grid device

and an authorized device or person); (3) the prevention of unauthorized modifications to

Smart Grid devices and the logging of all modifications made; (4) the physical protection

of Smart Grid devices; and (5) the potential impact of unauthorized use of these Smart

Grid devices on the bulk-power system.

31.   To the extent that any of the new Smart Grid standards or extensions to relevant

existing standards require adaptation or extension in order to address these security-

related concerns, such considerations should be given the highest priority. The Institute

has suggested that beyond the NERC Critical Infrastructure Protection Reliability

Standards, additional security standards to be investigated include ISA99/IEC 62443,

NIST Special Publication (SP) 800-53, and the work of AMI-SEC.[37] The Institute also

---

[37] ISA99/IEC 62443 represents a suite of standards for industrial automation and control system security. NIST Special Publication (SP) 800-53 involves security controls for federal agencies, including those who are part of the bulk-power system (e.g., Tennessee Valley Authority, Bonneville Power Authority). The Advanced Metering Infrastructure (AMI) Security Task Force (AMI-SEC), is defining common requirements and standardized specifications for securing AMI system elements.

Docket No. PL09-4-000   - 24 -

suggests examining harmonization of several of these standards in order to provide

additional protection to the bulk-power system. Commission staff will monitor Institute

activities with respect to Smart Grid cybersecurity and physical security in order to fully

coordinate the Commission's regulatory objectives and responsibilities in this arena. The

Commission seeks comments on this proposed approach to maintaining bulk-power

system reliability and security as smart grid technologies are deployed and integrated.

Communication

32.   The second cross-cutting issue is the need for a common semantic framework (i.e.,

agreement as to meaning) and software models for enabling effective communication and

coordination across inter-system interfaces. An interface is a point where two systems

need to exchange data with each other; effective communication and coordination occurs

when each of the systems understands and can respond to the data provided by the other

system, even if the internal workings of each system are quite different. A core group of

standards initiated by the Electric Power Research Institute provide the basis for

addressing this issue - these standards are IEC 61970 and IEC 61968 (together often

referred to as the "Common Information Model" standards) and IEC 61850. These

standards have been cited by both the Utility Perspective Paper, as well as the Institute's

recent Smart Grid Issues Summary.[38] This group of standards was designed to allow

---

[38] See infra n.32.

Docket No. PL09-4-000   - 25 -

different systems to talk to one another as well as to provide software development tools
for more efficient system integration. This suite of standards is already in use by a
number of utilities for enterprise system integration (enabling integration across "intra-
system" interfaces). Indeed, while additional work on these standards will also help
intra-system communication and coordination, we agree with the OpenSG Subcommittee
and the Institute that inter-system interfaces should be a priority.

33.   The Commission is not mandating that these particular standards be further
developed. Rather, we identify them here to establish priorities for further development
by the Institute and industry. The group of standards initiated by the Electric Power
Research Institute serves as a foundation for developing a complete set of
communications standards. These standards require some level of harmonization with
one another and other standards, and extensions to these standards will be required for
additional interoperability and functionality. Efforts to coordinate and/or harmonize
these standards with others intended to promote interoperability should be encouraged.
For example, ongoing efforts to coordinate IEC 61968 with "MultiSpeak" developed by
the National Rural Electrical Cooperative Association should be continued. But these
standards represent the best work to date and will be an essential building block in
realizing the most significant early benefits for the bulk-power system. These standards
are also key to the attainment of renewable power and climate policy goals and can help

Docket No. PL09-4-000 - 26 -

enable customers to manage their energy usage and cost. The Commission seeks comments on this proposed approach.

Four Priority Functionalities

34. In addition to the cross-cutting issues discussed above, the Commission seeks comments on the four Smart Grid functionalities that the Commission's preliminary analysis indicates will be most helpful in addressing the bulk-power system challenges and should be given priority in the standards development process.

Wide-area situational awareness

35. Wide-area situational awareness is the visual display of interconnection-wide system conditions in near real time at the reliability coordinator level and above. The wide-area situational awareness efforts, with appropriate cybersecurity protections, can rely on the NASPInet work undertaken by the North American SynchroPhasor Initiative (NASPI) and will require substantial communications and coordination across the RTO and utility interfaces. We encourage the RTOs to take a leadership role in coordinating the NASPI work with the member transmission operators.

36. Regarding the potential Smart Grid role in addressing transmission congestion and optimization of the system, increased deployment of advanced sensors like Phasor Measurement Units will give bulk-power system operators access to large volumes of high-quality information about the actual state of the electric system that should enable a more efficient use of the electric grid, for example through a switch from static to

Docket No. PL09-4-000   - 27 -

dynamic line ratings. However, such large volumes of data present challenges in the

form of information processing and management. Advanced software and systems will

be needed to manage, process, and render this data into a form suitable for human

operators and automated control systems. The Institute's process should strive to identify

the core requirements for such software and systems that would be most useful to system

operators in addressing transmission congestion and reliability.

Demand response

37.   Smart Grid-enabled demand response is a priority because of its potential to help

address several of the bulk-power system challenges identified above. Further

development of key standards would enhance interoperability and communications

between system operators, demand response resources, and the systems that support

them. In order to achieve an appropriate level of standardizations, a series of demand

response "use cases" should be developed using readily available tools.[39] In this regard,

we encourage a particular focus on use cases for the key demand response activities

discussed earlier: dispatchable demand response load reductions to address loss or

unavailability of variable resources and the potential for dispatchable demand response to

increase power consumption during over-generation situations.

---

[39] The "use case" is a concept from the software and systems engineering
communities whereby a developer, usually in concert with the end user, attempts to
identify all of the functional requirements of a system. Each "use case" essentially
describes how a user will interact with a system to achieve a specific goal.

Docket No. PL09-4-000   - 28 -

38.   It also appears that achieving such demand response capabilities will require

additional standardization of the interfaces between systems on the customer premises

and utility systems, including addressing data confidentiality issues. The Institute notes

that considerable work has been done to develop demand response standards. One

standard, Open Automated Demand Response (OpenADR) (developed for the interface

between the utility and large commercial customers) has already been referred to the

Organization for the Advancement of Structured Information Systems (OASIS).

OpenADR has been developed by the Lawrence Berkeley National Laboratory, and is

now going through a formal standards development process being coordinated between

OASIS and the Utility Communication Architecture International User Group.[40]

Accordingly, we would encourage a focus in this area as well.

39.   Specifications for customer meters are within the jurisdiction of the States, but it is

clear that communication and coordination across the interfaces between the utility and

its customers can have a significant impact on the bulk-power system, particularly as new

renewable power and climate policy initiatives introduce the need for more flexibility in

the electricity grid, which creates the need for increased reliance on demand response and

---

[40] The Utility Communication Architecture International User Group has also been developing OpenHAN, a specification for the energy services interface between the home area network (HAN) and the utility. Both OpenHan and OpenADR will benefit from the planned extensions of IEC 61850 and the common information model standards described above.

Docket No. PL09-4-000   - 29 -

electricity storage. A large portion of electricity storage may ultimately be located on

customer premises. As noted in the Institute's Smart Grid Issues Summary, an

appropriate starting point for further standards development would be the harmonization

of IEC Standard 61850 and several meter standards, namely ANSI C12.19 and C12.22,

and we encourage the Institute and industry to work together on this suggestion. The

Commission seeks comment from States and other parties on the optimal approach to

develop standards in this area, and we will pursue direct communications with the States

on this topic through the NARUC-FERC Smart Grid Collaborative and other NARUC

Committees.

Electric storage

40.   The third key grid functionality is electric storage. If electricity storage

technologies could be more widely deployed, they would present another important

means of addressing some of the difficult issues facing the electric industry. To date, the

only significant bulk electricity storage technology has been pumped storage

hydroelectric technology. However, we are aware that new types of storage technologies

are under development and in some cases are being deployed, and could also potentially

provide substantial value to the electric grid. While further research and development

appears necessary before any widespread deployment of such newer technologies can

take place, it may nevertheless be appropriate to encourage the identification and

standardization of all possible electricity storage use cases at an early stage. There are

Docket No. PL09-4-000   - 30 -

existing standards that can be the starting point for interoperability standards

development for DER. IEC 61850 addresses communications for DER, and IEEE 1547

has been designated as a federal standard for interconnection.[41]

Electric transportation

41.   The fourth key grid functionality is electric transportation. As indicated above, to

the extent that new electric transportation options become widely adopted in the near

future, maintaining the reliable operation of the bulk-power system will require some

level of control over when and how electric cars draw electricity off of the system. At the

most basic level, this could be accomplished by providing an ability for distribution

utilities to facilitate vehicle charging during off-peak periods so that this new electric

load would not increase peak loads and require the development of new peak generation,

demand response and/or more transmission to urban load centers that are being targeted

for these vehicles. A more advanced implementation could offer vehicle owners the

option to voluntarily limit their charging to times when variable renewable generation is

producing power or to permit utilities the limited use of the aggregated capabilities of

these vehicles for various grid-related purposes such as bulk power storage or ancillary

services.

---

[41] See Energy Policy Act of 2005, Pub. L. No. 109-58, sec. 1254, 110 Stat. 594, 970 (2005), adding a new subsection 111(d)(15) to the Public Utility Regulatory Policies Act of 1978 (16 U.S.C. 2621(d)).

Docket No. PL09-4-000   - 31 -

42.   Ultimately we would hope for a smarter grid to accommodate a wide array of advanced options for electric vehicle interaction with the grid, including full vehicle-to-grid capabilities. However, assuming full vehicle-to-grid capabilities cannot be achieved immediately, we would encourage the Institute's process to focus on the development of appropriate standards, or extensions to relevant existing standards, to provide at least the minimum communications and interoperability requirements that are necessary to permit some ability for distribution utilities to facilitate vehicle charging during off-peak load periods. The Institute's Smart Grid Issues Summary notes that the Society of Automotive Engineers (SAE) has developed two draft standards, SAE J2836 and SAE J2847, which address communications and price signals/demand response respectively. These standards are on the SAE 2009 Ballot. Looking forward to the potential provision of ancillary services to the grid by electric vehicles, electrical interconnection issues must be dealt with along with potential expansion of communications ability. To this end, we urge the SAE and the automobile industry to plan data communications systems between electric vehicles and the grid that are able to be upgraded. We also urge the Institute to include electric vehicles in its DER standards development.

43.   Several of the preceding paragraphs discuss the development of use cases or other standards that appear similar to business practice standards development, in order to help shape and identify the functional needs that the Institute's technical interoperability standards development process will address. Since the North American Energy

Docket No. PL09-4-000   - 32 -

Standards Board (NAESB) has a great deal of experience in helping the electric and

natural gas industries successfully negotiate business practice standards, it may be helpful

to the Institute to engage NAESB resources in the development of these use cases and

other business practice-like standards. We seek comment as to whether the Institute

would be helped by the incorporation of resources from other organizations such as

NAESB into the development of these various business practice-like standards.

44.   The Commission seeks comment on whether the priorities and reliability

principles articulated above are appropriate, and whether there are other priorities or

reliability principles that should be included in order to address potential challenges to

the operation of the bulk-power system.

C.   Interim Rate Policy: Guidance for Smart Grid-Related Filings by Jurisdictional Entities

45.   Given the trends discussed above, Smart Grid policies should encourage utilities to

deploy systems in the near term that advance efficiency, security, and interoperability in

order to address potential challenges to the bulk-power system. A key consideration for

utilities when determining whether to adopt such systems will be whether they are able to

recover the costs of these deployments in regulated rates. Another key consideration may

involve the potential for stranded costs associated with legacy systems that are replaced

by Smart Grid equipment. Additionally, as the electric system may require several of the

new capabilities of the Smart Grid before interoperability standards have been developed,

we recognize the need for guidance for jurisdictional entities. Thus, to offer some rate

Docket No. PL09-4-000   - 33 -

certainty and guidance regarding cost recovery issues, the Commission is proposing a

rate policy for the interim period until final interoperability standards are adopted.

46.   FPA section 205 requires that all rates for the transmission or sale of electric

energy subject to the Commission's jurisdiction be just and reasonable.[42] In evaluating

expenses for which cost recovery is appropriate, one of the criteria the Commission relies

on is whether the facilities are "used and useful."[43] Once interoperability standards are

completed, the Commission will consider making compliance with those standards a

mandatory condition for rate recovery of jurisdictional Smart Grid investments. For now,

we propose as an interim rate policy to accept rate filings, including single issue rate

filings, submitted under FPA section 205 by public utilities to recover the costs of Smart

Grid deployments involving jurisdictional facilities provided that certain showings are

made. In other words, we propose to consider Smart Grid devices and equipment,

including those used in a Smart Grid pilot program or demonstration project, to be used

and useful for purposes of cost recovery if an applicant makes the following showings.

47.   We propose that an applicant must show that the reliability and security of the

bulk-power system will not be adversely affected by the deployment at issue. Second,

---

[42] 16 U.S.C. 824d.

[43] The general rate-making principle is that expenditures for an item may be included in a public utility's rate base only when the item is "used and useful" in providing service. See NEPCO Municipal Rate Committee v. FERC, 668 F.2d 1327, 1333 (D.C. Cir. 1981).

Docket No. PL09-4-000 - 34 -

the filing must show that the applicant has minimized the possibility of stranded

investment in Smart Grid equipment by designing for the ability to be upgraded, in light

of the fact that such filings will predate adoption of interoperability standards. Finally

because it will be important for early Smart Grid deployments, particularly pilot and

demonstration projects, to provide feedback useful to the interoperability standards

development process, we propose to direct the applicant to share information with the

Department of Energy Smart Grid Clearinghouse, provided for in the ARRA.[44]

48. In order to satisfy our first concern about reliability and security, we propose that

applicants will be required to address the security concerns discussed in the previous

section on the development of key standards. Accordingly, an applicant must show how

its proposed deployment of smart grid equipment will maintain compliance with

Commission-approved reliability standards, such as the Critical Infrastructure Protection

Reliability Standards, during and after the installation and activation of Smart Grid

technologies so the reliability and security of the bulk-power system will not be

jeopardized. An applicant must also address: (1) the integrity of data communicated

(whether the data is correct); (2) the authentication of the communications (whether the

communication is between the intended Smart Grid device and an authorized device or

person); (3) the prevention of unauthorized modifications to Smart Grid devices and the

---

[44] ARRA sec. 405(3).

Docket No. PL09-4-000   - 35 -

logging of all modifications made; (4) the physical protection of Smart Grid devices; and

(5) the potential impact of unauthorized use of these Smart Grid devices on the bulk-

power system.

49.   Regarding the second concern about stranded smart grid investment, we propose

to require a showing that the applicants have made good faith efforts to adhere to the

vision of a Smart Grid described in Title XIII of the EISA, including optimizing asset

utilization and operating efficiency. In general, applicants should attempt to adhere to the

principles of the Gridwise Architecture Council Decision-Maker's Interoperability

Checklist.[45] In practice, we will place the most weight on an applicant's adherence to the

following principles: (1) reliance to the greatest extent practical on existing, widely

adopted and open[46] interoperability standards; and (2) where feasible, reliance on systems

---

[45] See Gridwise Architecture Council Decision-Maker's Interoperability Checklist
Draft Version 1.0, available at
http://www.gridwiseac.org/pdfs/gwac_decisionmakerchecklist.pdf (Interoperability
Checklist).

[46] An open architecture is publicly known, so any and all vendors can build
hardware or software that fits within that architecture, and the architecture stands outside
the control of any single individual or group of vendors. In contrast, a closed architecture
is vendor-specific and proprietary, and blocks other vendors from adoption. An open
architecture encourages multi-vendor competition because every vendor has the
opportunity to build interchangeable hardware or software that works with other elements
within the system. See Gridwise Architecture Council Decision-Maker's Interoperability
Checklist Draft Version 1.0, available at
http://www.gridwiseac.org/pdfs/gwac_decisionmakerchecklist.pdf. We note that
Congress recently made utilization of open protocols and standards, if available and
appropriate, a condition of receiving funding from the Department of Energy for

(continued...)

Docket No. PL09-4-000   - 36 -

and firmware that can be securely upgraded readily and quickly. Adherence to these two

key principles should minimize the possibility of stranded smart grid investment by

making it less likely that equipment replacement will be required once final standards are

approved.

50.   Regarding the information sharing concern, the following information should be

shared with the Department of Energy Smart Grid Clearinghouse: (1) any internal or

third party evaluations, ratings, and/or reviews including all primary source material used

in the evaluation; (2) detailed data and documentation explaining any improvement in the

accurate measurement of demand response resources; (3) detailed data and

documentation explaining the expansion of the quantity of demand response resources

that resulted from the project and the resulting economic effects; (4) detailed data and

documentation for any improvements in the ability to integrate variable renewable

generation resources; (5) detailed data and documentation that shows any achievement of

greater system efficiency through a reduction of transmission congestion and loop flow;

(6) detailed data and documentation showing how the information infrastructure supports

DER such as plug-in electric vehicles; and (7) detailed data and documentation that

shows how the project resulted in enhanced utilization of energy storage. To the extent

---

demonstration projects and grants pursuant to EISA sections 1304 and 1306. See
American Recovery and Reinvestment Act, Pub. L. No. 111-5, sec. 405(3) and 405(8), ___
Stat. ___, ___ (2009).

Docket No. PL09-4-000   - 37 -

that the Department of Energy specifies additional criteria for making grants under the ARRA for Smart Grid demonstration and pilot projects, the Applicant should agree to share information relevant to those criteria as well.

51.   Finally, consistent with the policy of supporting the modernization of the Nation's electric system announced in EISA section 1301, the Commission also proposes to permit applicants to file for recovery of the otherwise stranded costs of legacy systems that are to be replaced by smart grid equipment. However, an appropriate plan for the staged deployment of smart grid equipment, which could include appropriate upgrades to legacy systems where technically feasible and cost-effective, could help minimize the stranding of unamortized costs of legacy systems. Accordingly, we propose that any filing for the recovery of stranded legacy system costs must demonstrate that such a migration plan has been developed.

52.   The Commission will also entertain requests for rate treatments such as accelerated depreciation and abandonment authority (whereby an applicant is assured of recovery of abandoned plant costs if the project is abandoned for reasons outside the control of the public utility) specifically tied to Smart Grid deployments under our FPA section 205 authority. Any requests for such rate treatments for Smart Grid deployments will need to address all of the concerns discussed above for rate recovery and make the same showings described in that section. We would also consider applying these rate treatments to the portion of a smart grid pilot or demonstration project's cost that is not

Docket No. PL09-4-000 - 38 -

already paid for by Department of Energy funds, such as those authorized by EISA

sections 1304 and 1306.[47] To the extent that such showings are made, we propose to

consider permitting abandonment authority to apply to any Smart Grid investments that,

despite reasonable efforts, could not be made upgradeable and must ultimately be

replaced if found to conflict with the final standards to be approved under the Institute's

standards development process.

53. The Commission invites comments on all aspects of this proposed interim rate

policy.

III. Comment Procedures

54. The Commission invites comments on this proposed policy statement

[Insert_Date 45 days after publication in the **FEDERAL REGISTER**].

IV. Document Availability

55. In addition to publishing the full text of this document in the Federal Register, the

Commission provides all interested persons an opportunity to view and/or print the

contents of this document via the Internet through FERC's Home Page

(http://www.ferc.gov) and in FERC's Public Reference Room during normal business

hours (8:30 a.m. to 5:00 p.m. Eastern time) at 888 First Street, N.E., Room 2A,

Washington D.C. 20426.

---

[47] To be codified at 42 U.S.C. 17384, 17386.

Docket No. PL09-4-000   - 39 -

56.   From FERC's Home Page on the Internet, this information is available on

eLibrary. The full text of this document is available on eLibrary in PDF and Microsoft

Word format for viewing, printing, and/or downloading. To access this document in

eLibrary, type the docket number excluding the last three digits of this document in the

docket number field.

57.   User assistance is available for eLibrary and the FERC's website during normal

business hours from FERC Online Support at 202-502-6652 (toll free at 1-866-208-3676)

or email at ferconlinesupport@ferc.gov, or the Public Reference Room at (202) 502-

8371, TTY (202)502-8659. E-mail the Public Reference Room at

public.referenceroom@ferc.gov.

By the Commission.

(S E A L)

                                        Nathaniel J. Davis, Sr.,
                                        Deputy Secretary.

Docket No. PL09-4-000   - 40 -

Appendix A

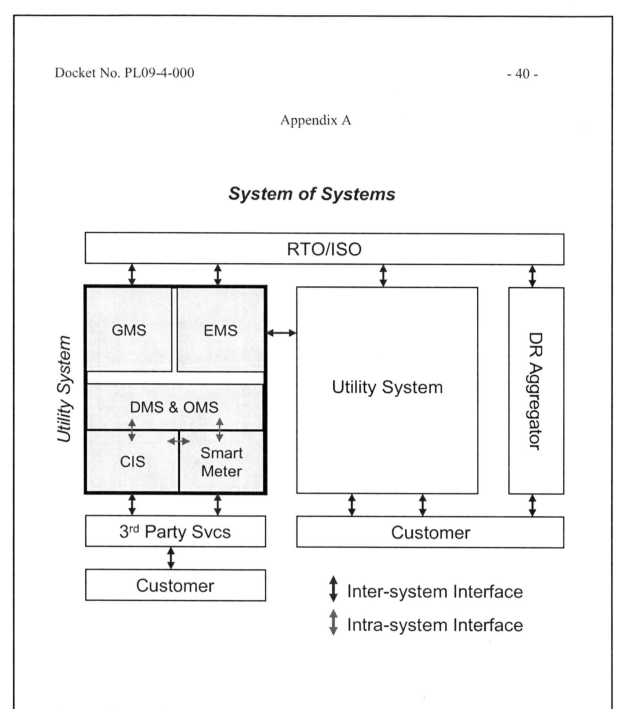

Source: <u>Smart Grid Standards Adoption: Utility Industry Perspective</u>, OpenSG Subcommittee of the Utility Communication Architecture International User Group, and Smart Grid Executive Working Group.

614

Other Resources
Internet Resources

- Energy Independence and Security Act of 2007, P.L. 110-140, Dec. 19, 2007
  <http://frwebgate.access.gpo.gov/cgi-bin/getdoc.cgi?dbname=110_cong_public_laws&docid=f:publ140.pdf>

- Smart Grid News.com
  <www.smartgridnews.com>

- National Institute of Standards and Technology Smartgrid
  <www.nist.gov/smartgrid>

- Federal Energy Regulatory Commission
  <www.ferc.gov>

- U.S. Department of Energy
  <www.doe.gov>

- National Council on Electricity Policy
  <www.ncouncil.org>

- Smart Grid: Smart Grid News—Grid Modernization and the Smart Grid
  <www.smartgridnews.com/artman/publish>

- SmartGridToday.com
  <www.smartgridtoday.com>

- The Smart Grid Security Blog
  <http://smartgridsecurity.blogspot.com>

- The Smart Grid Buzz
  <http://broadbandoverpowerlines.blogspot.com>

- Collaborative Energy—the Smart Grid and the End Node
  <www.newdaedalus.com>

- "Smart Electric Grid of the Future: A National 'Distributed Store-Gen' Test Bed" by Roger Anderson, Columbia University; Paul Chu, University of Houston; Ron Oligney, Texas Energy Center; and Rick Smalley, Rice University
  <www.ldeo.columbia.edu/res/pi/4d4/testbeds/Smart-Grid-White-Paper.pdf>

- Environmental & Energy Study Institute
  <www.eesi.org>

- Smart Grid Working Group
  <www.energyfuturecoalition.org/preview.cfm?catID=13>

- Listing and links to various reports
  <www.energyfuturecoalition.org/Resources/Energy-Efficiency-/-Smartgrid>

- "Challenge and Opportunity: Charting a New Energy Future", Energy Future Coalition
  <http://energyfuturecoalition.org/files/webfmuploads/EFC_Report/EFCReport.pdf>

- Edison Electric Institute
  <www.eei.org>

- 2009 National Electricity Delivery Forum
  <www.electricitydeliveryforum.org>

- National Electrical Manufacturers Association Seminar on
  "Defining Intelligence in the Intelligent Electricity Grid", January 18, 2008
  <www.nema.org/gov/energy/smartgrid/upload/Presentation-Smart-Grid.pdf>

- National Energy Technologies Laboratory -Presentation
  *<www.energetics.com/supercon07/pdfs/NETL_Synergies_
  of_the_SmartGrid_and_Superconducitivity_Pullins.pdf>*

- San Diego Smart Grid Study
  <www.sandiego.edu/epic/publications/documents/061017_SDSmartGridStudyFINAL.pdf>

- Electric Power Research Institute, EPRI Intelligrid
  <http://intelligrid.epri.com>

- Gridwise at Pacific Northwest National Laboratory (PNNL)
  <http://gridwise.pnl.gov>

- U.S. Department of Energy Smart Grid Task Force
  <www.oe.energy.gov/smartgrid_taskforce.htm>

- Smart Grid
  <www.oe.energy.gov/smartgrid.htm>

- GRIDWISE Alliance
  <www.gridwise.org>

- Grid Week
  <www.gridweek.com>

- IntelliGrid
  <http://intelligrid.epri.com>

- Galvin Electricity Initiative
  <www.galvinpower.org>

Books

- "Perfect Power: How the Microgrid Revolution Will Unleash Cleaner, Greener, More Abundant Energy," by Robert Galvin and Kurt Yeager (McGraw-Hill 2008), ISBN-10: 0071548823

- "The Green Guide to Power: Thinking Outside the Grid," by Ron Bowman (BookSurge 2008), ISBN-10: 1439207690

- "The Grid: A Journey Through the Heart of Our Electrified World," by Phillip F. Schewe (Joseph Henry Press 2007), ISBN-10: 030910260X

- "Understanding Today's Electricity Business," by Bob Shively and John Ferrare (Enerdynamics 2004), ISBN-10: 0974174416

- "Electric Power Industry in Nontechnical Language," by Denise Warkentin-Glenn (PennWell 2006), ISBN-10: 1593700679

- "Electricity Markets: Pricing, Structures and Economics," by Chris Harris (Wiley 2006), ISBN-10: 0470011580

- "Energy and Power Risk Management: New Developments in Modeling, Pricing and Hedging," by Alexander Eydeland and Krzysztof Wolyniec (Wiley 2002), ISBN-10: 0471104000

- "Electric Power System Basics for the Nonelectrical Professional," by Steven W. Blume (Wiley-IEEE 2007), ISBN-10: 0470129875

- "Electric Power Generation: A Nontechnical Guide," by Dave Barnett and Kirk Bjornsgaard (Pennwell 2000), ISBN-10: 0878147535

- "From Edison to Enron: The Business of Power and What It Means for the Future of Electricity," by Richard Munson (Praeger 2008), ISBN-10: 031336186X

- "Power Primer: A Nontechnical Guide from Generation to End Use," by Ann Chambers (Pennwell 1999), ISBN-10: 087814756X

- "Electric Power Generation, Transmission, and Distribution," by Leonard L. Grigsby (CRC 2007), ISBN-10: 0849392926

- "Electric Power Distribution Handbook," by Thomas Allen Short (CRC 2003), ISBN-10: 0849317916

- "Understanding Electric Utilities and De-Regulation," by Lorrin Philipson (CRC 2005), ISBN-10: 0824727738

- "Power System Economics: Designing Markets for Electricity," by Steven Stoft (Wiley-IEEE 2002), ISBN-10: 0471150401

- "Making Competition Work in Electricity," by Sally Hunt (Wiley 2002), ISBN-10: 0471220981

- "Market Operations in Electric Power Systems: Forecasting, Scheduling, and Risk Management," by M. Shahidehpour, H. Yamin, and Zuyi Li (Wiley-IEEE 2002), ISBN-10: 0471443379

- "The Electric Power Engineering Handbook, Second Edition," by Leonard L. Grigsby (CRC 2007), ISBN-10: 0849392934

- "America's Electric Utilities: Past, Present And Future," by Leonard S. Hyman, Andrew S. Hyman, and Robert C. Hyman (Public Utilities Reports 2005), ISBN-10: 0910325006

- "Electric Power Systems: A Conceptual Introduction," by Alexandra von Meier (Wiley-IEEE 2006), ISBN-10: 0471178594

- "Electric Power Planning for Regulated and Deregulated Markets," by Arthur Mazer (Wiley-IEEE 2007), ISBN-10: 0470118822

- "Economic Evaluation of Projects in the Electricity Supply Industry," by H. Khatib (The Institution of Engineering and Technology 2003), ISBN-10: 0863413048

- "Electricity Economics: Regulation and Deregulation," by Geoffrey Rothwell and Tomás Gómez (Wiley-IEEE 2003), ISBN-10: 0471234370

617

- "Fundamentals of Power System Economics," by Daniel S. Kirschen and Goran Strbac (Wiley 2004), ISBN-10: 0470845724

- "Distributed Generation: A Nontechnical Guide," by Ann Chambers, Barry Schnoor, and Stephanie Hamilton (Pennwell 2001), ISBN-10: 0878147896

- "Power Industry Dictionary," by Ann Chambers and Susan D. Kerr (Pennwell 1996), ISBN-10: 0878146059

- "Electrical Power System Essentials," by Pieter Schavemaker and Lou van der Sluis (Wiley 2008), ISBN-10: 0470510277

- "Power System Engineering: Planning, Design, and Operation of Power Systems and Equipment," by Juergen Schlabbach and Karl-Heinz Rofalski (Wiley-VCH 2008), ISBN-10: 3527407596

- "Power System Operation," by Robert Miller and James Malinowski (McGraw-Hill 1994), ISBN-10: 0070419779

- "Power System Stability and Control," by Prabha Kundur (McGraw-Hill 1994), ISBN-10: 007035958X

- "Power Generation Handbook: Selection, Applications, Operation, Maintenance," by Philip Kiameh (McGraw-Hill 2002), ISBN-10: 0071396047

- "Electric Power Substations Engineering, Second Edition," by John D. McDonald (CRC 2007), ISBN-10: 0849373832

- "Modeling and Forecasting Electricity Loads and Prices: A Statistical Approach," by Rafal Weron (Wiley 2006), ISBN-10: 047005753X

- "Electric Power Distribution System Engineering, Second Edition," by Turan Gonen (CRC 2007), ISBN-10: 142006200X

- "Electricity Demystified," by Stan Gibilisco (McGraw-Hill 2005), ISBN-10: 0071439250

- "Understanding Today's Electricity Business," by Bob Shively and John Ferrare (Enerdynamics 2004), ISBN-10: 0974174416

- "Transmission and Distribution Electrical Engineering, Third Edition," by Colin Bayliss and Brian Hardy (Newnes 2007), ISBN-10: 0750666730

- "Power Distribution Planning Reference Book, Second Edition," by H. Lee Willis (CRC 2004), ISBN-10: 0824748751

- "Guide to Electrical Power Distribution Systems, Sixth Edition," by Anthony J. Pansini (CRC 2005), ISBN-10: 084933666X

- "Electric Utility Systems and Practices," by Homer M. Rustebakke (Wiley-Interscience 1983), ISBN-10: 0471048909

- "Understanding Electric Power Systems: An Overview of the Technology and the Marketplace," by Jack Casazza and Frank Delea (Wiley-IEEE 2003), ISBN-10: 0471446521

Other Resources from TheCapitol.Net

Capitol Learning Audio Courses™
<www.CapitolLearning.com>

- Congress and Its Role in Policymaking
  ISBN: 158733061X

- Understanding the Regulatory Process, A Five Course Series
  ISBN 13: 9781587331398

- Historical Overview of the Federal Regulatory Process
  ISBN 13: 9781587331466

- Overview of the Rulemaking Process
  ISBN 13: 9781587331466

- OMB's Role in the Regulatory Process and Pertinent Executive Orders
  ISBN 13: 9781587331480

- How to Read and Comment on a Proposed Rule
  ISBN 13: 9781587331541

- Devising a Policy and Issue Management Strategy
  ISBN 13: 9781587331558

- Preparing for Congressional Oversight and Investigation
  ISBN: 1587330644

Live Training

- Understanding the Regulatory Process:
  Working with Federal Regulatory Agencies
  <www.RegulatoryProcess.com>

- Understanding Congressional Budgeting and Appropriations
  <www.CongressionalBudgeting.com>

- Advanced Federal Budget Process
  <www.BudgetProcess.com>

- The President's Budget
  <www.PresidentsBudget.com>

- Capitol Hill Workshop
  <www.CapitolHillWorkshop.com>

About TheCapitol.Net

We help you understand Washington and Congress.™

For over 30 years, TheCapitol.Net and its predecessor, Congressional Quarterly Executive Conferences, have been training professionals from government, military, business, and NGOs on the dynamics and operations of the legislative and executive branches and how to work with them.

Our training and publications include congressional operations, legislative and budget process, communication and advocacy, media and public relations, research, business etiquette, and more.

TheCapitol.Net is a non-partisan firm.

TheCapitol.Net encompasses a dynamic team of more than 150 faculty members and authors, all of whom are independent subject matter experts and veterans in their fields. Faculty and authors include senior government executives, former Members of Congress, Hill and agency staff, editors and journalists, lobbyists, lawyers, nonprofit executives and scholars.

All courses, seminars and workshops can be tailored to align with your organization's educational objectives and presented on-site at your location. We've worked with hundreds of clients to develop and produce a wide variety of custom, on-site training.

Our practitioner books and publications are written by leading subject matter experts.

TheCapitol.Net has more than 2,000 clients representing congressional offices, federal and state agencies, military branches, corporations, associations, news media and NGOs nationwide.

Our blog: Hobnob Blog—hit or miss ... give or take ... this or that ...

TheCapitol.Net is on Yelp.

TheCapitol.Net supports the
TC Williams Debate Society, Wikimedia Foundation
and the Sunlight Foundation

Non-partisan training and publications that show how Washington works.™

PO Box 25706, Alexandria, VA 22313-5706   703-739-3790   www.TheCapitol.Net